Family Farming and Rural Economic Development

NIPA® GENX ELECTRONIC RESOURCES & SOLUTIONS P. LTD.
New Delhi-110 034

Family Farming and Rural Economic Development

Edited by

M.L. Choudhary

Vice-Chancellor

Bihar Agricultural University

Sabour, Bhagalpur, Bihar

Aditya

Assistant Professor

Department of Extension Education

Bihar Agricultural University

Sabour, Bhagalpur, Bihar

NIPA® GENX ELECTRONIC RESOURCES & SOLUTIONS P. LTD.

New Delhi-110 034

NIPA® GENX ELECTRONIC
RESOURCES & SOLUTIONS P. LTD.

101,103, Vikas Surya Plaza, CU Block
L.S.C. Market, Pitam Pura, New Delhi-110 034
Ph : +91 11 27341616, 27341717, 27341718
E-mail: newindiapublishingagency@gmail.com
Web: www.nipabooks.com

For customer assistance, please contact
Phone: + 91-11-27 34 17 17
Fax: + 91-11- 27 34 16 16
E-Mail: feedbacks@nipabooks.com

ISBN: 978-81-19103-94-2

Composed and Designed by NIPA®.

Preface

Family farming is the predominant form of agriculture in the food production sectors in both developing and developed countries. It mainly includes all family-based agricultural activities, and it is linked to several areas of rural and social development. It is understood as a means of organizing agricultural, forestry, fisheries, pastoral and aquaculture production which is managed and operated by a family and increasingly relevant on family labour of both men and women. In developing countries like India, a number of factors are key for the successful development of family farming, such as: agro-ecological conditions and territorial characteristics; access to markets; policy environment; access to land and natural resources; access to technology and extension services; access to finance; demographic, economic and socio-cultural conditions; availability of specialized education among a few. The United Nations has designated the year 2014 as the International Year of Family Farming at the 66th session of the United Nations General Assembly recognizing the increasing role of family farming in poverty alleviation and rural development in particular.

The concept of family farming incorporates various elements. Considering the sociological perspective, family farming is associated with family values, such as solidarity, continuity and commitment/ bonding; in economic terms, family farming is synonymous with specialized entrepreneurial skills, business ownership and management, choice and risk behaviour, sustainability, resilience and individual achievement. Family farming can be considered more than a professional occupation because it reflects a lifestyle based on beliefs and traditions about living and work.

Smallholder and family farming are means to reduce poverty and hunger in the rural areas of the developing world. The contribution of small farmers to total farm output in India exceeds 50%, while they cultivate 44% of land. They have lesser capital but require higher use of labour and other family-owned inputs, and usually have a higher index of cropping intensity and diversification. They grow a variety of cultivars, many of which are landraces. These landraces are genetically more heterogeneous than modern varieties, and thus offer greater resilience against vulnerability and enhance harvest security in the midst of diseases, pests, droughts and other stresses. The diversity in farming, crops and livestock, often results in higher productivity than the large farms practicing usually monoculture. The livestock forms are an integral component of sustenance for small and marginal farmers, especially in difficult agro-climatic regions. In arid areas where droughts are common, the livestock lend sustainability to rural economy. The diverse activities contribute to food basket, nutritional security and household income of farmers and play a significant role in generating gainful employment in rural areas, particularly among the landless, small and marginal farmers including women.

The content of the book offers a diverse selection of chapters that address issues of importance to those in the agriculture industry, researchers, faculty, and others. The chapters focus on the role and aspirations of youth in family farming, economic opportunities for youth in farming that provide alternatives to migration to urban areas, challenges and mitigation strategies to the problems faced by youth in family farming, role of Information and Communication Technologies (ICTs)/ recent innovations in contributing to better family farming, cases of innovative ideas/ strategies/ business proposals for improving family farm's productivity and production efficiencies, advances in research methods vis-à-vis participation of youth in family farming.

The book will provide valuable inputs on the diverse topics covered under family farming and is an excellent resource for researchers; instructors; students in agriculture, horticulture, environmental science, and other allied subjects; together with content of interest to policymakers.

We express our sincere gratitude to all the authors for their professional commitments and patience during the review and publication process.We would also like to thank the publishers which were extremely efficient, thorough, and professional in their outlook. Their concern for producing a high quality book was evident at every stage. Last, but never least, the Bihar Agricultural University, Sabour and its officials deserve special praise and recognition for their enduring support, trust and benevolent funding for the cause of family farming in the country.

December, 2014 **Editors**

Sabour

Contents

CHAPTER - 1

Involving Rural Youths in Farming – A Suggested Roadmap

P P Pal, A K Singh and H K De
ICAR-Zonal Project Directorate, Zone II
Kolkata 700097

ABSTRACT

Rural youths represent both a challenge to and an opportunity for reducing rural poverty. To overcome this challenge and seize the opportunity concerted efforts are needed to generate decent employment opportunities. A complementary set of awareness generation, entrepreneurship development, enabling policy, capacity building and technical support activities need to be designed for creating an enabling environment that meets the multi-dimensional needs of rural youths. Sustainable poverty reduction and rural economic development can only be achieved when national enabling environments facilitates and empower young people to achieve their full potential. Agriculture being the core of rural activity, potentiality of farming sector to create employment opportunity both in primary as well as in secondary agriculture needs to be highlighted with proven success before the rural youths. Enabling environment in all aspect needs to be created to motivate the educated youths to live in villages and launch a new agricultural movement based on integrated application of science and social wisdom. The untapped demographic dividend will become our greatest strength when the aspiration turns into reality.

Key words : Poverty, challenge, opportunities, youth, employment

INTRODUCTION

In India, every third person is a youth. By the year 2020, the median individual in India will be 29 years making it the youngest country in the world. India is set to experience a dynamic transformation as the population burden of the past turns into a demographic dividend though it is apprehended that the benefit will be tempered with social and spatial inequalities. However, we are deriving very little demographic dividend in agriculture at present. Farming not being so intellectually satisfying and economically benefitting, rural youths are very often averse to take up farming as a livelihood option. On the contrary, population pressure on land is steadily increasing, average size of farm holding is going down to below 1 hectare and the farmers are getting indented and tempted to selling prime farm land for non-farm purposes due to steep rise in the price of land. Collapse of millions of livelihoods in agriculture and its related occupations has led to deepening agrarian crisis resulted into suicide of more than 2, 40,000 farmers, mostly broken by debt, between 1995 and 2009 (BBC News; South Asia 2011). Hence, motivating rural youths towards farming in its present dimension is quite difficult and needs well-thought multipronged strategy to create enabling environment so as to make farming as lucrative as other sectors to attract and retain the youths on a sustained basis.

Rural Youth: Challenge and Opportunity

The global scenario of youth population indicates that nearly 1.2 billion youths live in the world today comprising over 14.0 percent of the global population (UNDESA, 2013). Almost 88 percent of these young people are in developing countries – a figure that is expected to increase over the coming decades. Asia alone is home to 60 percent of the world's youth. The majority of the youths live and work in rural areas where productive resources are lacking and the youths rely on their own labour to earn a living. As the rural labour markets do not provide sufficient decent work opportunities for young men and women, they are to compromise their ability to lead productive lives. Major constraints that youth encounter are - weak enforcement of labour legislation, insecure and low income, poor health, safety and environmental conditions,gender inequality in pay and opportunities, the exploitation of migrant workers, inadequate social protection and weak social dialogue. These factors collectively accounts for high levels of unemployment, under employment and poverty among young people.

Such poor prospects for young people originate in part from the stagnation of rural economy which can only be tackled through promotion of identified activities that bring greater economic diversity and dynamism to rural areas. This will also provide employment and income-generating opportunities for young rural people across the socio-economic ladder. Tapping into the economic potential of the youth force will ensure a tremendous opportunity to help drive this rural transformation and reduce rural poverty among youths and adults alike.

Women Spearhead Agri Business

Smt. Basanti Panna is a progressive farmer from Khuntkewal in Chatra district of Jharkhand. She possesses a farm with land holding of 2.5 ha. She is a well known personality in the field of agriculture and social work. In the year 2006-07, she received training in 'Vermicompost making and integrated farming system' from KVK, Chatra. She established the units with the support of line departments and since then no looking back. At present she is engaged in commercial vermi compost production and also imparts training to women SHGs. Annually she earns around Rs 150000.00 from vermi compost. She has also adopted integrated farming system (cow-piggery-duckery-fishery-crop production) and running it profitably.

Smt. Chhabita Pramanik, a farm woman of Bankura district of West Bengal took initiative in the year 2001 to form a self-help group with 14 women. Rangamatichar Bhachati Swanirbhar Dal, her SHG was involved in the activity of parboiling and frying of rice. Later she came in contact with KVK, DRDC and other govt. organizations to get training, inputs for vegetable cultivation and preservation, farm machinery, etc. In the year 2009 she formed a cluster of 10 self-help groups under a single umbrella to motivate them to go for dairy, goatery, mushroom, fruit and vegetable cultivation alongwith household preservation. From 2010 onwards she intensified agricultural activities and with the support of KVK Bankura she successfully cultivated paddy through SRI method, groundnut and kharif onion. She also was involved in MGNREGA programme to produce 4.0 lakh saplings of forest trees under the supervision of KVK. Her effort created 4.5 lakh mandays under MGNREGA cell and earning of Rs.4.0 lakh as profit for her groups. Her contribution has been recognized by Govt. of West Bengal, in awarding with Krishak Samman Puruskar, to the only woman farmer of the state and Mahindra and Mahindra for Mahindra Samriddhi India Agri. Award.

Youth Employment, Poverty Reduction and Food Security – the Intermingled Linkage

The prospect of finding gainful work is often limited for the youths living in rural areas. With little or no access to land, market, finance and education, the rural youths struggle to engage in profitable agricultural activities. The resulting unemployment, deficit in decent work and working poverty among youths pose a serious threat to their livelihood and food security. Moreover, migration of rural youths to urban areas in search of work is making farm population ageing to aggravate the situation further. Without increased involvement of youth in agriculture, long-term shortage in skilled agricultural labour will inevitably occur. Hence, the need of the hour is to arrest/reverse the trend to avoid negative effect on agricultural productivity, output and food supply, which in turn may undermine household and national food security.

For providing much-needed employment opportunities for rural youths, promoting and incentivizing youth participation in agricultural sector are pre-requisites. Apart from addressing unemployment and income-generating issues, it will also fortify food security at the household and national levels. As labour is often the only asset rural youths possess, their participation in the labour market is one of the principal determinants of their well-being. Youth will continue to struggle if employment opportunities are limited and of poor quality. In contrary, decent work empowers youths and allows them to make the necessary investments – in skill acquisition, their health aspect and educational front – to break the vicious cycle of rural poverty.

However, knowledge and capacity are somehow lacking at the Apex level to effectively generate decent employment opportunities for rural youths. Policy framework at national level should account for the creation of both sufficient and decent employment opportunities in rural areas. Organizational and administrative capacities must be strengthened to ensure effective implementation of youth employment activities. On the other hand, youths themselves must be better equipped to make a meaningful contribution to agriculture and rural development. Development of youths should be considered as catalyst for reducing poverty and improving food security.

Reaching Out The Goal

For a country like India to prosper in agriculture, demographic dividend in terms of vast youth energy must be leveraged. Across the entire agricultural value chain-from farming to research, innovation, product development and market participation – human asset should be utilized with creation of decent job opportunities, entrepreneurial opportunities in agri-business and weather creation. In attaining the objective to attract and retain youths in agriculture, education, financial norms, legislation and even intellectual property need to re-orient to convince the youths to looking towards agriculture with pride, profit and prestige.

The following suggested measures would ensure youths involvement in Indian agriculture:

Education

- Value based education on the potential in the agricultural value chain should be the core of agricultural education system rather than to overhaul the pedagogic education curricula in inculcating agriculture.
- Addressing investment in social capital should be another area of

focus. Comparatively low education and literacy among rural youths especially women prevents to exploit the available opportunities for gainful employment in farming and non-farming rural agricultural livelihood.

- Creation of altruism of agriculture from early childhood from simple understanding that food does not come from supermarket but from the farm and the parents should encourage the children to take up farming rather than migrating to urban areas to take up daily labour job.
- Sustain and grow agri-business at the rural sector to empower the youths psychologically and economically. Youths with potentiality may be encouraged to concentrate on off-farm business opportunities at the cost of agricultural produce. Greater involvement of rural youths in agri-business will lead to enhanced participation of youths in farming to transform agriculture into self-sustaining livelihood option.
- Vocational education on agriculture needs re-orientation to address development of skill, creation of awareness on potential of agriculture, development of skill and capacities to sustain and grow the young people's efforts in engagement of agriculture.
- Opening up avenues to obtain financial education, capital and profitability for young people. It has often been observed that rural financial institutions demand collateral (usually land and assets) that young people do not have. Changing the paradigm of access to credit in rural financial institutions will certainly encourage the rural youths to take up agri-based enterprises for their livelihood.

Intellectual Property

A strategic way needs to work out to equip young inventors and innovators to make them understand their rights and that wealth creation starts from ideas and its ownership. Opportunity for the rural youths to enjoy such institutional covering will follow it. This will eventually lead to creation of enabling environment of moving an idea into a 'million dollar' institution.

- Attracting human resource pool of capital will have to focus on market research linking ideas and content rural young people have within the agricultural value chain into products that can be delivered to an already demanding market. This includes a focus on improved rural circles, additional investment, lending, franchise ownership of brands and agro-processing trendy market oriented products.

Enacting Rules and Regulations

To safeguard the involvement of rural youths in a high-risk-oriented sector like agriculture, enactment of rules and regulations is a pre-requisite. This will lead to:

- Ensuring youths have legal redress, a regulatory framework that cushions the risk of engaging in agriculture.
- Minimize restriction of age, formalities of land ownership and use
- Rural youth shows the way

Shri Laxman Das, a farmer of North Andaman through innovative method of bee rearing has enabled a large number of islands families to earn sizeable income from bee keeping. He has been successful in domesticating *dammer*bee. He has made an artificial bee hive with the help of a split bamboo pol. The bees are allowed to build the colony in the bamboo structure. In the colony multilayer arrangements including pollen chamber, food chamber, honey storage chamber and brood rearing chamber can be distinctly observed. A single colony can produce 400-500 gm of honey per year. The indigenous method of domesticating the wild stingless bee assumes significance as such honey is sold @ Rs.1500/- per kg as the honey is having higher medicinal quality.

Sri Kamal Mandal of Baruipur, south 24 Paraganas, West Bengal inherited small farm of 5bighas of land. After finishing his education, he spent a few years in search of employment. He came in contact with a large vegetable farm and was encouraged to start growing cowpea, brinjal, chilli etc. Vegetable has huge demand in Kolkata metro city. He has been engaged in vegetable cultivation for last 3 years in over 3 bighas of land. He cultivates different types of vegetables with integrated approach. He adopted improved package of practices to fetch more income. By doing so, he has been benefited financially as is reflected from his annual return of Rs.1, 98, 000/- from cultivation of vegetables year round as compared to Rs. 1,20,000/- from traditional paddy cultivation. He is being assisted by Sasyashyamala KrishiVigyan Kendra in growing capsicum, broccoli etc.

Sri Amitabh Anand, an educated unemployed youth from Bhagalpur, Bihar is a self made entrepreneur. He is a post graduate (M.A.) and also did diploma in rural development. He has 10 acre of land and grew wheat, paddy, chickpea, mung and vegetables. After completing his post graduation, he started cultivation in his land by own and saw yield was not good as well as input cost was more due to improper management of time and application of inputs. Meanwhile, he came to know about KVK, Bhagalpur and discussed his problems with KVK experts. KVK experts advised him to participate in its training programmes. He participated in a training programme on 'Repairing, maintenance and management of agricultural implements' and learned various agricultural technologies and advanced practices. In 2011, he started his KrishiYantra Bank namely **Vishwakarma Krishi Yantra Bank** in his village Mohanpur and started to provide all agricultural machineries and farm implements to the farmers. In Bihar, mechanization among small and marginal farmers has always been a big challenge, and keeping this in view his work in increasing farm mechanization level is commendable. He also got 'innovative farmer award – 2012 by Bihar Agricultural University, Sabour, Bhagalpur in recognition of his work.

Inclusion of youth as key actors, beneficiaries and equal stakeholders in policy discussion and instituted policies on the future of food, farming and development.

Rebranding and Repackaging of Agriculture

Farming to make attractive, rebranding and repackaging should be done at the basic level for which the following measures need to be considered:

- Appreciating young agripreneurs through award scheme to add impetus to their already existing efforts
- The secondary agriculture would be providing value addition to agricultural products, creating facilities for primary processing and stress management in agriculture.
- Creating awareness and building momentum on enhancing a support framework able to introduce innovation into the agriculture sector. It will enhance a positive attitude among rural youths that business opportunities do exist in agriculture.
- Introducing marketing as short production cycle theory of agriculture that will work as catalyst to ascertain that the agriculture value chain rises/surpasses beyond planting and harvesting.
- Proving agriculture has high fiscal value through the assets approach and also emphasizing the regular income potential for young people who want to settle down and have families and be rest assured that they can provide financially for their households.

Tapping Opportunities in Non-farm Sectors

- Custom hiring and service centres are fast becoming career options for rural youths. With increasing demand for farm machineries, its sale and maintenance makes a good business sense. Besides, transportation, marketing of inputs, disposal of produce etc. are some other areas that offer scope for gainful employment. In fisheries and aquaculture 'Aquaculture Service Centres' offering soil and water testing, PCR analysis, feed testing etc. have become quite popular.

- Post harvest value addition of agricultural commodities; its trading through 'e-chaupal' and other means offers attractive opportunities for youth to venture into.
- Entrepreneurship development of youth in primary as well as secondary agriculture would go a long way in ensuring that the youths get self employment.

CONCLUSION

If agriculture to prosper and sustain in long run, it needs involvement of rural youths for carrying forward the noble goal to feed the country. However, present mind-set towards agriculture as a profession hardly presents any lucrative future – both socially and economically – to the rural youths. With the narrowing scope to absorb vast majority of rural youth in non-agricultural sectors, alternate way out must be devised to tap the enormous youth energy into a productive chain. Starting from re-look at the present course curricula at primary, secondary and graduation level, more and more opportunities need to be created for self-employment. At the later stage in order to motivate young people into agri-business and retain those who are already in, more attention by the government must be paid to provision of agricultural insurance package for young investors. Likewise, easy access to start-up capital has to be provided by the financial institutions to encourage entrepreneurship development among rural youths. Changes in existing rules and regulation of ownership of land will add further impetus to the rural youths to shape their own agriculture with zeal, enthusiasm and confidence. Modernizing agriculture by employment of hi-tech equipment and machinery in production of crops and rearing of livestock, processing and marketing would act as incentive to young people. Utilizing computer and ICT tools in the farm operations would force the society to change its outlook and perceive that rural youths engaged in farming are no different from others engaged in elsewhere. Lessons learnt from previously launched programmes like agri-clinics and agri-business centres will help in renewed/modified approach in creating young entrepreneurs. Enabling policy framework and its implementation with dedication can only attract and retain the rural youths in farm sector.

REFERENCES

1. BBC News: South Asia 2011
2. ILO (2013). ILO Global Employment Trends for Youth 2013. Geneva. ILO

3. Information material October 2013. FAO promoting decent employment opportunities for rural youth. Retrieved from http://www.fao.org/docrep/018/i2976e/i29768
4. Lawrence Ndambuki Muli (2014). Youth, the Green and Blue Revolutions and Agriculture Transformation: African Observatory for Policy Practice and Youth Studies. Retrieved from http//allafrica.com/stories/201409041270.html
5. UNDESA (2013). UNDESA World Population Prospects, the 2012 revision Key findings and advance tables. New York. UNDESA
6. Swaminathan M S (2013). Harnessing the demographic dividend for agricultural rejuvenation. Retrieved from http://www.fao.org/fieldnim/templates/ERP/docs 2010/MSSwaminathan.doc etc.

CHAPTER - 2

Role of Youth in Family Farming

Vijay Kumar[1], [2]Jyoti Manjusha, [3]Ranjana Sachan and Y.K Verma[4]
[1]T.V.O, Jharkhand Government; [2,3,4]Ph.D Scholar, IVRI, Bareilly.

ABSTRACT

The family farm exists as one of the most important factors in food production in India. Rural areas in the country are home to 73 per cent of India's population, most of whom rely upon agriculture for their livelihood. Whole families run the farms and, because of the rural location, are less educated than their urban countrymen. Youth are increasingly losing interest in agriculture and migrate to urban areas in search for job opportunities in other sectors, yet in some countries opposite trends were observed recently. Access to adequate productive resources, goods and services, and job opportunities would convince youth to stay closer to the family farm. Access to education would be essential for youth to be able and willing to stay in rural areas and engage in agriculture. Young Farmers' Organizations can effectively represent the needs and interest of rural/agricultural youth in policy making processes. Women are the backbone of family farming. They play multiple roles, undertaking both productive and reproductive activities in agricultural/rural households. Women's large contribution of work in family farms is not recognized, in terms of income earned and access to productive resources and assets. Families in agriculture have multiple vulnerabilities. Both women and men need adequate access to productive resources to make rural societies more resilient. If women get equal access to resources, they could improve their yields which would reduce food insecurity of agricultural/rural households. Government should develop policies, programs and projects that target specifically young farmers and provide them with privileged access to land, credit and technical knowledge. Improve educational instruments and support alternative models for knowledge transfer in rural areas, e.g: agro-ecological education in local schools. Youth should actively participate in farmers' organizations in order to let their voice heard and better access agricultural resources - including credit and finance. Young farmers should pay special attention to the conservation of natural resources. Governments should invest in young farmers for tomorrow's food security.

Key words: Youth, family farming, credit, women, agriculture

INTRODUCTION

Family farming includes all family-based agricultural activities, and it is linked to several areas of rural development. Indeed, 70% of global food supply comes from more than 500 million family farms, which are supporting rural employment in many countries (FAO, 2014). 2014 has been proclaimed as the International Year of Family Farming (IYFF). Up to 40% of people worldwide work in agriculture and the vast majority of them are family farmers – smallholder or subsistence farmers predominantly producing food to eat themselves, as well as to sell. Collectively, these farmers are responsible for producing over 70% of the world's food, so they play a central role in food security. And as the world population increases, the reliance on smallholder farming is set to grow. Agriculture has an image problem. There is a decreasing interest among youth in entering agricultural related fields due to the persistent perception of agriculture as an outdated field with minimal financial returns. But it remains true that youth face large challenges accessing land and finance without collateral. Youth make up about one fifth of the population of developing and emerging economies and face global unemployment levels from 10-28%. The number of young people of working age is increasing while this same group continues to shy away from careers in agriculture. Over 60 per cent of the world's rural population is made of youth, with half of them being young women and girls. Due to their limited access to assets (in particular land), markets, finance and education and skills training, rural youth are often unemployed or work informally – often in unpaid, very low-skilled, insecure and sometimes hazardous jobs. If young people living in rural areas do not find enough incentives, profitable economic opportunities and attractive environments in which to live and work, they will continue to migrate to cities. Young people have enormous potential for the innovation and risk-taking that is often at the core of growth and development in rural areas, particularly in smallholder agriculture. Young farmers and producers often have greater capacity for innovation and entrepreneurship than older adults. This capacity may better equip them to address the emerging requirements of agriculture and the rural non-farm economy.

12 August 2014 is International Youth Day and, coincidentally, 2014 is the UN's International Year of Family Farming. This constitutes an opportune time to focus world attention on the central role that family farming plays in eradicating hunger and poverty, and the equally important role that young people play in the

establishment of a thriving and sustainable food security system. When it comes to food security, one must think about sustainability not only in terms of the environment but also people – creating the next generation of farmers. Yet the family farm, in developing countries especially, is under threat.

Youth support family farming as well and generally carry out gender-specific activities as well. Young women help their mothers in production and care-giving tasks, whereas young men help with or take charge of some of the activities left undone by the absent father. In both cases, they also seek paid employment to contribute toward the family's income without abandoning the domestic chores or other tasks at the farm.

The work of each member of the family is therefore crucial. Gender specific activities are determined by the roles that the rural society considers as either masculine or feminine. Men normally spend more time on farm activities, crop and equipment maintenance, animal herding, the purchase of supplies and sale of farm products. Consequently, they are normally considered the head of the household.

Today's youth are tomorrow's family farmers, and maintaining interest in farming as a profession is vital to future food security. Youth make up roughly one fifth of the population of developing and emerging economies and face global unemployment levels from 10 to 28 percent. The number of young people of working age is increasing while this same group typically rejects careers in the agriculture and food system. But we can change that and cultivate the next generation of agricultural leaders—not just farmers, but food entrepreneurs, scientists, agronomists, extension agents, union and government leaders. Governments, particularly in developing countries, but also elsewhere need to invest in policies and practices that provide access to land, credit and banking services, education and knowledge, and technical skills for young farmers. And these governments need to ensure that young people have access to markets, goods and service, employment opportunities, and leisure so that they want to stay on the farm.

One of the reasons for this is youth migration. More and more young people are attracted away from rural communities into cities by the promise of jobs, better housing and access to hospitals, shops, even entertainment. While a career in the city is commonly embraced as a step forward, life in agriculture is often perceived as a step back. Ever fewer youth work in family-run smallholder farms. The UN forecasts that by 2030, 60% of the global population will live in cities and up to 60% these people will be under the age of 18.

Although young people may believe there's no future in smallholder farming, quite the opposite is true. Global food demand is set to increase by 60% in 2050 (based on 2005 levels), making effective, efficient and sustainable food production more important than ever. Perhaps the greatest challenge we face, therefore, is educating young people about the opportunities that exist in agriculture and giving them the necessary education skills and tools to make a successful future for themselves. In a nutshell, we need to make youth and agriculture a priority as a part of tackling food security.

To be successful, farming must also be as science-based and technology-focused as work in the city. Even in remote regions, many families now own a mobile phone. Mobile delivery of agricultural information to farmers – often using voice messages – can help smallholders grow better crops to feed their families or to sell, for example.

Youth are increasingly losing interest in agriculture and migrate to urban areas in search for job opportunities in other sectors, yet in some countries opposite trends were observed recently.

- Access to adequate productive resources, goods and services, and job opportunities would convince youth to stay closer to the family farm.
- Access to education would be essential for youth to be able and willing to stay in rural areas and engage in agriculture.
- Young Farmers' Organizations can effectively represent the needs and interest of rural/agricultural youth in policy making processes.

RECOMMENDATIONS

- Develop policies, programs and projects that target specifically young farmers and provide them with privileged access to land, credit and technical knowledge.
- Improve educational instruments and support alternative models for knowledge transfer in rural areas.
- Youth should actively participate in farmers' organizations in order to let their voice heard and better access agricultural resources - including credit and finance.
- Young farmers should pay special attention to the conservation of natural resources.
- Governments should invest in young farmers for tomorrow's food security

CONCLUSION

Farming will not be sustainable if the next generation does not want to do it. To feed nine billion people in 2050, we have to adopt a long-term approach and ask ourselves how to attract young people into a profession that they may consider outdated and unprofitable, starting with the family farm. To ignore this issue is to ignore a hugely important aspect of food security. The major problems that inhibit youth participation in agricultural activities were lack of commitment, lack of logistic support and lack of land ownership. It was recommended that a robust relationship between agencies interested in encouraging youth involvement in agriculture should be evolved through legislation and implementation of policies to guarantee training programmes, credit facilities and land accessibility to youths at the identified rural youth organizations. This will enhance youth involvement and catalyze agricultural development.

REFERENCES

http://www.fao.org

http://www.economistinsights.com

Contd...

CHAPTER - 3

Role and Aspirations of Youth in Family Farming

M. Bhavya Manjari
Research Associate, Krishi Vigyan Kendra, Rudrur, Nizamabad District, Telangana

ABSTRACT

Agriculture is an important sector of Indian economy that contributes about 17% to the total GDP and provides employment to over 60% of the population. But, the villages are becoming poorer and poorer and only older generations are left there to do the agricultural operations. During the past two years, cost of food items like grains, pulses, oil and vegetables are increasing, month after month. Family farming is the best approach to produce agricultural products economically, utilizing locally available resources and crop residues and manpower very efficiently and sustainably. Family farming is a way of life" - there are unique advantages - the security of access and control over resources, meeting multiple needs of the family – food and income, the relationships with nature, the bonding with the past, present and future; ensuring freedom from external exploitation as well as freedom to do your own way. Family farms as multipurpose enterprises have been serving as buffers in times of disasters, as glue of social bonding, often functioning as a 'value web' between farming families and communities. The United Nations defines youth as young men and women from age 15 to 24. The present situation of rural youth in developing countries can generally be characterized by illiteracy, lack of educational and training opportunities, unemployment, under-employment and the flight of large numbers of rural youth to the cities. There are serious challenges too – stifling and non supportive policy environment, lack of extension system to handle the farmer's distinct needs and capacities, and lastly 'converting' family farms to be 'business like' models. However, successful pilots can become big impact programmes, for instance, as in Andhra Pradesh, through NGO-GO collaboration. Hence, rural youth need to be prepared in: (a) improving their capabilities to produce food and to conserve productive resources in the rural environment, (b) improving skills and abilities in undertaking income-generating activities in rural areas, (c) relieving population pressure and improving nutrition and the well-being of farming families and (d) developing leadership and the ability to work well with others in group and community situations.

Key words: Aspirations, family farming, resources, business, youth

INTRODUCTION

In her review of the literature on rural youth in the US, MacBrayne (1987:1) defines aspiration as: "an individual's desire to obtain a status object or goal such as a particular occupation or level of education", while expectations are: "the individual's estimation of the likelihood of attaining those goals, plans, ambitions or dreams". Attainment is then described as "the behaviouralrealisation of the goal" (MacBrayne, 1987:1).

Globalization and WTO have resulted in the youth of our country losing interest in agriculture and going into cities to earn more money. The villages are becoming poorer and poorer and only old people are left there to do the agricultural operations. During the past two years, cost of food items like grains, pulses, oil and vegetables are increasing, month after month. China which has the second highest rate of economic growth and the first highest population becoming consumerist may need huge amounts of food and other items for which India has to pay a huge price to provide food security for its population. Hence, it is very important for the administration and the people to give top most importance for agriculture production. But this does not mean that huge external input oriented agriculture production.

Family Farming

Family farming is the best approach for any country and particularly for India, to produce agricultural products economically, utilizing locally available resources and crop residues and manpower very efficiently and sustainably. But all of us think that industrial type of agriculture could produce food economically. In reality, external input agriculture needs and consumes fossil fuels which are very scarce and very soon may not be available.Again use of any external inputs causes global warming due to long distance transport. Hence, younger generation has to think twice before going away to cities to get higher wages. If integrated system of agriculture is adapted like tree cropping, animal husbandry, agro forestry and food production, which are interdependent and self supporting, farming, can be economically viable and sustainable. Because of "green revolution" technology lands have been degraded due to loss of humus, resulting in poor soil water holding capacity and the destruction of soil organisms and soil regenerating processes, demanding more and more agro chemicals. Punjab and Haryana are the examples of destructive external input agriculture. Fortunately, youth are getting into agriculture in a small way. But, for them it is very important that small scale farming or family farming is very important to be free from external inputs and in getting their soils become more productive year after year. Agriculture should not be compared with other professions only on economic gains.

Rural Youth

In the less developed regions, the rural sector makes up approximately two-thirds of the population. More than half are children and youth. Current estimates indicate that the youth age 15 to 24 comprise 19 percent of the rural population. When 10 to 14 year olds are included in the youth category, this percentage increases to 32 percent of the rural population.

How to Help Youth Thrive in Staple Agriculture

i. Comprehensive & full cycle capacity building for target youth (needs assessment, trainings, awareness creation, implementation, monitoring & evaluation). The content must be relevant, up to date and practical.

ii. Diversification of youth involvement in the Sector (Look at the Staple Food Value Chain).

iii. Prioritize socio-economic benefits and strategies to achieve them (It is about their aspirations) and therefore there must be a win - win situation.

iv. Training & Implementation Support even on pilot basis (Technical & Financial Support).

v. Social interaction & entertainment should be mainstreamed (music, skits,drama, & sports as tools for agricultural extension services to farmers).

vi. Comprehensive policy that would address full participation by youthin the entire process, training government officers to specialize in working with youth, collaboration with other youth service providers (NGOs, Academic & Research Institutions and Private Sector) adequate funding must be secured for both short & long term programmes.

vii. Creativity, initiativeness, responsibility, self-reliance and involvement of other key segment of the society is important (parents & community leaders).

viii. Benefits & Incentives to motivate youth and their leaders.

ix. Affirmative action for disadvantaged youth (out of school youth)such as specialized training packages and access to financial and marketing services.

x. Increased & Diversification of funding for youth project.

In the area of empowerment of family farmers, equal attention should be paid to the women and men in the farm family. Women play a critical role in all aspects of agriculture, but invariably their intellectual role and managerial skills remain unrecognised. The IYFF affords a unique opportunity to engender all agricultural policies and programmes. To provide a new deal to family farmers, we need to attend to the following four areas of importance to sustainable food security and elimination of hunger.

(A) Conservation

Family farmers have been at the forefront of bio-resources conservation. In crops like rice and wheat, they have conserved large numbers of local varieties and land races for public good at personal cost. For example, there are now over 150,000 varieties of rice in the world. Much of this genetic diversity constitutes the contributions of women and men operating small holdings. The conservation ethos of family farmers is based upon criteria such as cultural diversity, culinary diversity, curative diversity and ecosystem diversity. The FAO Treaty emphasises the need to recognise farmers' rights in the field of genetic resources conservation and enhancement. The government has also introduced an integrated legislation titled Protection of Plant Varieties and Farmers' Rights Act (PPVFRA) thereby emphasising the need for concurrent attention to breeders' and farmers' rights. The PPVFRA of India also confers the genome saviour award on rural and tribal communities who have conserved important genetic material. Similarly, Biodiversity Authority of India has instituted breed saviour awards to recognise and reward the contributions of those who are conserving local breeds of farm animals. We should do everything possible during the IYFF to recognise the contributions of family farmers to genetic resources conservation and strengthen their tradition in the areas of in-situ on-farm conservation as well as ex-situ measures like sacred groves.

(B) Cultivation

Family farmers are usually eco-sensitive and also take steps to adopt low risk agronomy. It is important for small farm families to adopt the evergreen revolution pathway of agriculture improvement, that is, increase in productivity in perpetuity without ecological harm. The smaller the farm, the greater is the need for marketable surplus. Organic farming can also be promoted wherever efficient methods of soil health and plant health management can be developed and introduced.

(C) Consumption

In the area of consumption, there is great opportunity for introducing agricultural and horticultural remedies for the prevailing nutritional maladies. Naturally occurring biofortified crops can be promoted. Greater attention is needed for designing farming systems based on nutrition sensitive agriculture. Our aim should be to make every family farm a biofortified farm. The UN should also declare one of the remaining years of this decade as the international year of underutilised and biofortified crops. A food-based approach should be adopted for overcoming micronutrient malnutrition.

(D) Commerce

Ultimately, farmers' interest in farming is based on opportunities for assured and remunerative marketing. India made huge progress in agriculture from what was known as a "ship to mouth" existence to the status of conferring the right to food with homegrown food. This was possible because of public procurement at a remunerative price. Unfortunately, there is a controversy in the World Trade Organisation about the support extended to Indian farmers for ensuring the achievement of Mahatma Gandhi's vision of a "hunger free India". Providing family farmers with adequate financial and scientific support to ensure food security should be the bottom line of all food and agriculture policies of developing countries. Food security should remain non-negotiable, especially in countries where more than 50 per cent of the population depends on agriculture for their livelihood. In countries like India, agriculture is a backbone of the rural livelihood security system. This position should be recognised and respected in WTO negotiations. While trade distorting subsidies and support should be discouraged, life saving support should be encouraged.

The zero hunger vision can be converted into reality, provided family farmers are enabled to sustain their livelihoods based on the principles of ecology, economics, equity and employment. Family farms should be enabled to become both climate smart and nutrition sensitive. With increasing pressure of population on land and water resources and prospect of adverse changes in climate, decentralised family farming based on gender and nutrition sensitive agriculture is the pathway for food for all and forever.

Chapter - 4

Role and Aspiration of Rural Youth in Family Farming

Akanchha Singh
Institute of Agricultural Sciences –Banaras Hindu University -221005

ABSTRACT

Family farming is the blending of experience, skill, managerial qualities, business strategies and finally economic independence because this type of farming involves the blending of different minds of the family members. There is not any doubt in saying that youth are the of future. This is the time between childhood and adulthood. It is the time when a young individual get out of their dreamy world of childhood and face some of the most complicated realities of the world.Family farming is still facing the problem of urbanization and unemployment. But youth of present scenario are the solution of every problem because skilled youth of rural background can drive future economic growth as they are innovative in bringing and applying new ideas on a large scale. Rural youth are enthusiastic and they can act as good leaders as their orientation is scientific. Youth support family farming as well as perform gender specific activities.They are said to be the key to sustainable and innovative development. Investment in rural youth is the key to country's progress and boost economic opportunities. Rural youth of today want that their capabilities should be utilized for producing food and for conserving productive resources in rural environment. Their mental skills make them different from others so Govt. should adopt such policies that mark the correct value of their skill.This research paper emphasizes the role of rural youth in agriculture and family farming and what are their aspirations from this sector.

Key words: Family farming, Urbanization, Innovative development, Aspirations

INTRODUCTION

It is rightly said that the real India lies in its village,agriculture is the soul and farmers are the backbone of those villages.Young brains are the most important connecting link between innovative ideas and the traditional thought. Young persons in a family welcome innovative ideas and they always make themselves ready in order to apply that new idea effectively. Since family farming covers various elements from differentpoints of view like from sociological perspective family farming is associated with family values such as solidarity, continuity and commitment, in economic terms,family farming is identified with specific entrepreneurial skills, business ownership and management, choice and risk behavior, resilience and individual achievement. So a rural youth play an important role in each and every perspective of family farming. These papers are a review of a decade and highlight some of the important points related to the aspirations of rural youth and how they can be efficiently utilized in family farming.

Rural Youth and Family Farming

Youth is often understood to be the period of transition from childhood to adult hood. Government of India officially defines youth as persons between the ages of 13 & 35 years.Rural youth differs from urban youth in many aspects. Rural youth have limited access to educational programs that are geared to their situation and need not surprisingly, many rural people drop out of school at an early age. Hence most of the rural youth are not attracted towards school and involve themselves in agriculture. Family farming offers an effective and economic solution to help meet the challenges of making sure that each person has access not just to calories but to nutritious food, also. Traditionally familyfarming is practiced in India but the real concept of family farming are not getting its way. Family farming demands young and catalytic brains but these brains are shifting toward urban areas for doing under rated jobs. One very big and alarming question for all of us that why rural youth are not opting family farming and why they are moving toward certain under-rated jobs.

Current Situation of Rural Youth

India is having world's largest youth population and it is a positive point for any country if that country is utilizing the human resources efficiently. Young employment and underemployment is prevalent around the world because young people lack skills, work experience , job search abilities and financial resources to find employment (United Nations, 2003). Although a large proportion of young rural women are employed in agriculture,rural males are increasingly turning to the non – farm sectors (Mitra, Arup;Verick Sher, March 2013). There are some factors which are deviating our rural youth from agriculture to another sector.

1. Poverty–Young Indian face major problem of poverty. Even after more than 50 years of independence India still has the world's largest no. of poor people in a single country. Of its nearly 1 billion inhabitants, an estimated 260.3 million are below the poverty line of which 193.2 million are in the rural areas. More than 75 % of poor people reside in villages (www.azadindia.org). Poverty makes them incapable to utilize their full potential.

2. Lack of productive employment opportunities - Unemployment is a less affordable option for people living in rural areas, where most young workers have to accept any job in order to survive (www.fao-ilo.org). This situation put a question mark in front of all our rural people that whether they get the value of their talent and skill.

3. Inadequacy of social protection schemes – Young people in Indian economies have little financial support outside their family and friends. Simply copying publicly provided and financed program from rich countries will not do the trick in many cases. It requires a more comprehensive approach which draws attention to many more risks, and which possess many more instruments of dealing with diverse risks, than by social protection. Social protection schemes play a very important role in capturing rural youth attention and make them able to do their jobs even in tough situations.

4. Inadequacy of active labor market policies in rural areas– The main aim of active labor market policies are developing employment related skills with measures such as training, providing efficient labor market with employment services, job matching and counseling. But in Indian condition such policies are lacking and this is one of the biggest reasons of the migration of rural youth from rural to urban areas.

5. Farm–related safety hazards – Each year, more than 2 million youth under the age of 20 are exposed to farm related safety hazard in USA however in Indian condition this problem is still not prevalent because like America our youth are not that much technically upgraded.

6. Illiteracy –Illiteracy is common among rural youth. However, two kinds of people are found in rural areas one who are achievers because they want to achieve high in this field and another are stayers whose achievement motivation is quite low in comparison to achievers. But the population of stayers is more in comparison to achievers. Because of the financial constraints and other facilities they do not take interest

in getting higher education and secondly enhancing illiteracy rate of India.

7. Customary social system – Since head of the family take the necessary decision, a rural youth can not take independent decisions and most of the times their ideas can not be efficiently utilized. Social systems have rigid rules and boundaries which always hinders a rural youth's aspiration in a negative way.
8. Lack of recognition and awards – Recognition and awards are the source of extrinsic motivation but rural society or rural people lack these kinds of motivation.This is also one of the most prominent reason why an individual does not want to stay in family farming.
9. Migration – Rural youth desires and aspirations compel them to move toward urban areas. Opportunities in urban areas for employment, education, etchave been a pull factor attracting migrants from rural to urban areas. Migration stream out of rural to urban areas is approximately 20 million. (www.censusindia.gov.in) according to 2001 census. Rising disenchantment with the profession pushes them out of agriculture while opportunities in other sector of the booming economy pull them out of agriculture (Sharma, Amrita; Bhaduri, Anik).
10. Fast moving media and Communication Technology - Modern era is the era of ICT and related technologies. ICT play an important role in bringing technological revolutions. Rural areas are still devoid of such innovative technologies so as rural youth. A rural youth always aspire for such technological upgradation and revolutions.

METHODOLOGY

This paper is a literature review of a decade and unfolds some of the major challenges of rural youth and what are the expectations and aspirations of rural youth from this sector.

RESULTS AND DISCUSSION

This review brings out some of the hidden and most important points related to role and aspirations of rural youth by understanding the current situation of rural youth. Research evidence have pointed out that the aspiration is one of the psychological factor that influence the youth's decision for a career in the future (Rojewsji, 1999; Sherwood, 1989). Rural youth are important for India because their aspirations and opportunities are also the indicators of the characteristics of adults in near future .We cannot think about the future of family farming without taking on board the perspective of the youth. Rural youth are innovative skilled, enthusiastic and can support family farming efficiently. But due to above mentioned situations their talent and skill cannot be efficiently utilized. The greatest challenge we facetherefore is educating young people about the opportunities that exist in agriculture and giving them the necessary education skills and tools to make a successful future for themselves.

1. **Education** - A rural youth first of all aspire for a good education, various constraints make them unable to find their appropriate goal, and one important factor is good education. An educated rural youth can bring innovative idea as well as work on that. For improving rural social condition and for breaking traditional trends in family farming an educated rural youth is the need of the hour.
2. **Occupation** - A good and reputed occupation is the dream of every individual and the rural youth is not a different case. Family farming is not preferred by rural youth because it lack glamour and recognition and a youth is highly motivated for recognition. So in order to make them stay in this field agriculture should come at first and foremost bench.
3. **Economic Gardening / Local work opportunities** -A rural youth always seek opportunities.One approachto rural economic development is "economic gardening", where local entrepreneurial talent is nurtured and supported by investment and gifts of land on which to build a new business. (Howley, Caithelin; Hambrick, Kimbley 2011). Thus by giving this opportunity their talent and skill can be efficiently utilized.
4. **Flexible and supporting policies** – A rural youth aspire for such policies which motivate them to choose agriculture as career and make them sustain there and at the same time solve their problem of unemployment.
5. **Commercialized Farming** -Agriculture is still used as a means of subsistence. There is a need to commercialize this sector so that every farmer related to this sector gets their share as well as the correct value of their hardwork.Comprehensive agribusiness education programmer are scarce and not easily accessible for rural youth and promoting agricultural opportunities relate not only to economic growth but also to overall domestic productivity and stability in agricultural based countries. This will definitely promote our rural youth and ensure their better condition.
6. **Rewards and recognition** –Rewards and recognition are the source of extrinsic motivation so a rural youth should be awarded for their

activities and works. A reward of any kind makes an individual feel better and enhances his/her confidence level.

7. **Independence in terms of finance and rights** – An individual always aspire for independence in terms of finance because financial independence make them able to do their job efficiently while independence in terms of rights make them able to take decision efficiently along with full self confidence.

8. **Technological Revolution**– A rural youth wants to be technically upgraded because the present as well as the future world is the world of technology.ICT is now recognized as a technological tool which can serve as a catalytic intervention in respect of transforming the lives and livelihoods of rural families. The economic and income divides between urban and rural areas can be overcome only by technological upgradation of rural professions.

RECOMMENDATIONS

This review flag out some of the necessary recommendations which are as follows

1. Policies considering needs and importance of rural youth should be made so that a rural youth get their basic needs as well as they aspire for other higher opportunities.

2. Social protection schemes should also be promoted so that a rural youth get necessary support even in difficult situations and they do not feel inferior in family.

3. Good education is the pre–requisite of social development so education is the first and foremost right of any individual. An educated person knows their right better in comparison to other illiterate persons.

4. Local work opportunities or economic gardening should be promoted so that economic opportunities can be created even at local level. This will stop the migration of rural youth to urban areas as well as provide various economic opportunities at local level.

5. There should be active labor market policies which can improve the condition of rural youth. Market is an interface between money and produce so active labor market policies make them alive in terms of their value in the market.

6. Commercialized farming should be promoted so that the farmer get appropriate share of their efforts. Commercialized farming means every aspect is dealt for commercial purpose i.e. for profit motive.

7. Rewards and recognition programmes should be conducted periodically so that a youth aspire for better and best opportunities because rewards and recognition are the source of extrinsic motivation.
8. Technology should be in reach of rural youth so that they can tune with the rapid and latest technologies and the related change.

CONCLUSION

India is having world's largest population of magnificent youth. These human resources should be efficiently utilized for economic betterment.Government policies need to address the issues of low productivity employment among the rural youth. The challenge is both in terms of job creation andpreparation of young people for labor market. Some efforts should be done by rural youth and some efforts should be left for Government. The only thing is to rejuvenate rural youth with confidence in their abilities and the possibilities of their sustenance in rural India.So a youth can bring change and not only change but can create history if better facilities and opportunities are provided.

REFERENCES

1. Howley, Caithelin, Kimberly Hambrich. 2011. Attachment and Aspiration: What Influences Rural Youth's Educational and Residential Plans? ICF International
2. http://www.azadindia.org
3. http://www.censusindia.gov/indiastats2015.htm
4. http://www.fao-ilo.org
5. Mitra, A. and Verick, S.2013. Youth Employment and Unemployment: An Indian Perspective, International Labor Office, Geneva.
6. Sharma, Amrita; BhaduriAnik: The "Tippling Point in Indian Agriculture, Understanding the Withdrawal of Indian Rural Youth.
7. United Nations. 2003. World Youth Report 2003: The Global Situation of Young People, New York.

CHAPTER - 5

Exploring Youth Aspirations for Agriculture

A.S. Tigga
Asst. Prof.-cum-Jr. Scientist, Department of Extension Education
Bihar Agricultural College, Sabour, Bhagalpur

INTRODUCTION

It is becoming obvious that the interest of youth in agriculture is declining and this trend is likely to continue if we do not look into this seriously. A large number of young people are migrating to urban centers in the hope of getting employment and for a better life. This is due to lack of jobs and opportunities in rural areas, and a view that working in farming is low status and does not offer good future prospects. Young people play an important role and are required in the agriculture sector. But education or training programs are not always readily available, technologies are not accessible, inputs may be scarce, and other vital services are not accessible. Young people now-a-days are becoming less connected with their food, often making little or no association between the food they eat, who is behind it and it is produced. They may not see beyond the store or market, nor make the connection between the food they are eating and the efforts made by farmers to grow it. This growing disconnection with agriculture implies an uncertain future for sustainable food production to feed the growing population. It is important to make the link between food and agriculture visible. The important role young people play in farming needs to be understood as agriculture is sustainable if it can attract future generation farmers. In fact the young people are the future farmers. Efforts are required to discourage the overwhelming tendency of young people to move out of farming in search of a more comfortable life and better income. It is true that there are major challenges, farming is not an easy or a more glamorous option, but we should be exploring ways of doing agriculture in a sustainable and rewarding manner.

Conditions are increasingly challenging. There is less land, markets and other facilities may not be readily available. So it is a really important challenge for us to find a space in the agricultural sector for young people. There is a general perception that farming is for rural people, non-educated people, people who are poor or aged or who do not really have any other option except to do farming work. But we need to work out how best to change this mindset and create an environment to attract young people into this sector.We are of the view that a central part of this process is to interrogate young people's aspirations and expectations for the future. Aspirations play an important role in influencing how young people make life choices, how they think and feel about themselves and their livesand ultimately their life outcomes. The aspirations of rural youth and the factors contributing to their formation should therefore be of interest not only to the young people themselves, their families and communities, but to all those with an interest in agricultural and rural policy and development. The formation of aspiration – determinants and correlates of aspirations from the literature it is clear that aspirations are personal and dynamic: aspirations may mean different things to different people and they are formed and develop in response to different environments and circumstances. The formation of aspirations tends to begin early in childhood and aspirations are adapted and changed in light of new experiences, choices and information, including an individual's awareness of their own abilities and the opportunities open to them. Social class, socio-economic status and income have been identified as correlates of aspirations, with higher levels related to higher aspirations, although these can be over-ridden by the effects of attitudes and beliefs (Morrison Gutman and Akerman, 2008). In UK studies, social class has been found to have both direct and indirect effects on aspirations, the latter through academic attainment. For young people with a high level of confidence in their academic ability, aspirations remain high irrespective of social class (Furlong and Biggart, 1999). Similarly, young people with a high level of self-belief in their ability to achieve and who attribute success to hard work rather than luck tend to have higher aspirations than their peers. Attainment and aspiration are thus mutually reinforcing and it can be difficult to state with certainty the direction of causality (Morrison Gutman and Akerman, 2008). Gender is another key determinant of aspirations, both in the form they take and how they evolve over time. However, some studies have found girls tend to be less circumscribed by gender stereotypes than boys; girls are also consistently found to have higher educational and career aspirations than boys, in spite of the reality that men continue to achieve higher occupational status, career enhancement and financial rewards (Morrison Gutman and Akerman, 2008). Clearly these patterns are context specific, with outcomes in other socio-cultural settings likely to be different depending on the context of gender norms and gender relations.

Fulfillment of Aspirations and The Role of Expectations

The previous section highlighted the tendency for aspirations to be formed early in childhood but shaped and modified over time by environment and experience. Inirrespective of race and class, and suggesting they were identifying outside their historical and socio-political context. These points to the need to explore cultural variation in the development of aspirations and in the relationship between aspirations and expectations through empirical work in different contexts. The importance of prestige in the formation and fulfillment of aspirations is illustrated very well with respect to migration where, along with economic security and social mobility, the potential for status enhancement via occupation and income can be considerable. This provides compelling reasons why agriculture, for example, may be unappealing as it may not bring status regardless of the economic outcomes (Rao, 2009). In this regard, it is important to note that aspirations are not just about economic opportunities. Degrading work, such as manual labour, performed in other locations does not affect status in the same way as when it is performed in a person's home location, and can allow acquisition of life skills and funds for self-employment and social and kinship contributions that confer higher status. Gender and inter-generational relationships have a critical influence on the acquisition of status and prestige via different occupational routes. In Bangladesh, for instance, women's migration increases their autonomy and confidence but this, and the type of work they do in towns and cities, is not valued by older generations and therefore their families try to restrict their migration in order to retain their own status. This example underlines the need to disentangle the multiple and sometimes contradictory influences on aspirations, including through interrogating the aspirations of parents and peers.

Aspirations and Outcomes

We now turn to the relationship between aspirations and outcomes. There are two important questions here: First, to what extent are aspirations related to eventual outcomes? And, following from this, what are the implications of the aspirations – outcomes link for current policies and practices? Numerous studies from the US, UK and Australia point to strong relationships between young people's educational aspirations and future educational attainment; and occupational aspirations and eventual career choices.

Aspirations and Agriculture

The impetus for developing a stream of research on youth aspirations is the observation that young are reluctant to pursue an agriculture-based livelihood, which may have implications for continent-wide initiatives to revitalize the agriculture sector. In this section we look at the (limited) literature which speaks to youth aspirations in relation to agriculture. Despite being an agricultural country with, at the time, a strong 'back to the land' campaign by government, students expressed no desire to farm and the stated preference for any involvement in agriculture was as an agricultural officer. This has resulted to the fact that agriculture is regarded as an employer of the last resort to young people.

Emerging Hypotheses

There are strong links between aspirations and expectations of young people (and their parents) and their outcomes. These are influenced by the environment close to the individual and the broader societal context, including socio-cultural factors such as gender and social-class, as well as geography. Further, occupational status is an important factor for young people in making life choices. Rural young people tend to have lower aspirations and expectations compared to their urban counterparts and this is reflected in lower levels of attainment. The relative isolation of rural communities and the tendency towards higher levels of poverty and lower socio-economic status play a significant role in this. Supply-side issues in rural education and limited rural labour markets are also importantbarriers both to high aspiration formation and higherlevels of achievement. However, young people can often face an 'aspiration-attainment' gap, where there is a mismatch between what young people strive towards and what their realities (socio-cultural and economic) can offer them. Rural youth are certainly being brought closer to the 'outside world' through increased globalization, urbanization and migration but the interplay between these major forces and the socio-economic and cultural realities of rural life, and the impact this will have on agriculture and the rural economy, are yet to be explored in relation to aspiration formation.

The Review Suggests the Following Facts

- Aspirations are formed against a broader, changing social context and wider changes in society. Both the environment close to the individual and the broader societal context therefore influence aspiration formation.
- Young people's aspirations and expectations in relation to economic outcomes are strongly related to sociocultural influences and the degree of social-embededness, which can exacerbate/ perpetuate entrenched inequalities along the lines of gender.
- Social influences on aspirations tend to be stronger in rural areas because of smaller population sizes, relative isolation and community culture, resulting in social pressures that encourage uniformity and limit student achievement regardless of aspiration and motivation to

succeed.

- There is a tendency for rural young people's educational expectations to be lower than those of their urban counterparts.
- Higher poverty rates and lower socio-economic statusin rural communities negatively impact on the aspiration levels of young people.
- Aspirations are not just about economic opportunity – status is important: agriculture is unappealing to young people because it does not bring status regardless of economic outcomes.
- There is an increasing gap between these aspirations and rural young people's expectations about the style of life and level of livelihood that local rural environments and economies can support.
- Young people in areas of high agricultural growth are likely to be more interested in making farming a central element of their livelihood than those living in low growth areas.
- Young people's interest in making farming an important element of their livelihood will likely be positively related to their ability to put together or gain access to the resources needed to farm on a "commercial" basis (i.e. land, credit, labour etc.)

REFERENCES

1. Furlong, A. and Biggart, A. (1999) Framing 'choices': a longitudinal study of occupational aspirations among 13- to16-year-olds. Journal of Education and Work, 12 (1): 21-35
2. http://www.future-agricultures.org.
3. Morrison Gutman, L. and Akerman, R.(2008) Determinants of Aspirations. Centre for Research on the Wider Benefits of Learning, Research Report 27. Institute of Education: London.
4. Rao, N. (2009) Gender differences in migration opportunities, educational choices and wellbeing outcomes. Sussex: Development Research Centre on Migration, Globalisation and Poverty.

CHAPTER - 6

Youth Involvement in Family Farming

Sneha Singh[1], **Raghuveer Singh Meena**[2] **and O.P. Mishra**[3]
[1,2]Research Scholars [3]Associate Professor
Department of Extension Education, Institute of Agricultural Sciences
Banaras Hindu University, Varanasi- 221005

ABSTRACT

India is a developing country and agriculture is at the core of this development. Youth participation in agriculture for its sustained contribution to a nation's development is crucial but this is not ascertained. FAO has declared year 2014 as the International Year of Family Farming (IYFF). Today's youth are tomorrow's family farmers, and maintaining interest in farming as a profession is vital to future food security. The number of young people of working age is increasing while the same group is typically rejecting careers in agriculture and food system. This scenario can be changed by cultivating the next generation of agricultural leaders—not just as farmers but as food entrepreneurs, scientists, agronomists, extension agents, union and government leaders etc. Governments, particularly in developing countries, need to invest in policies and practices that provide access to land, credit and banking services, education and knowledge, and technical skills for young farmers. And these governments need to ensure that young people have access to markets, goods and service, employment opportunities, and leisure so that they prefer to stay on the farm.

Key words: IYFF, Family Farming, FAO, Youth Participation.

INTRODUCTION

2014 has been proclaimed as the International Year of Family Farming (IYFF). Family farming includes all family-based agricultural activities, and it is linked to several areas of rural development. A family farm is one where the family owns or has land tenure, makes the decisions and takes the risks, and provides much of the labour.

Since the beginning of agriculture, family farming has been contributing significantly to food and nutrition security globally. Indeed, 70% of global food supply comes from more than 500 million family farms, which are supporting rural employment in many countries (FAO, 2014). There are around 55 crore farmers in India (NSSO-2009). Agriculture is the largest unorganized sector in India, with 57 per cent of India's total population and 73 per cent of India's total employment.

Family farms all over the world can be identified by their unique socio-cultural and economic attributes. Today's youth are tomorrow's family farmers, and maintaining interest in farming as a profession is vital to future food security. Youth make up roughly one third of the population of India and emerging economies and face unemployment levels. The number of young people of working age is increasing while this same group typically rejects careers in the agriculture.Young people perceive agriculture as a profession of intense labour, not profitable and unable to support their livelihood compared to white collar jobs offer. They think agriculture would not afford them to enjoy the pleasures of owning a beautiful home, fast cars, the latest gadgets and mobile phones like what their colleagues in white collar jobs have access to.

Against the background of challenges facing family farms, farm youths like other family farm's natural and economic resources, must be preserved. They are the future generation, who will work the land in the twenty-first century. Without their involvement, agricultural development efforts will not succeed. Although youths are noted for their active involvement in family farming (Ugwoke, 2005),

There is a general lack of interest of youths in farming, In some countries, many youths wonder about the viability of a career in agriculture, and yet others perceive farming as less dignifying (Russell, 1993; Malumfashi, 1999). But it can be change the next generation of agricultural leaders—not just farmers, but food entrepreneurs, scientists, agronomists, extension agents, union and government leaders. Governments need to invest in policies and practices that provide access to land, credit and banking services, education and knowledge, and technical skills for young farmers. And these governments need to ensure that young people have access to markets, goods and service, employment opportunities, and leisure so that they want to stay on the farm.

Youths play an integral role in the development of any country and cognizant of the fact, much emphasis continues to be placed on areas youths can further excel, especially in Agriculture.

Increasing Interest of Youth in Family Farming

Access to adequate productive resources, goods and services, and job opportunities would convince youth to stay closer to the family farm. Access to education would be essential for youth to be able and willing to stay in rural areas and engage in agriculture. Young Farmers' Organizations can effectively represent the needs and interest of rural/agricultural youth in policy making processes. Governments should take the necessary steps to make sure that arable government land is only used for agricultural purpose, fairly distributed among youth farmers and that government should put in place mechanisms to help youth entering sustainably into farming. Governments should invest in water conservation/harvesting and irrigation systems in arable areas. Governments should support local seed banks to facilitate access of youth farmers to seeds and seedlings.

A Future for Family Farming and Youth

It was generally agreed that youth have a positive role to play as the future of family farming young people currently operate in a difficult environment with lacks of vital resources coupled with a negative perception, they are increasingly tackling these issues with advocacy, use of technology as well as with their energy and various innovative ideas. We cannot think about the future of family without taking on board the perspective of the youth.

Towards Youth's Forward Thinking

The question is what could be the role of youth in the futures of agriculture. The agricultural development field, are engaging the right stakeholders to drive key policies that not only make agriculture more appealing to the youth, but also strive to give more visibility and recognition to the positive and important roles of youth to the futures of farming, food security and sustainability.

Identifying the alternative futures of agriculture is crucial to recognizing and pinpointing the areas of maximum impact where young people could contribute to a food secure world. Family farms face around the world and on the ways to make research more responsive to family farming.

CONCLUSION

Therefore the relationship between the socio-economic characteristics of the youths and their attitude towards family farming needs to be exploited and understood. Rural youths in family farms require. Technical support and an enabling environment in the form of increased access to agricultural research and extension services, entrepreneurship skills, inputs, credit and market (both local, regional and international). Finally there is need for a deeper and sounder research on this problem to explore it on a wider scale.

REFERENCES

Russell, E. B. (1993). Attracting Youth to Agriculture, Journal of Extension, 13:4. URL: http://www.joe.org/joe/1993winter/a2.html

Ugwoke FO (2005). Youths Participation in Farming Activities in Rural Areas of Imo State, Nigeria. Implications for Extension. J. Agric. Ext. 8: 136-142.

CHAPTER - 7

Rural Youth: The Future of Family Farming

Mukesh Kumar, Aparna Radhakrishanan and Minu Singh
Ph.D Scholar, Dairy Extension Division, National Dairy Research Institute, Karnal Haryana

ABSTRACT

Youth denotes-"Young, Original, Unique, Talented, and Hopeful". Youths represent the most active segment of the population and the engine that do most productive work of the society (Adesope, 1996). Youths are a formidable force in the agricultural production, process, constituting a sizeable proportion of future progressive farmers and better citizens, especially in the rural areas. Fifty percent of Indian population has average age group 25 and more than 65 percent below the age of 35 years therefore United Nations (2005) defines youth as 'all individuals aged between 15 and 24 year' and declared 2014 as the "International Year of Family Farming". Family farming is a means of organizing agricultural, forestry, fisheries, pastoral and aquaculture production which is managed and operated by a family and predominantly reliant on family labour, including both of women and men.

Key words: Youth, family farming, agriculture, age

INTRODUCTION

According to United Nation's General Assembly, "Youth are the persons falling between the ages of 15 and 24 years inclusive". The age-structure figures reveal a tendency towards a young population, where more than 50% population falls below the age of 25. The Census 2011 data reveals an increase from 353 million people in 2001 to 430 million people in 2011. In fact, as predictions go, the youth population of India is set to reach 464 million by 2021. By 2020, a young working population will make up a substantial 64% of India's population in the

Census 2011 data revealed. More than two-thirds of the total population (68.8%) resides in villages while the rest (31.2%) live in urban India. Both in developing and developed countries, family farming is the predominant form of agriculture in the food production sector. Rural youth act as a centre of agricultural, environmental and socio-economical policies in the national agendas by identified gaps and opportunities to promote a shift towards a more equal and balanced development through family planning that aims to raise the profile of family farming and smallholder farming by focusing world attention on its significant role in eradicating hunger and poverty, providing food security and nutrition, improving livelihoods, managing natural resources, protecting the environment, and achieving sustainable development in rural areas.

Agricultural production in India is affected by a number of factors including lack of continuity in agricultural policies, politicized environment in agricultural support institutions, isolation of agricultural education, research and extension wings, unfavourable prices, buyer's and middleman "Mafia", absence of necessary infrastructure for farm exports, deficient management and marketing skills, a large number of small operators and unproductive tenancy systems, etc. These problems may be over come with the help of youth-serving farming and the practitioners.Rural Development in India is one of the most important factors for the growth of the Indian economy. India is primarily an agriculture-based country. Agriculture contributes nearly one-fifth of the gross domestic product in India. In order to increase the growth of agriculture, the Government has planned several programs pertaining to Rural Development in India. An innovative national pilot project to attract and retain rural youth in agriculture, in a bid to tap the youth potential in farming and also prevent their migration to urban areas, is in the pipeline." The prospects of agriculture are getting affected because of the youth shunning farming and migrating to cities in search of menial jobs. The only way of attracting them to agriculture is to turn farming into a profitable venture. The proposed project aims at not only making farming a profitable venture, but also increasing the farmers' incomes by three folds in five years.

According to him, efforts will be made to increase the farmers' incomes under the proposed project through various technological interventions depending upon the local requirements and bringing about convergence among the local research institutes, administrative bodies and farmers. This would also involve organising the youth on the basis of crops being grown by them and promotion of integrated farming.To begin with, about 500 farm youth would be chosen from each centre, which would comprise a cluster of three to four village panchayats. They would be given training in various aspects of farming including hi-tech farming and value addition to crops. The project would also link them with banks to get loans for setting up processing units and custom-hiring services related to farm machinery.

Importance of Family Farming

- Rural youth in Family farming preserves traditional food products, while contributing to a balanced diet and safeguarding the world's agro-biodiversity and sustainable use of natural resources.
- Rural youth in Family farming represents an opportunity to boost local economies, especially when combined with specific policies aimed at social protection and well being of communities.
- It generates food and income for hundreds of millions of rural people, including the poor and marginalized.
- Family farming creates jobs for women, men and young people, both within their family farms and in related enterprises along food and agricultural value chains.
- It provides models of adaptability and resilience for more sustainable food production.

Challenges Faced by Family Farmers

- Climate change and climate variability.
- Limited access to financial resources, inputs, technology, training, research and advisory services and education.
- Lack of tenure security in a context of increasing competition for land and water and inadequate governance of land tenure.
- Price volatility and limited access to market.

Strengths of Youth involvement in Family Farming Agriculture

- Youth having latent energy, capacity and ability to produce
- Excellent source of new ideas and innovation
- Youth creating market for food consumption and knowledge of their own eating peculiarities
- Significant impact on public opinion, policy and action

Opportunities for Youth Involvement in Family Farming Agriculture

- In current scenario vibrant and expanding market for primary and secondary agricultural commodities offers tremendous opportunities for young people to earn good income from agricultural activities

- There are existing incentives and policies that target youth involvement in agriculture which encourages young persons.
- Many tertiary training opportunities exist for youth interested in agriculture

CONCLUSION

The challenge now is to initiate action and strengthen collaborative efforts that bring together the disparate lines of thinking and policy discourse and to enhance the knowledge base on youth and farming in an era of rapid change. This must include the need to develop clear evidence based country-specific strategies for the future of small-scale farming built on inclusive debate on agricultural pathways and long-term choices. Given the present and anticipated future role of agriculture in employment and the sheer number of youth in rural areas today and anticipated into the future, new models to enhance decent employment and livelihood in the agriculture sector must be developed, including support to employment opportunities along the entire agri-food market chain and the associated service sectors.

REFERENCES

http://www.nddb.org/English/Pages/default.aspx (Accessed on 10 October 2014).

United Nations (2011). World Youth Report [online]. UNWorldYouthReport.org. (Accessed on 6 October 2014).

Adesope, O.M(2007). Agricultural Youth Organizations. Introductory Concepts. 2nd edtion.

CHAPTER - 8

Role, Challenges and Opportunities for Agricultural Youths in Family Farming

Sarvesh Kumar and B.K. Mandal
Scientist (Agricultural Extension), JNKVV, Krishi Vigyan Kendra-Harda (MP)
Programme Coordinator, KVK-Saharsa, Bihar Agricultural University, Sabour (Bihar)

ABSTRACT

The United Nations has declared 2014 as the International Year of Family Farming. The central role of increased participation by youth in agriculture is necessary and vital to facilitate food and nutrition security. Rural Family farming generates food and income for hundreds of millions of rural people, including the poor and marginalized fostering jobs for men, women and young people, both within their family farms and in related enterprises along food and agricultural value chains. It provides models of adaptability and resilience for more sustainable food production and is socially and culturally part of communities and territories constituting regions and as such underpins national food security in most countries safeguarding and protecting the environmental assets and natural resources, biodiversity and cultural heritage. Under the right conditions and with supportive policies, family farmers particularly the youth and women have proven to be innovative, productive and resilient, especially if they are supported to use agro-ecological practices that are based on their own resources and priorities. This makes youth in family farming key players in tackling the pressing issues of our time, such as hunger, climate change and biodiversity loss. Family farming includes all family-based agricultural activities, and it is linked to several areas of rural development. Family farming is a means of organizing agricultural, forestry, fisheries, pastoral and aquaculture production which is managed and operated by a family and predominantly reliant on family labour, including both men and women. Rural youth, smallholder and family farmers face numerous challenges in the prevailing times such as climate change and climate variability; lack of tenure security in a context of increasing competition for land and water (population growth, urbanization) and inadequate governance of land tenure; limited access to financial resources, inputs, technology, training, research and advisory services, and education, price volatility and limited access to markets among a few. As an opportunity the availability of real income generating options in agriculture can offer rehabilitating young persons whose alternative options are neither socially acceptable nor responsible. Simultaneously, the vibrant and expanding market for primary and secondary agricultural commodities offers tremendous opportunities for young people to earn good income from agricultural activities.

Key words : Employment, Opportunities, Family Farming, Food Security, Rural Youth

INTRODUCTION

The quest for agricultural and rural development has rural sector in general depends to a large extent on pre- occupation of both academics and policy youth and adult farmers' adoption of improved agricultural practices. The youth have greatly contributed to agricultural and rural development. Their areas of contributions may be on agricultural productions, promotion of people's cultural heritage, effective mobilizations of the people on national issues as well as infrastructural development. Farming youths greatly contribute to agricultural development (provision of farm labour, direct production of crops and livestock, agricultural information to farmers) and rural development. Despite their significant effort in community development, major factors constraining the effective performance of farming youth may be faced, inadequate fund, lack social amenities, non-recognition of youths as partners in rural development (Umeh and Odom, 2011).

Over 60 per cent of the world's rural population is made of youth, with half of them being young women and girls. Due to their limited access to assets like land, markets, finance, education and skills training, rural youth are often unemployed or informally unpaid, very low-skilled and insecure. If young people living in rural areas do not find enough incentives, profitable economic opportunities and attractive environments in which to live and work, they will continue to migrate to cities. This trend would not only contribute to the mega urbanization and growing urban unemployment that is already under way, but is expected to affect global food production and arise a central challenge that who will then feed the global population that is projected to reach 9.2 billion people by 2050.[3]

There is need to invest in "The rural youth of today, the farmers of tomorrow". Investing in young people living in rural areas is key to enhancing agricultural productivity and food security, boosting rural economies, and reducing rural-to-urban migration. Young people have enormous potential for the innovation and risk-taking that is often at the core of growth and development in rural areas, particularly in smallholder agriculture. Young farmers and producers often have greater capacity for innovation and entrepreneurship than older adults. This capacity may better equip them to address the emerging requirements of agriculture and the rural non-farm economy (Nwanze, 2012).

Rural youth do not perceive agriculture as a remunerative or prestigious profession, and until they find meaningful economic opportunities and attractive environments in rural areas, they will continue to migrate to cities. [4]

By 2050 an estimated seven out of ten people worldwide will live in low-income food-deficit countries (Oxfam, 2011).

A majority of rural youths preferred living in urban areas because of economic push factors included poor physical infrastructure and social amenities in the rural areas, search for education and skills acquisition, and the absence of desirable job opportunities. Other factors include a general dislike of village life or expulsion from rural communities resulting from the commitment of an offense. These economic factors are the dominant reason for rural youths increased involvement in non-farm activities and migrating to urban areas (Olayiwola and Echebiri, 2005).

Youth make up approximately one-fifth of the total population in many countries in developing and emerging economy regions. The prospects of rural youth finding decent work in many countries of the developing and emerging economy worlds (Lucchesi, 2012).

The United Nations has declared 2014 as the International Year of Family Farming. This is a very important recognition for the multiple social, economic, environmental and cultural functions of family farming. Rural Family farming generates food and income for hundreds of millions of rural people, including the poor and marginalized fostering jobs for men, women and young people, both within their family farms and in related enterprises along food and agricultural value chains. It provides models of adaptability and resilience for more sustainable food production and is socially and culturally part of communities and territories constituting regions and as such underpins national food security in most countries safeguarding and protecting the environmental assets and natural resources, biodiversity and cultural heritage. Under the right conditions and with supportive policies, family farmers particularly the youth and women have proven to be innovative, productive and resilient, especially if they are supported to use agro-ecological practices that are based on their own resources and priorities. This makes youth in family farming key players in tackling the pressing issues of our time, such as hunger, climate change and biodiversity loss. Family farming includes all family-based agricultural activities, and it is linked to several areas of rural development. Family farming is a means of organizing agricultural, forestry, fisheries, pastoral and aquaculture production which is managed and operated by a family and predominantly reliant on family labour, including both men and women. Rural youth, smallholder and family farmers face numerous challenges in the prevailing times such as climate change and climate variability; lack of tenure security in a context of increasing competition for land and water (population growth, urbanization) and inadequate governance of land tenure; limited access to financial resources, inputs, technology, training, research and advisory services, and education, price volatility (energy, food, etc.) and limited access to markets among a few.

Role of Rural Youths in Family Farming

The increased participation by youth in agriculture and family farming as well is being necessary and vital to facilitate food and nutrition security because;

1. If young farmers do not replace the old age producers the production of food within the country will be seriously compromised in the next 10-15 years. The young farmers have much potential to regulate modern agricultural practices and handling of innovative agricultural implements and tools which confirms the high production and productivity both to feed the mounting population of the developing countries.
2. The youth population is strategic to the success of these efforts to boost regional food production.
3. The poor image of persons involved in agriculture needs to be changed and the young people are the ideal catalysts for such change given their greater tendency and willingness to adopt new ideas, concepts and technology which are all critical to changing the way agriculture is practiced and perceived.
4. Farming youth involvement in agriculture may increase employment particularly rural employment as required. Given the many opportunities available in food production, agriculture can play a significant role in impacting youth unemployment and by extension, unemployment generally in the region or Nation.
5. Increased productivity in the agricultural sector depends on youths who comprise above 50% of the population in the country. The energy that youths possess and their significance in the regional population may provide tremendous opportunities for increasing agricultural productivity.

6. The engagement of rural youths in family farming may reduce crime and other social problems with productive alternatives also. The availability of real income generating options in agriculture can offer opportunities preventing and indeed rehabilitating young persons whose alternative options are neither socially acceptable nor responsible.

Strengths of Youth Involvement in Agriculture or Family Farming

- They are the latent energy, capacity and ability to produce remarkable and targeted.
- Youths are the excellent source of ideas and innovation and areas of interest in the agricultural and allied sector for informative and technical back stopping.
- They are the main market for food consumption and knowledge of their own eating peculiarities
- Youth put significant impact on public opinion, policy implementation and action orientation.

Weakness of Youth Involvement in Family Farming

- Vulnerability to marketing and sales promotion and unavailability of strong marketing facilities at grass root level that sometime de-motivate rural youths to involve in family arming at priority level.
- Large sections of the rural youth's population are untrained whereas modern agriculture requires highly trained youth involvement.
- Sometimes the control into their roles, responsibilities and functions places the youth at the disadvantage because they have to wait for the interventions of adults or old aged involved in agricultural pursuits.
- The negative stigma of pursuing agriculture as a career influences youth to stay away from it and involvement of rural youths at great extent.

Opportunities for Youth Involvement in Family Farming

- There are tremendous opportunities for young people to earn good income from agricultural activities through vibrant and expanding market for primary and secondary agricultural commodities.
- There are existing incentives and policies that target youth involvement in agriculture which encourages young persons.
- Many free training institutes like KVKs and other tertiary training

opportunities exist for young persons interested in agriculture.

Threats to Youth Involvement in Family Farming

- Abnormal behavior may be seen (Crime and violence, drugs abuse etc.)
- Better and luring opportunities outside of the agricultural sector
- Negative social impacts generally seen as failure and dishonored
- Lack of motivation and the feelings of depression
- Poverty and Unemployment
- Lack of education

Recommendations for Increasing Youth Involvement in Family Farming

1. Improved access to training and capacity development

- Improved training at the primary and secondary school level
- Linking School Agricultural Operations with General Curriculum
- More opportunities for on farm training for youth
- Enterprise development training particularly in value added activities such as food processing and packaging

2. Improved and easier access to resources such as:

- Land
- Capital
- Technology and information
- Facilitating access to and encouraging the use of appropriate technology
- Developing a data base of agencies that can provide access to youth on information in agriculture

3. Facilitating market opportunities

- Providing targeted marketing opportunities for primary and value-added agricultural products produced by young entrepreneurs via special arrangements with schools, hotels etc.

4. Providing motivation and improving the image of agriculture

5. Facilitating networking among youth etc.

These interventions will assist with motivating and rewarding youths involved in agriculture which is critical to their involvement while unearthing the many innovations that are certain to surface.

CONCLUSION

To stimulate the interest and encourage the participation of youth in agriculture which is necessary to ensure the sustainability of agricultural production and realize the goal of food and nutrition security. As an opportunity the availability of real income generating options in agriculture can offer rehabilitating young persons whose alternative options are neither socially acceptable nor responsible. Simultaneously, the vibrant and expanding market for primary and secondary agricultural commodities offers tremendous opportunities for young people to earn good income from agricultural activities. Young people have enormous potential for the innovation and risk-taking that is often at the core of growth and development in rural areas. Young farmers and producers often have greater capacity for innovation and entrepreneurship than older adults. This capacity may better equip them to address the emerging requirements of agriculture. There is need of hour to invest in "The rural youth of today, the farmers of tomorrow". Investing in young people living in rural areas is key to enhancing agricultural productivity and food security, boosting rural economies and reducing rural-to-urban migration.

REFERENCES

1. Echebiri, N., (2005) Characteristics and Determinants of Urban Youth Unemployment in Umuhia, Nigeria. A Paper Presented at World Bank Conference on Share Growth in Africa held at Accra, Ghana.
2. G.N. Umeh and C.N. Odom (2011) Role and Constraints of Youth Associations in Agricultural and Rural Development, World Journal of Agricultural Sciences, Vol. 7(5): 515-519
3. http://ifad.org/farmer/2012/youth/index.htm
4. http://www.fao.org/docrep/017/ap668e/ap668e.pdf
5. Kanayo F. Nwanze (2012) Youth in Agriculture, Paper Presented in Conference in Special Session on Farmers' Forum, International Fund for Agricultural Development IFAD Headquarter, Via Paolo di Dono, 4400142, Rome, Italy.
6. Olayiwola, O.F. (2005) Perception of Rural-Urban Migration in Selected Rural Communities in Ondo State, Nigeria. A Publication from Department of Sociology, Adekunle Ajasin University Akungba, Nigeria.
7. Oxfam (2011) Growing a Better Future. Food Justice in a Resource-Constrained World. Available at: www.oxfam.org/grow.
8. V. Lucchesi (2012) Small-scale Farming and Youth in an Era of Rapid Rural Change, IIED/HIVOS, London/The Hague.

CHAPTER - 9

Challenges Faced by Rural Youth in Choice of Career in Agriculture

Renu Gangwar[1] and V.L.V. Kameswari[2]
[1]Ph.D Student (Agricultural Extension and Communication)
[2]Professor,Department of Agricultural Communication
G B Pant University of Agriculture and Technology, Pantnagar, Uttarakhand

ABSTRACT

Agriculture and allied sectors is the largest livelihood providers in rural India, still Indian farmers are among the poorest in the World. Agriculture sector is potentially largest employment providing sectors in the country. Despite this, unemployment is high among educated rural youth, and they prefer to migrate to urban areas to take up low paying jobs. The present study was carried out mainly to find out the challenges faced by rural youth in choice of agriculture as a means of livelihood. The study was conducted in four villages in Udham Singh Nagar district in Uttarakhand. Data was collected from 115 youth using pre-tested interview schedule. It was found that maximum number of respondents had education up to higher secondary level, majority were students and had medium level of mass media exposure and change proneness. The study indicates that limited access to new agricultural technologies, limited access to farm machinery, perception regarding suitability to agriculture, damage of crops from pests and insects, and lack of information on agricultural inputs and government policies were the main challenges faced by rural youth in choosing agriculture as a means of livelihood. The study indicates that government has developed policy framework for involving youth in agriculture. This study can be carried on training needs of youth in agriculture and allied sectors.

Key words: Agriculture, Livelihood, Unemployment, Uttarakhand

INTRODUCTION

Across the world, there were one billion youth falling in the age group of 15 to 24 year in 2000, and this increased to 1.2 billion youth in 2011, Out of these, 754 million (85 per cent) live in developing countries and approximately 60 per cent live in Asia alone, so that United Nations declared 2011 as the 'International Year of Youth. The youth population growth rate is generally higher in the poor countries. And their number has tripled since 1950s. By the year 2025, the number of youth living in developing countries is likely to reach 89.5 per cent. Rural youth account for around 55 per cent of the total world youth population. They are among the most disadvantaged groups. They have limited access to educational programmes that are geared to their situation and needs (World youth report, 2012). For the development of the nation as well as youth it is necessary to involve youth in agriculture. Their active participation is possible if they are imparted relevant concepts and issues related to agriculture. This is important because awareness about agriculture is one of the major prerequisites for the willingness to participate in the process of agricultural development and choose agriculture as a career.

In India, About 70 per cent population is below the age of 35 years, that makes India the youngest nation in the World. According to some estimates, the proportion of population under 25 years in India is 51 per cent. Some experts refer to this large proportion of youth as "demographic dividend" because greater proportion of the population is young and in the working age-group which can lead to economic growth. However, out of the youth population of 460 million, only 333 million youth in India are literate and unemployment rate is highest (10.6 per cent) among youth. By 2020, the average Indian will be only 29 year old, whereas in China and the United States of America the average age is estimated to be 37 years. This high percentage of youth can be utilized for taking Indian agriculture to new heights by channelizing the creative energies of the youth through development of appropriate skills, knowledge and attitude. This predominance of youth in the population is expected to last until 2050. Hence, the National Youth Policy (2012) emphasizes youth empowerment in different spheres of national life. There are several advantages of involvement of youth in agriculture as they have the latent energy, capacity and ability to produce, propensity to learn and grasp new ideas or technologies faster and they are excellent source of ideas and innovations (Government of India, Ministry of Agriculture, 2012).

Agriculture sector contributes a major portion to gross production of India. And its role remains significant in the nation's economy. Agriculture and agri-enterprises are the high employment potential sector in India. Hence, Agriculture remains unattractive to the youth leading to their movement to other sectors of the economy to grab better life. In recent times, migration of young people in urban areas in search of job has reduced availability of labour force for agricultural production. Tackling the factors that negate youth interest in agriculture would ensure improvement in the agricultural production which is essential for food security of the nation. If agriculture is profitable, then it can serve as a source of gainful employment for the youth. This will go a long way to solve the problem of migration of youth arising from under or disguised unemployment. In view of the above, the present study has been designed to find out the constraints faced by rural youth in choice of agriculture as a means of livelihood. Specifically, the objectives of the study were, to study the socio-economic, communication and psychological characteristics of rural youth and to identify the challenges faced by rural youth in choosing agriculture as a means of livelihood.

METHODOLOGY

The study was conducted in two blocks namely Kichha and Sitarganj of Udham Singh Nagar district of Uttarakhand state. These were selected purposively as these blocks have maximum population as well as maximum cultivable land in the district Udham Singh Nagar. Two villages Ajeetpur and Bara from Kichha and Sisaiya and Barri from Sitarganj block were selected purposively as these villages have large population. Thirty per cent youth in the age group of 15-24 years belonging to the households owning agricultural land and residing permanently in the village were selected randomly. Total 115 respondents were selected through pretested structured interview schedule.

RESULTS AND DISCUSSION

Sampling Characteristics

Findings of study revealed that maximum number of respondents (31.30%) had educational qualification up to higher secondary level it indicates that youth and their families were aware of the importance of education as well as its role in their future development. Majority of the respondents were students and unemployed the figure of unemployed youth in the study area is higher than the national figure of unemployed youth. Majority of the respondents belonged to medium size family (68.7%) and from joint family (54.79%) This is similar to the findings of Hosenally (2012), where it was found that maximum number of youth belonged to medium size family, vast majority of the respondents (86.08%) were unmarried, this high percentage of unmarried youth is due to delay in age of marriage even in rural areas because of economic, socio cultural and legal reasons. This is similar to findings from other studies where it was found that majority of the rural youth were unmarried (Mustapha, 2007 and Abdulla, 2010). (64.50%) youth belonged to middle caste. Further, it was found that majority relied on agriculture (68.70%) as a means of livelihood it is interested to note that in households where the head of the family was engaged in agriculture, (53%) respondents had land up to one acre this data is very similar which observed in rest of the country. Majority of the respondents belonged from APL family (71.30%). It was also found that majority of respondents had medium level of media ownership (85.20%) it can be shows that the youth were using various media for a variety of purposes including education, information seeking and entertainment, mass media exposure (72.17%) It was observed that most of the school going youth were using mobile phone for various purposes whether they owned or not. Television and newspaper were the other frequently used mass media whereas the reach of computer and landline phone was limited to few households. Majority (76.52 per cent) respondents had medium level of extension agency contact it can be concluded that neither the government nor the private extension agencies rarely visit the study area to give agricultural information or the youth in the area do not proactively contact with these agencies. 73.10 per cent of respondents had medium level of cosmopoliteness it can be concluded that very few youth in the study area were in touch with outside environment/sources, 70 per cent respondents had medium level of achievement motivation this indicates that the rural youth in the study area may not have a high urge to do things solely for their own satisfaction or to prove their worth and majority (58.30 per cent) of respondents had medium level of change proneness, it shows that youth were enthusiastic to accept change and new practices and had a very high propensity towards change.

Table 1: Characteristics of the respondents

S. No.	Variables	Frequency	Percentage
1.	**Age**		
	Very young (up to17 years)	24	20.87
	Young (18-23 years)	77	66.96
	Mature (above 23 years)	14	12.17
2.	**Education**		
	Illiterate	3	2.60
	Can read only	0	0
	Can write only	0	0
	Can read and write	0	0
	Primary education	10	8.70
	Secondary education	27	23.48
	Higher secondary education	36	31.30

Contd...

	Diploma	10	8.70
	Graduate	27	23.48
	Post-graduation	2	1.74
3.	**Present employment status**		
	Employed	22	19.54
	Unemployed	26	22.91
	Student	54	46.96
	Farmer	12	10.59
4.	**Family size**		
	Small (up to 4)	23	20
	Medium (5-8)	79	68.70
	Large (above 8)	13	11.30
5.	**Family type**		
	Nuclear	46	40
	Joint	63	54.79
	Extended	6	5.21
6.	**Marital status**		
	Unmarried	99	86.08
	Married	16	13.92
7.	**Caste**		
	General caste	21	19.50
	Other backward caste	74	64.50
	Schedule caste and Schedule tribes	18	16
8.	**Occupation of head of the household**		
	Labour	10	8.70
	Caste occupation	3	2.60
	Business	12	10.44
	Independent profession	0	0
	Cultivation/farming	79	68.70
	Service	11	9.56
9.	**Main source of income of family**		
	Labour	9	7.90
	Caste occupation	4	3.50

Contd...

	Business	13	11.20
	Farming	79	68.70
	Service	10	8.70
10.	**Size of land holding**		
	Up to 1 acre	61	53
	Up to 5 acres	36	31.30
	Up to10 acres	10	8.70
	Up to 15 acres	8	7
11.	**Total family income**		
	APL (more than Rs. 27,000 p.a)	82	71.30
	BPL (less than Rs. 27,000 p.a)	33	28.70
12.	**Media ownership**		
	Low (up to 1)	12	10.40
	Medium (2-3)	98	85.20
	High (above 3)	5	4.40
13.	**Mass media exposure**		
	Low	15	13.05
	Medium	83	72.17
	High	17	14.88
14.	**Extension contact**		
	Low (up to 1)	19	16.53
	Medium (2-5)	88	76.52
	High (6-9)	8	6.95
15.	**Cosmopoliteness**		
	Low	30	26.10
	Medium	84	73.10
	High	1	0.80
16.	**Achievement motivation**		
	Low	32	28
	Medium	81	70
	High	2	2
17.	**Change proneness**		
	Low	1	0.90
	Medium	67	58.30
	High	47	40.80

Challenges Faced by Youth in Choice of Career in Agriculture

The probable problems which youth could face in choosing agriculture as a career identified from the literature. The list of statements was translated into local language (Hindi) and the youth were asked to identify three major constraints in decreasing order of significance. The challenges identified by the youth were ranked on the basis of weighted mean score. The limited access to new agricultural technology was ranked first by the respondents. Limited access to farm machinery was ranked second, perception that he is better suited for job and living in urban areas was ranked third, damage of crops from pests and insects ranked fourth and lack of information about agricultural inputs and government policies were ranked fifth. This is similar to finding of Butt (2011), where it was found that the government should take serious steps for developing policies and government should establish technical and vocational institutes in villages for the training of rural youth.

Table 2: Challenges faced by youth in choice of career in agriculture

S. No.	Challenges	Weighted mean score	Rank
1.	Limited access to new agricultural technologies (seeds, practices, etc.)	4.32	I
2.	Limited access to farm machinery	3.44	II
3.	Perception that the person is better suited for job and living in urban areas	2.51	III
4.	Damage to crops from pests and insects leading to low returns	2.50	IV
5.	Lack of information about agricultural inputs (seed, pesticide, etc. availability) and government policies	2.32	V

CONCLUSION

The government has emphasized the need to involve youth in agriculture and has accordingly framed several policies. Despite these efforts, youth do not have a positive attitude towards agriculture as a means of livelihood. Hence, a study to find out various challenges faced by rural youth in choice of career in agriculture such as limited access to new agricultural technology was considered as major constraint in choosing agriculture as a means of livelihood by the respondents, limited access to farm machinery, perception that he is better suited for job and living in urban areas damage of crops from pests and insects, and lack of information about agricultural inputs and government policies were other challenges cited by youth in choice of career in agriculture.

REFERENCES

Abdullahi, Y.M., Gidado, A.S. and Jibril, S.A. (2010) Attitude of Rural Youth Towards Family Farming and The Implications for Policy.Journal of Agricultural Extension. 14 (2): 24-26.

Butt, T. M., Hassan, M. J. Y. and Sahi, S. T. (2011) Role of Rural Youth in Agricultural and Rural Development: A Self-Perceived Case Study. Journal of Agricultural Extension and Rural Development. 3(2): 23-28.

Government of India.Ministry of Youth Affairs and Sports. Draft National Youth Policy. 2012.

Hosenally, N. (2012) Current and Emerging Youth Policies and Initiatives with a Special Focus on Links to Agriculture: A Case Study. Food, Agriculture and Natural Resources Policy Analysis Network. South Africa.

Mustafa (2007) Evaluation of Youth Empowerment Scheme of the National Poverty Eradication Programme on Agriculture. Thesis, M.Sc. Department of Agricultural Economics and Extension, University of Maiduguri, Nigeria, pp.10-12

CHAPTER - 10

Problems of Rural Youth in Family Farming: Challenges and Mitigation Strategies

O.P. Mishra
Associate Professor, Department of Extension Education
Institute of Agricultural Sciences
Banaras Hindu University, Varanasi – 221 005 (Uttar Pradesh)

It is agony of the situation that even after attaining food security after a wonderful revolution, we are at a verge of complex combination of several crisis. This notably includes lack of nutritional security, deteriorating genetic base, loss of biological-diversity etc.Therefore at the 66th session of general assembly of the United Nation 2014 was formally declared to be the "International Year of Family Farming" to reposition family farming at the centre of discussion to address multiple problems existing in agriculture (www.fao.org/family-farming-214/home/what-is-famly-farming/en/).

Family is the basic unit of the society. A self reliant and prosperous society needs self sustaining families which in turn require equitable and creative participation by all members of the family. Family farming is the backbone of agriculture where a rural youth effuse and rejuvenate this backbone through effective participation and obeisance of the tradition through which indigenous technical knowledge is transferred.

Family farming is the same participatory phenomenon, employed from generations immemorial where each and every member of the family manage, actuate and link different agricultural activities to not only ensure nutritional security to its family as well as nation but also provide farm animals with sufficient amount of fodder throughout the year. Besides it, rural youth involved in family farming also maintains and learns the indigenous technical knowledge (ITK). They transfer this value and natural germplasm preservation to next generation.

Family farming utilizes family labour as an effective tool in accomplishment of family objectives.

Inspite of being conservatory of indigenous technical knowledge and maintainer of environmental hygiene and agro biodiversity, the body of family farming is sobbing due to crest fallen rural youth whose non participation tells many gloomy things beneath the glittering facade of pictures showing them in smiles.

Lack of effective policies and remunerations, less farm mechanization, absence of global market in its real sense, often brings disinterestedness in place of interest. It's a fact that rural youth is the engine that takes the compartment of tradition inform of indigenous technical knowledge, culture of integrated faming predominantly existing in family farming etc, to the next generation and non participation of rural youth brings immediate concern from the world to address their problems if we want a more sustainable and a diverse agricultural ecosystem.

Challenges

The concept of rural development on a global level first originated in the mid of 17th century in England when a group of people known as 'Quakers' organized themselves on the principles of selfless sacrifices. This philosophy propagated throughout the world. In India, rural development is considered as "sine qua non" for faster economic development and betterment of rural life, but the question remains is that- Are rural youth at par with the youth group of urban part of our country? Well the challenges faced by the rural youth tell us the story of disparity of rural and urban youth (Maulick, 2013).

1. **Limited non-farm sector:** One of the problems that could lead to dissatisfaction in rural youth is limited development of non farm sector, which leads to the poverty, unemployment and underemployment. In the present scenario the fact is that most of the highly productive sectors of employment are located in urban areas. People of the rural area have no option but to move towards town and cities to have an income to sustain their family by grabbing the opportunity of diversified livelihood. The fact is that rural youth still consider migration as one of the most reliable way to sustain their needs. This happens with the rural youth because the elders of the family have to remain with family and farming and it's the youth of the family who have to leave the family and migrate towards urban area to look for further opportunities.
2. **Agriculture is considered as underprivileged:** It is said that all that glitters is not gold but the fact is that everyone is eventually

attracted towards glittering object, same happens with the rural youth. Agriculture work, being a farmer do not set up well with the mental set up of new generation of rural youth, they want to explore the ocean of unlimited opportunities. Most of them are attracted towards jobs in urban area because society considers the farming work as underprivileged. Even the smaller of the jobs that are offered in the urban areas are considered superior to the agricultural work. This leads to a feeling of dissatisfaction among the rural youth. According to a survey conducted National Sample Survey Organization (2003) it is revealed that 41% of the farmers are ready to quit agricultural activity provided they have opportunity on offer. Because of involvement of whole family in agriculture rural youth have no option to remain attach with the farming.

3. **Division of land:** Another problem that stands in front of rural youth is that they are not able to get sufficient land to cultivate so that they can sustain their livelihood. In European countries it is a tradition that elder son is the successor of the entire property of his parents. But in India the situation is different; here all the children of a father have equal share in the property of their father. Families in rural areas are generally bigger than that of urban areas. More are the children; more is division of land holding, thus the land inherited to successive generations become smaller and smaller. Even at the time when India got freedom the land holdings were smaller as compared to other European or most of the western countries, but thereafter the population grew and became almost three fold but the total cultivable land not changed much. So this led to division of land in subsequent generation. In many of the farms, land holdings have become such small that they are not economical now. This is such a problem that is bound to grow in the times to come.

4. **Irregular income and no year round employment:** Regular income prospective and year round employment is another big challenge in front of the rural youth. Many don't get employment round the year in the rural area and commute between rural and urban areas. After the harvest of crop they migrate to the urban area before the beginning of the next cropping season. They live in the slums of the cities and work as a rag picker or may go for being a casual construction worker.

   This was about employment which is not offered year round in rural areas but in work of rural areas income regularity cannot be assured. As compared to jobs in the urban areas where wages or salaries are fixed, farmers working on their agricultural land are never sure about what will be the net income of the season. Their income can be massively affected if there is incidence disease or unfavorable weather condition. A young farmer who invests money at the start of the season can only get money back after the season is over and crop is harvested. The period between sowing to harvest they are left with no source of income to get their pockets full; even if they face some financial crisis in the middle of season they have no source but to borrow money from money lenders at a heavy interest rate.

5. **Youth are not involved in any decision making of farm:** Due to involvement of whole family in farming practices, rural youth are not involved in decision making except to follow the decisions taken by non progressive elder members of the family. Sometimes these kinds of constrains hurt the agriculture production and family income because rural youth have more access to mass media and research institutions, so new beneficial techniques or methods which they think can be beneficial for farm production,theyare not allowed to adopt in the field without satisfying other members of the family involved in family farming.

6. **Higher education is not really an option:** Youth of the family that is involved in family farming have to somewhat sacrifice their educational prospect of life. During the course of primary education they are directly or indirectly involved in the agricultural activities along with the family members. So for them the value of education is not that highas the value of agriculture activities. Higher education is not really an option for most of them.A few of them those who want to go for further studies generally don't go far away from their village because general view of family is that their children complete their education as early as possible and start helping in agriculture production practices.

7. **No future for their next generation in village:** This is an era in which everyone is concerned about what they will offer to their family and children; view of present generation of the rural people i.e. rural youth is different from generation preceeding them. They think about the education that will be offered to their offspring. Lack of quality education facility and infrastructure in rural area is a hindrance for their buds to grow and blossom. They often face a dilemma to support their family in farming or to move toward urban area so that future of their offspring is brighter; so either they sacrifice their family farming or they sacrifice their offspring's future.

8. **Role of female youth is a big question:** In families involved in farming, participation of female youth in any decision making process is almost zero. They hardly get any opportunity to put forward their opinion on any matter involved in the farming. In their parents' home they remain below the ideology of their parents and after marriage under their in-laws. They are hardly allowed to express like female youth in the urban areas do. Females are not considered to have any effective income generation capacity and hence they are often restricted to household activities or they go out as a cheap labour. Their capacity to do work is never considered. Without any power in home to express, without independence to work and to have an income opportunity their voices are nearly silenced.

Mitigation Strategies

Finding a method to remove or mitigate the non participation and disinterestedness of the rural youth in family farming is not simple and straight forward solution. Rather have to evolve a balanced strategy like a balanced diet which must address social, economical and political issues. The strategy should focus on the way to increase the income of the farm families, this alone can solve many problems that are faced by rural youth in family farming. Some of the strategies that can be adopted have been discussed below.

1. **Farm mechanization can be an option:** India is a country in which agriculture has high proportion of labour involved and is less dependent upon mechanization. So option of farm mechanization by involvement of modern equipments, machineries, gadgets could contribute to agriculture. This could increase profit of farmers. Small machineries can be used by farmers as they have less landholding, new machineries can also be used by the farmers practicing agriculture at terraces and hills to maximize their profitability. By this way farmers will be able to decrease the cost involved in the production thereby increasing their overall productivity and profit. Farm mechanization can help15-20 per cent savings in seeds, 15-20 percent savings in fertilizers, 5-20 percent increase in cropping intensity, 20-30 percent savings in time, 20-30 percent reduction in manual labour and 10-15 percent overall increase in farm productivity. This can some what address to less land holding and low income problem of rural youth (Gautamand Kumar, 2014).

2. **Refining indigenous techniques of farmers:** It should be a high priority to evolve location specific technologies, which is economically viable for the farmers and which should increase their income.

It is important to give value to indigenous technologies of farmer and refine it rather than replacing it. These farmers are conservator of huge biodiversity so it is necessary to work with them in a participatory mode.

3. **Investment in agriculture sector to boost infrastructure:** If it is said that agriculture sector is starved of capital then it will not be wrong. Investment in agriculture is decreasing over the period of time. For settlement of disparity, development of off farm infrastructure and improving connectivity of villages it is necessary to increase investment in agriculture. This will increase opportunities of jobs for youth.

4. **Safeguard from climatic changes to assure regular income:** Indian agriculture is suffering with a syndrome of gambling with the monsoon. Changing climate has its effect all over the world but Indian agriculture is vulnerable to this change more than others. This is because of the fact that majority of farming community in Indian agriculture are family farmers and they lack sufficient mechanism to cope up with the change in the environment. So climate resilient technique is required, rather it is the need of the hour because agriculture is the main occupation and livelihood of large part of our population, this technology will ensure regular income to rural youth (Kumar, 2014).

5. **Agriculture for agriculture:** Linking agriculture students with the rural youth engaged in agriculture will not be a bad idea. Same age of people can go along well; this will help them to understand each other's perspective which will help in growth of agriculture by mutual cooperation.

A lot of fund is allocated to various rural development programmes but identification of beneficiaries is a big problem. Mis-appropriation of subsidy in resources by non target people living in rural areas will create wastage of money and time of government and most importantly will lead to frustration in the minds of people who should have been real beneficiaries but in reality they are not so.

If one sees the present scenario of agriculture and people involved in agriculture then it can be observed that there is a wide diversion between people of this sector. Rural youth contribution in family farming is decreasing and they are moving away from the strict boundaries of family farming.On the other hand we have agriculture graduates who instead of contributing to the rural life and rural people prefer to go for safer career options. So we have to find a way to link these two communities to improve the situation of Indian agriculture.

Programmes such as a ***compulsory internship*** or ***krishi panchayat*** can be adopted. This can be an innovative platform which will enable proper linkage of rural youth with agriculture students. If agriculture students don't come in front for development of rural background of country then their studying agriculture is waste. This will also help agriculture graduates in their future because they will learn from real life experience of rural people. In turn this willfurther help them to become proper scientist who has practical knowledge of situation. The panchayat level committee would have 3 groups:

Group 1: One group consisting of people of SHG, rural youth and agriculture graduates.

Group 2: This group could have elderly members of the village.

Group 3: this group could have government officials involved in development of rural area.

6. **Employing linkage of rural youth and agriculture students:** Very often we see that officials and primary teachers are given the task of identifying the beneficiaries and they are given extra payment for this task. It would not be wrong if this job is performed by rural youth along with the help of agriculture students. This will not only increase the income and satisfaction in the mind of rural youth but also assure effective participation of agriculture students.

7. **Panchayat level planning committee:** This committee can be spectacle through which we can see and plan different agriculture activities at the grass root level. SHG can be created in panchayat, thereafter these SHG can look after

   (a) Performance of different farms and farming activities in village and provide them with necessary knowledge input and new technologies.

   (b) Proper utilization of the subsidy.

   (c) Promote use of conservative practice and organic fertilizer.

8. **Strengthening the primary school infrastructure:** Future of coming generation is future of all whether it is rural or urban youth. So by strengthening the primary school infrastructure will guarantee better future for their future generation. This way they can get confidence of staying in rural area and continuing with agricultural work.

9. **Providing option of work in idle time:** During the period between sowing to harvest when youth members are generally free and most

of the works like weeding, hoeing is done by female. Depending upon the situation of different villages old practices of growing vegetables can be adopted in form of kitchen garden. Use of roof, wall of house, backyard of house can be done for this purpose. This strategy will ensure work for youth and will provide additional income along with nutritional security. By this way integration of different enterprises will increase farm income and youth would not be forced to migrate to urban area for search of work. This is especially important for villages which are poverty stricken,having less income opportunities, poor market access and where migration is a regular phenomenon.

10. **Organizing social fellowship program for rural development:** "These issues are debated profusely, triggers circular blame games, with a general conclusion that the responsibility lies in the hands of the country's youth to put things right. Yet every year we see the educated youth overlooking this sector which is crying out loud for their attention" says Dr. G.N.Hariharan,Principle Coordinator, M.S. Swaminathan Research Foundation (MSSRF), Chennai (Prabu, 2011)

    Youngsters or young professionals take part in Youth for India Fellowship of State Bank of India (SBI). This program ties up with well established NGO's like MSSRF, BAIF and SEVA Mandir, allowing the fellows to work on various rural projects for a year in different states and union territories. In this fellowship many graduates, post graduates including alumni of eminent institutions such as IIT's and IIM's take part. This was a well blend of youth from background of engineering, management, biotechnology, urban planning, law, mathematics and agricultural sciences. This was a unique platform to bring together people of different background to work on similar topic.

    One of the young professionals Shuvajit Payne said "basically though we might have been financially secure working in big companies, at one point of time we started developing a feeling that apart from earning money there must be something that's as individuals we must achieve to the rural poor's".

    This can be an innovative approach which could lead achievement of new heights in the field of agriculture and various problems can be addressed.Initiatives of this sort must be taken by ICAR and other SAUs to give a boost to the promotion of vibrant youth in this sector.

11. **Appreciation in the form of award and reward:** Youth of any kind want recognition and appreciation for their work. So appreciation in the form of award and reward act as a source of extrinsic motivation and this will boost their work and efficiency.

12. **Interface between rural youth and agricultural professionals:** Organizing interface will lead to the exposure of rural youth towards latest technologies and trends and on the other hand agricultural professionals will get the understanding of grassroots problems of field.

CONCLUSION

On the whole it can be said that Rural Youth play an important role in family farming to ensure food and nutritional security along with income security. However, they face lot of problems in family farming. Problems such as poor employment, lack of income security throughout the year, low prestige occupation, poor facilities for their children, etc. are responsible for low motivation of the rural youth to work in the family farming.Mitigation Strategies such as Farm mechanization, Investment in agriculture, Agriculture for agriculture, Organizing social fellowship program, Appreciation in the form of award and reward, Interface between rural youth and agricultural professionals etc. may work to overcome the problems of rural youth and inject enthusiasm in rural youth to work in family farming.

REFERENCES

1. Gautam, H.R. and Kumar, R. (2014) Agricultural Development-The Road Ahead. Kurukshetra, 62(8):3-6.
2. Kumar, P. (2014) Technologies to Boast Agriculture Production. Kurukshetra, 62(8):16-18.
3. Maulick, B.G. (2013) Rural Credit-A Growth Catalyst. Kurukshetra, 62(2):21-24.
4. Prabu, M.J. (2011) Tackling Rural India's Problems is the Mission of 28 Youngsters. The Hindu, August 18, (Also available at www.thehindu.com/todays-paper/tp-features/tp-features/tp-sci-tech-and-agri/tackling-rural-indians-problem-is-the-mission-of-28youngester/article2367493.ece)
5. www.fao.org/family-farming-214/home/what-is-famly-farming/en/

CHAPTER - 11

Assessment of DSR and MTR as A Source of Livelihood for Rural Youths in Northern Bihar

Ram Pal and Devendra Mandal
Subject Matter Specialist,Krishi Vigyan Kendra, Rohtas, Bikramganj Bihar-801222

ABSTRACT

Assessment of technical and financial feasibility of two resource conservation technologies suitable for rice establishment1. Zero Tillage Machine (ZTM) for direct seeded rice (DSR) and; 2. Paddy transplanter machine (PTM) for mechanical transplanted rice(MTR) in Rohtas district was carried out at the demonstration farm of Krishi Vigyan Kendra (KVK), Rohtas during Kharif-2012 and 2013 in a view to reduce the drudgery and cost of cultivation. An eight row self-propelled PTM (Model 2ZT-238-8) and 9 tined National Make ZTM were used for the purpose. Technical performance of these resource conservation technologies was found satisfactory. The field capacity, field efficiency, fuel consumption and cost of establishment of rice by these technologies were evaluated for the purpose. For the assessment of financial feasibility; benefit-cost ratio, internal rate of return, pay-back period and net present worth of these technologies were evaluated and demonstrated to rural youths for their quick adoption and get rid of labour-shortage and delayed transplanting of rice crop and hence saving of significant loss of Rabi crops that happens due to delayed harvesting of rice crop.

Key words: Paddy mechanization, DSR, ZTM, MTR, paddy transplanter, field capacity, field efficiency, economic feasibility

INTRODUCTION

Rice is grown on all the continents of the world, except Antarctica. As a global food, it has a large influence on human nutrition and food security all over the world. It is the staple food for over half of the world population. In Asia alone, about more than two billion obtains 60-70 per cent of their calories intake from rice and its derivatives. Almost a billion households in Asia, Africa and Americas depend on rice for their main source of employment and livelihood. Rice is therefore on the frontline in the fight against world hunger and poverty and is also symbol of both cultural identity and global unity.Worldwide, India stands first in the rice area and second in rice production, after China. It contributes about 21.5 per cent of global rice production.[2] Within the country, rice occupied one quarter of total cropped area, contributes about 40-43 per cent of total food grain production and continues to play a vital role in national food and livelihood security.[4] Bihar stands on 8th position among rice producing states of country and contributes about 5.34 per cent in national rice production. About 62 per cent of cultivated area is covered with paddy. Paddy is main crop of Bihar and is being cultivated in an area of about 3.6 Mha with an average productivity of 14.19 q/ha while its productivity is about 50.0 q/ha in Son command area.[1,3] Instead of being a major crop, it is also has the distinction of highest energy conversion ratio, labour intensive i.e. employee generation at village level also. Due to more pressure on natural resources like soil-water, migration of workforce from agriculture and frequent occurrence of draught and flood, traditional way of rice cultivation became obsolete and loss making business.[8]

Paddy cultivation is a tedious process in which mechanization of various operations is not only necessary to reduce the production cost but also provide a huge employment opportunity to rural youths without migration. Transplanting of paddy seedlings from nursery to main field is one of the similar operations which are tedious and time bound. Manual transplanting requires 250-350 man hr/ha that is 25% of the total labour requirement of the crop.[6,7,8] Manual transplanting also fails to meet the agronomical requirements like plant population, uniform row and hill spacing, thereby limiting the use of agricultural implements and hand tools for inter-cultural operations. Labour availability becomes a severe problem at that time because most of the farmers wish to finish transplanting at the same time. In these circumstances, self propelled rice transplanter is a boon for farmers and unemployed rural youths as well. It reduces drudgery involved in transplanting and generates huge employment opportunity for rural youths.[6,8] Transplanted rice has deleterious effects on soil environment and nearly 30% of the total water used (1400 to 1800 mm) in rice culture is consumed mainly during puddling and transplanting operation. Puddling requires lot of scarce water at a time when there little water available, high evapo-transpiration and lower ground water table, destroy soil structure and adversely affects soil productivity.Therefore, a key concern is how the water requirement of rice cultivation can be reduced and how farmers can avoid puddling and transplanting operations without yield penalty. DSR which removes puddling and drudgery of transplanting the young rice seedling provides as option to resolve the adaphic conflict and enhance the sustainability of rice and subsequent cropping system. DRS overcome the problems of seasonality in labour requirements for rice nursery raising and transplanting operations. DRS facilitates timely establishment of rice and succeeding crops. Scarcity of water is on the increasing and the excessive dependence on groundwater and poor irrigation systems are leading to declining of water table resulting in higher cost of pumping aggravating energy crisis. Continued puddling over decade's and led to deterioration in soil physical properties, and transplanting operations are usually performed by migratory labour, which has an element of seasonality and thud increasingly becomes a serious concern for the timely transplantingof rice. Looming water crisis, water-intensive nature of rice cultivation and escalating labour costs drive the search for alternative management methods to increase water productivities in rice cultivation. DSR has received much attention because of its low-input demand. It involves sowing pre-germinated seed into a puddled soil surface (wet seeding), standing water (water seeding), or dry seeding into a prepared seedbed (dry seeding). [8]

Keeping above problems in mind,two prominent resource conservation technologies; PTM and ZTM are assessed to minimize the labours shortage and both water and labours availability during transplanting, respectively. The focus of the assessment was centred for getting clear cut data base for counselling unemployed rural youths to opt these technologies as a source of their livelihood at village level and their contribution in nation building.

MATERIAL AND METHODS

Assessment of technical and financial feasibility of self propelled PTMand DRS sown by ZTM were carried out at the demonstration farm of KVK, Rohtas. Rohtas has the distinction of highest rice productivity,assured canal irrigated area, number of modern rice mills and number of ZTM, mechanized district and food security index in Bihar state (Food Security Atlas of Rural Bihar).An eight row self propelled PTM (Model 2ZT-238-8) and 9 tined National Make ZTM were used for the purpose, specification of these machineries are given in Table-1 and Table-2, respectively. Field data like depth of planting, number of plant per hill, Number of missing hills, number of floating hills, float sinkage, field capacity and efficiency, fuel consumption etc were recorded during operation of paddy transplanter and field capacity, fuel consumption etc for ZTM for technical feasibility assessment. Financial analysis such as Net Present Value (NPV),

Internal Rate of Return (IRR), Benefit–Cost ratio (BCR) and Payback period were calculated by following the standard methods and prevailing rates in the districts.

RESULTS AND DISCUSSION

Technical performance of machines were measured and evaluated during the operation and just after the transplanting/sowing. The observed details are given in Table 4. As the PTM was riding type, number of buried hills was nil and damaged hills were almost negligible. Due to use of rotavator and providing enough time for settlement of puddle soil the efficiency of machine was recorded very satisfactory. Float provided in the machine works as leveller and helped in significant reduction in number of missing and floating hills. In DSR, ZTM was used as seed drill, i.e. field was well prepared before using the machine. Seed drill or ZTM might be used for the purpose depending on the availability, but if a person is willing to purchase a new machine one should select a ZTM as it may be used in wheat sowing in Rabi season and enhance the financial feasibility of machine by increasing the working hours. Use of ZTM for DSR further increase the adoptability of machine as the machine used to keep ideal after the wheat sowing season and proved a dead asset for farmers who owned the machine earlier.

Financial feasibility of these machines were analysed on the basis of five parameters, viz; payback period, internal rate of return, net present worth, benefit and cost ratio and breakeven point and is shown in Table-4.The very short payback period attracts prospective rural youths who are interested to adopt farm machinery as their source of livelihood and the same to financial institution to finance them as sown in Table-4. The break-even- point indicates that if the machine has possibility to use even for 12.8 h/year and 19.18 ha/year for PTM and ZTM respectively, one can invest in these. Higher benefit cost ratio and IRR coupled with positive NPV further enforce that these machinesare very much suitable for taken as an entrepreneurship mode.PTM would not work like other farm machinery as owner works in the fields of farmers on rental basis without any prior agreement. Since raising of mat nursery is a very skilful work, entrepreneur should contact with farmers in advance and raise mat nursery in his own supervision and do the transplanting. Keeping all in view it may say that eight-row paddy tans planter and ZTM are suitable option for upcoming entrepreneurs if taken as a source of livelihood.

Table 1 : Specification and working features of PTM

Sl. No.	Particulars	Descriptions
1.	Model	2ZT-328-8
2.	Name of Company	VST Tillers & Tractors Ltd. Banglore, India
3.	External dimensions L x W x H (mm)	2410 x 2131 x 1300
4.	Type	Single Wheal driven
5.	Engine	4-Hp air cooled diesel engine
6.	No. of rows	8
7.	Row spacing (mm)	238 mm
8.	Hill to hill spacing (mm)	140 & 170
9.	Frequency of transplanting	263
10.	Working speed (km/hr)	1.57 to 1.97
11.	Speed of travel on highway (km/hr)	8.24
12.	Field capacity (ha/hr)	0.13 to 0.20
13.	Weight (kg)	320
14.	Cost (Rs')	2,25,000.00

Table 2 : Specification and working features of ZTM

Sl. No.	Particulars	Descriptions
1.	Name of Company	National Agro Industries Link Road, Industrial Area A, Ludhiana-141003.
2.	Length xWidth x Height,mm	1940x970x1070
3.	Seed metering mechanism	Fluted roller
4.	Fertilizer metering mechanism	Corrugated roller type
5.	Power transmission	Through chain and sprockets
6.	Number of furrow opener	9
7.	Size of feed shaft, mm	16
8.	Size (diameter) and number of flutes	42.7mm and 9
9.	Diameter of fluted roller, mm	42.7
10.	Size of fertilizer shaft, mm	20
11.	Seed box capacity, cm^3	66241
12.	Fertilizer box capacity, cm^3	63310
13.	Ground wheel diameter, mm	390
14.	Weight, kg	224
15.	Cost (Rs')	40,000.00

Table 3 : Technical features of PT and ZTM

Sl. No.	Particulars	PTM	ZTM
1.	Depth of planting, mm	28.19	28
2.	Number of plant per hill	6.70	-
3.	Missing hills,%	0.70	-
4.	Floating hills, %	1.64	-
5.	Damaged hills/seeds, %	0.17	12.5
6.	Buried hills,%	nil	-
7.	Float sinkage, mm	17.95	-
8.	Field capacity, m^2/hr	1875.5	2245
9.	Efficiency, %	93.73	86.3
10.	Fuel consumption, l/hr	1.06	3.45

Table 4 : Annual Cost Calculation

S.No.	Particulars	Amount(Rs)	
(A)	**Fixed cost**	**PTM**	**ZTM**
1.	Depreciation	17500.00	3000.00
2.	Interest @12% of average cost	16500.00	2640.00
3.	Insurance and taxes@2% of average cost	2750.00	440.00
4.	Housing cost @ 1.5 of average cost	2063.00	330.00
	Total fixed cost(Rs/annum)	38813.00	6410.00
(B) Operating costs (Rs)			
1.	Fuel consumption, lt/hr	1.06	3.45
2.	Lubricating oil consumption, lt/hr	0.053	0.1725
3.	Maintenance cost, 5% of fuel	0.053	0.1725
4.	Labour charge 1.	Driver, 12.	Labour, 3 250/ 8 hr450 /8 hr 250/ 8 hr150 /8 hr
	Total Operating cost (Rs/hr)	151.63	328.58
5.	No. of working hours/ annum	480	400
6.	Annual operating cost ,Rs	72782.40	131432.00
7.	Tractor cost, Rs	-	54000.00
8.	**Total Annual cost (Rs)**	**111595.40**	**191842.00**

Contd...

(C) Annual Return

1.	Area covered in a year,ha	90 .0	110.00
2	Annual return (@ Rs3450/ha transplanting cost	310500.00	275000.00
3.	Net profit	198904.60	83158.00
(D)	**Financial feasibility indicators**		
1.	NPV	+ve(701692.2)	+ve(386226)
2.	IRR	24.21%	19.08%
3.	B:C ratio	1.96	1.36
4.	Breakeven point	12.8 ha	19.18 ha
5.	Payback period	1.31 year	1.51 year

ACKNOWLEDGEMENT

The authors pay their heartfelt thanks to Director Extension Education, Bihar Agricultural University, Sabour, Bhagalpur, and Programme Coordinator, Krishi Vigyan Kendra; Rohtas for providing necessary facilities and moral support as and when required during carrying out the field work and permission to publish this data. We also acknowledge the help and support received from scientists and Farm Manager of KVK, Rohtas for carrying out the experiment and data recording.

REFERENCES

1. Anonymous, Annual Report 2013-14 of KrishiVigyan Kendra, Rohtas, Bikramganj.
2. Anonymous, Direct Seeded Rice (SEMIDRY), Adoption Technology for Climate Change in Semi Arid Region. Published by International Water Management Institute, ICRISAT centre, Hydrabad, India.
3. Anonymous, Economic Survey 2010-11.Finance Department, Govt of Bihar.
4. Anonymous, State of Indian Agriculture 2012-13. Ministry of Agriculture and Cooperation, Govt of India.
5. Anonymous, The Food Security Atlas of Rural Bihar produced by the Institute for Human Development (IHD) and the UN World Food Programme (WFP).
6. Chaudhury, V.P.; B.P.Varshney and S.K.Singh. 2005. Self propelled rice transplanter- A better alternative than manual transplanting, Agricultural Engineering Today, Vol. 29(5-6): 32-37
7. Kumar, V and Ladha, J.K.(2011). Recent development in direct seeded rice. Advances in Agronomy, Vol.111, Elsevier Inc.
8. Singh Y, Singh VP, Chauhan B, Orr A, Mortimer AM, Johnson DE, Hardy B, editors. 2008. Direct seeding of rice and weed management in the irrigated rice-wheat crop- ping system of the Indo-Gangetic Plains. Los Baños (Philippines): International Rice Research Institute, and Pantnagar (India): Directorate of Experiment Station, G.B. Pant University of Agriculture and Technology. 272 p.

CHAPTER - 12

Exploring Awareness Among NREGA Beneficiaries Under The National Rural Employment Guarantee Act 2005

A.S. Tigga
Asst. Prof.-cum-Jr. Scientist, Department of Extension Education
Bihar Agricultural College,Sabour, Bhagalpur,Bihar

INTRODUCTION

The National Rural Employment Guarantee Act 2005(NREGA) was brought into force by the Union Government in February 2006 which was renamed on 2nd October 2009 as MGNREGA i.e. Mahatma Gandhi National Rural Employment Guarantee Act. It is the first nation-wide employment scheme that guarantees employment legally to India's rural population. According to the Union Rural Development Ministry's figures, more than 1.36 crore people of the country's total population have been given jobs under the Act. It guarantees 100 days of unskilled jobs per rural household. This is meant to regenerate the rural natural resource base, which in turn, will result in sustainable livelihoods for residents. It is with this background the present article has been planned with the specific objectives to explore the extent of awareness about NREGA among the target group of the Act. Ray (2008) told that, director, Development, Allahabad University included NREGA in the curriculum of rural sociology and has suggested that seminar; workshops and university publications can be used for awareness generation.UNIFEM Workshop (2009) concluded that, in case of Kerala, Kudumbashree and Panchayat have been very vigilant in spreading awareness about NREGA through radio, poster, television, notice board on gram Sabhas, etc.

METHODOLOGY

There are altogether 20 blocks in Samastipur district. Out of these Pusa block has purposively been selected for this study. Further, two panchayats from this Pusa block have been chosen by following the random sampling technique. Then again from each Panchayat, two villages selected following the same technique. In this way, the locales of the study were four villages selected from two Panchayat of Pusa block of Samastipurdistrict.For selection of respondents, simple random sampling techniques have been followed. From each selected village, a total of 30 respondents consisting of 15 male card holders and 15 female card holders selected from the list of NREGA card holder by following the random table. In this way the sample respondent of the study was consist of 60 male and 60 female NREGA beneficiaries/card holders.Karl Pearson's coefficient of correlation and multiple regression analysis, besides frequencies and percentage, were used to analyze the data. In order to measure the awareness of beneficiaries about the NREGA, and its different provisions, which might affect the benefit of NREGA, to its beneficiaries, awareness test was developed and comprehensive score was obtained by the respondents on the question framed for the purpose.

RESULT AND DISCUSSION

Awareness of the different component/provision of the Act was measured on the basis of awareness index which included all the components/provisions which was calculated with respect to each respondent. The NREGA beneficiaries were classified into, low, medium and high level of awareness on the basis of the score achieved by them.

Table 1 : Level of awareness of NREGA beneficiaries

S.N.	Level of awareness	Distribution of Respondents					
		Male(N=60)		Female(N=60)		Total (N=120)	
		F	%	f	%	f	%
(1)	Low	–	–	1	2	1	1
(2)	Medium	9	15	21	35	30	25
(3)	High	51	85	38	63	89	74

It is evident from the table 1 that as regards the male beneficiaries, majority of them (85.0%), possessed high level of awareness about the NREGA guidelines. Another 15 per cent beneficiaries were found to have medium level of awareness. It was good to notice that none of the beneficiaries were having low level of awareness. In comparison to male, 63 per cent of female beneficiaries possessed high level of awareness followed by medium level of awareness (35

%) and low level of awareness (2 %). The present findings were in line with the findings of Menon (2010), who also found that there is high awareness about the different provision and component of the Act among the beneficiaries. Present finding also supported by the observation made by Mourlin (2007).

The awareness of different component/provision of the Act is affected by various socio-economic characteristics of beneficiaries. Therefore an attempt has been made in this section to explore the relationship, between the selected socio-economic variables and the awareness of different component/provision of the Act.The Pearson's correlation co-efficient was computed to know the relationship of the awareness of different component/provision of the Act with the related variables viz., Age (X_1), Caste (X_2), education (X_3), family education status (X_4), type of family (X_5), family size (X_6), land holding (X_7), family income (X_8), constraints (X_9) and migration (X_{10}).The results of the correlation co-efficient between awareness of the Act and socio-economic variable of male NREGA beneficiaries have been presented in table 2.

Table 2 : The co-efficient of correlation between awareness of the Act and socio-economic variables of the male NREGA beneficiaries

S. N.	Independent variables	Value of co-efficient of correlation
1.	Age (X_1)	-.398*
2.	Caste (X_2)	-.790**
3.	Education (X_3)	.735**
4.	Family Education status (X_4)	.140
5.	Type of Family (X_5)	-.283*
6.	Family size (X_6)	-.424**
7.	Land holding (X_7)	-.298*
8.	Family income (X_8)	-.343**
9.	Constraints (X_9)	-.427**
10	Migration (X_{10})	-.461**

* Significant at 5% level of probability, **Significant at 1% level of probability, NS-Non significant

Table 2 revealed that, the variables viz., age, caste, education, type of family, family size, land holding, family income, constraints and migration were found to be negatively and significantly correlated with the awareness of male NREGA beneficiaries about the different provisions under the Act. However the variables, land holding size and education, were found positively but significantly correlated with the impact of the provision/Act. There was only one variable viz., family education status which was not having significant correlation with awareness of the Act. Thus, except family education status, all the remaining variables had significance correlation with awareness of the Act.

CONCLUSION

On the basis of above discussion it could be concluded that both male and female NREGA beneficiaries are having high level of awareness about the different provisions and components of the NREGA in totality with the 74 percent. The awareness of different component/provision of the Act can be increased substantially among the male NREGA beneficiaries, with an increase in education. The awareness of Act will decrease with the increase of age, caste, type of family, family size, land holding, family income, constraints and migration.

REFERENCES

1. Mourlin, K. (2007) NREGA-A Key to Sustainable Rural Development: An Empirical Evidence from Betul district. Vikas Vani Journal, Vol. 1 (4) 15-18.
2. Menon, S. (2010) Right to Information Act and NREGA: Reflections on Rajasthan. http:/ /mpra.ub.uni-muenchen.de/7351/MPRA Paper No. 7351, posted 27. February 2008 / 05:29
3. Ray, B.B. (2008) Minutes of the Round Table on NREGA with IITS, IIMS, Agricultural Universities and Other Professional Institutions held on August, 1, 2008, Vigyan Bhavan, New Delhi., www.nrega. nic.in/minuts-round-table.pdf
4. UNIFEM Workshop (2009).Engendering NREGA: Women's Empowerment through Guaranteed Employment, http://www.unifem.org.in/PDF/Engendering%20 NREGA% 20-% 20 Workshop % 2031 % 20August%202009.pdf.

Chapter - 13

Economic Opportunities for Youth in Farming that Provide Alternative to Migration in Urban Areas

S.K. Sinha
Jute Research Station, Katihar

During the last visit of Barak Obama in the year 2011, he pointed out that India is fortunate to have a youth population with a half of the total population being under the age of 30-35 years. Out of 600 million young people, over 60% live in villages and majority of them are educated. Mahatma Gandhi considered that migration of rural youth from villages to town and cities as the most serious form of brain drain affecting adversely rural India's development. The National Commission on Farmers stressed the need for attracting and retaining rural youth in farming. Recently, National Sample Survey has indicated that over 45% people want to quit farming under these condition there is need to persuade educated youth including farm graduates to stay in villages and take up agriculture as a profession.There are various economic opportunities for farm graduates which are discussed below:

Agriclinic and Agribusiness as An Opportunity for Farm Graduates

As an opportunity for self employment of farm graduates to start agri-clinic and agri-business center, the government of India launched a programme for enabling farm graduates to start agri-clinic and agribusiness centers in rural areas. This programme was administrated by NABARD and MANAGE located in Hyderabad is yet to attract the interest of educated youth. It is hence, time of re-structuring of the programmeso that farm graduates may be willing to take-up advantages of the programme. This programme has been launched with the concept that ideally a group of five farm graduates who have specialized in agricultural, animal husbandry, fisheries, agri-business management and home science could jointly launch an agriclinic and agribusiness centre in every block in the country, Agriclinic will provide the services needed during the production phase of the farming while the agribusiness centres would cater to the needs of farm families during the post-harvest phase of agriculture. The farm women and men can be assisted during the entire cropping cycle starting with sowing and extending up to value addition and marketing.There are several possible ventures which can be effectively introduced to develop rural entrepreneurship through agri-based and allied activities under agriclinic and agribusiness centres. Some of these are soil and water testing lab, pest surveillance, agri service centres, agri insurance scheme, agro processing, maintenance of agri-implements, micro propogation, beekeeping, hatcheries etc. All these ventures are economically viable and could be started with low investment. The service sector designed to meet the demand driven needs of farm families and an important one is soil testing work which need to go village to village in the area of their domain and issue them soil health card to each and every family of his area. This could be done on permanent basis.

Economic Opportunities in Production of Biofertilizer

Youth can also help women's self help group s to manufacture and sell the biofertilizer, bio-pesticides and vermicompost. The technology for manufacturing these items is available. The demand of biofertilizer is increasing day by day as increasing farmers' consciousness towards deteriorating soil fertility. The production of bio fertilizer need low investment and less man power.

Economic Opportunities in Production of Blue Green Algae (BGA)

Rice is grown in about 43 million hectare of land in India. With the development of high yielding varieties, the demand for fertilizer–N is very high.To partially meet this demand, blue green algae bio-fertilizer plays an important role. In addition to providing improved biologically fixed nitrogen, the fertilizer use efficiency and improvement in physic-chemical properties of soil is also accomplished with the use of bio-fertiliser. The biofertilizer provides about 25 kg N/hectare per crop in term of urea. For commercial production of this bio-fertilizer, it is very beneficial venture and technology of production is available. This involves low capital investment and maximum use of local resources.

Setting of Village Knowledge Centres

The establishment of village knowledge centre could be done by young and educated rural youth. Such centre will be based on the integrated use of internet, FM radio and mobile telephony. They can disseminate the agricultural information on payment basis. The village knowledge centre will be helpful in motivating the farmers for initiating steps for setting up agricultural based ventures by providing technical knowledge.

Setting up Climate Risk Management Centres

Young and educated farmers can operate climate risk management centre which will help farmers tomaximize the benefits of good monsoon and minimize the adverse impact of unfavorable weather conditions. Sensitization of farming community with the environmental concern needs high priority. Balanced use of biomass, organic and in organic fertilizer and controlled use of agro chemical through INM & IPM needs to be permitted to achieve the sustainable increase in agricultural production. The young farmers can undertake this task by setting up risk management centrein each and every block.

Micro-Propagation

Micro propagation through tissue culture has been standardized and is being extensively used for sugarcane and envisaged to be useful for banana as well. Sugarcane is a vegetative propagated crops with low seed multiplication rate.The availability of quality seed of newly released varieties is very low. Micro-propagation is the best method for the production of quality seed free of disease and pest infestation.

Mushroom Cultivation

Mushroom is a macro fungus with a distinct fruiting body which is used as human food. It is the richest source of minerals and can cover up for the mineral vitamin deficiencies in our food. However, all mushroom are not edible and some of them are poisonous and harmful.At present, three types namely button mushroom, oyster and paddy straw mushroom is widely cultivated in India. These mushroom help to utilize the farm waste and crop residues, for example, one kg of straw can be converted into about 320 gram of mushroom.Thus, mushroom cultivation is successful agro based industry. Recently, a variety of mushroom products have been manufactured to be used in nutritional and pharmaceutical company.

Besides these ventures there are several other ventures like agro forestry, sericulture, hatchery, poultry, piggery, goatery and many other ventures available which can be started by young farmers as profession for their economic stability.Looking these opportunities, it can be concluded that rural youth have several opportunities in villages itself and theydon't need to migrate for earning their livelihood.

REFERENCES

1. Venkateswarla, B. andRamarao,C.A (2011).Yojana. January 2011
2. Swaminathan, M.S. (2011). Yojana. 2011
3. Gupta VP, Summer School on Agriculture-Based Entrepreneurship Development,RAU Bihar, 2001

CHAPTER - 14

Rural Youth Migration and Measures to Combat it Efficiently through Agriculture

Rekha Balodi[1], Lajja Vati[2], Sunaina Bisht[1] and Abhijeet Ghatak[3]
[1] Plant Pathology, COA, Govind Ballabh Pant University of Agriculture and Technology, Pantnagar (263 145), U.S. Nagar, Uttarakhand
[2] Plant Breeding and Genetics, BAC, Bihar Agricultural University, Sabour Bhagalpur 813 210, Bihar
[3] Plant Pathology, BAC, Bihar Agricultural University, Sabour Bhagalpur 813 210, Bihar

ABSTRACT

Agriculture is an important economic activity in India, contributing around 50% for the employment generation and 14% to the total gross domestic product (GDP) of the nation. Being a subsistence activity, agriculture shares maximum contribution in the national economy. However, the income generated from agriculture is not sufficient to engage the rural youth in farming profession. Lack of economic incentives coupled with other factors like lack of infrastructure for education, and opportunities for employment lead to the migration of the rural youth to the urban areas. Migration of human population can be both interstate as well as intrastate to establish a new permanent or semi-permanent residence. According to the 2001 census, total number of migrants is 314 million by the last residence. Of these 85% of the migration is intrastate while 13% migration is interstate. Approximately 20 million people have migrated from rural areas to urban areas; the reason for migration was predominantly marriage for females, and employment and education for males. This phenomenon has resulted in decrease in the work force in the villages leading over labour population in urban areas – those results in low wages due to competition and wide spread poverty. The situation becomes worse when the youth drop out from the formal education system and move to the urban areas in search of better employment opportunities. This unskilled labour constitute the labour class in the urban areas, and due to the low wages received by them, they remain economically vulnerable. This situation is aggravated by the biased government policies for the rural youth, lack of awareness and ever spreading poverty. Thus, it is important to understand the factors responsible for the youth migration from rural to urban areas and effect of this migration on both rural and urban economy. This paper highlights the issue of youth migration from rural areas to urban areas in search of better education and employment opportunities and its consequences. Further, the policies are described in brief for the involvement of the rural youth in rural economy.

Key words: Agriculture, employment, migration, unskilled labour, youth

INTRODUCTION

Meeting the demand of food along with nutrition supply is a great challenge for current policy makers and for the scientists. This brings agriculture to the centre-stage worldwide for mitigating the increasing demands of increasing world population, changing food habits and rise in the standard of living. On the one hand,the world has seen an upsurge in demands of food commodities, along with greater awareness for the nutritional security, which has led to the increase in demand of quality food;and on the other hand, agriculture is facing multitude of constraints due to changing climate change, land utilization and decrease in workforce in agriculture. The contemporary issues in agriculture have led to the emergence of the concept of "Sustainable Agriculture", which forms the basis of formulation of all activities and policies related to the agriculture globally.

Indian agriculture is depending on active rural population engaged in farming profession. Alongside, a balance of urban population is always needed to sort-out the problems associated with poor infrastructure at rural context (Fig. 1). Agriculture in India is an important economic activity, contributing a major part of the total Gross Domestic Product (GDP, approximately 14%) of the nation. Along with the economic inputs provided by the agriculture at national level, it supports the rural economy and livelihood by providing rural people with the food, feed, fuel and other inputs. Most of the rural households are dependent on the agricultural activities for their sustenance directly or indirectly. Whole family remains involve in agricultural activities, thus family farming is an important component of rural architecture in India. Among the challenges faced by the Indian agriculture, the migration of the rural youth to the urban areas and declining interest of the youth in agricultural activities is one of the major concerns. According to 2001 census the total population of India was 1,028 million comprising of 532 million males and 496 million females. This census reported that 315 million persons have migrated based on the place of last residence, i.e.

31% of the total population of India is internally migrated from the place of their residence[12]. Out of this 315 million migrant population about 30% are youth. With a high percentage of its population below the age of 30 and a projected labor force expansion of 95 million by 2020 (a 12% gain from 2010), India is uniquely positioned to reap the benefits from its demographic dividend[10]. Rapid urbanization coupled with high decadal growth (31.8%) is the driving factors for the migration of the population. And it is estimated that by 2011, 33% of the total population will be migrated from their native residence[6]. It is important to understand the driving forces for this migration and evaluate its impact on the rural economy, as this migration will lead to the drain of workforce from the rural areas to urban areas (Fig. 2).

The reason for this migration is attractive employment opportunities or facilities for higher education in the urban areas which are lacking in the villages/ remote areas. These opportunities lure the rural youth out of the farming profession. This problem becomes a social concern when the drop outs from the schools and colleges migrate to the cities in search of job opportunities, but they find it difficult to sustain their families with the meager salaries they earn in the cities. This trend cause a two-fold burden on the economy as on the one hand the migration of the youth from the rural areas lead to decrease in the workforce available in the rural areas, and at the same time this unskilled youth causes a burden on the urban economy[1].Thus, it is important to understand the causes of migration of the rural youth to urban areas for drafting effective strategies for their involvement in the farming and other rural activities. At the same time there is a need to increase awareness in the rural youth about the employment opportunities generated by the farming profession, which is required for the retention of theyouth in the villages and their involvement in farming profession.

Migration of Rural Youth to Cities

In India, distribution of population in rural and urban area shows a ratio of 1:2.5 (Fig. 1). Migration of the rural youth to cities is directly affecting the involvement of the youth in farming. Additionally, this is a social concern as this causes an imbalance in the workforce in rural and the urban areas. This issue becomes more conspicuous when the migrated population is unskilled and drop out from the education system. This migration of the human capital from one location to the other with in the country is known as internal migration, and this type of migration is facilitated by high wages and employment opportunities in cities compared to rural areas (Fig. 3)[7].

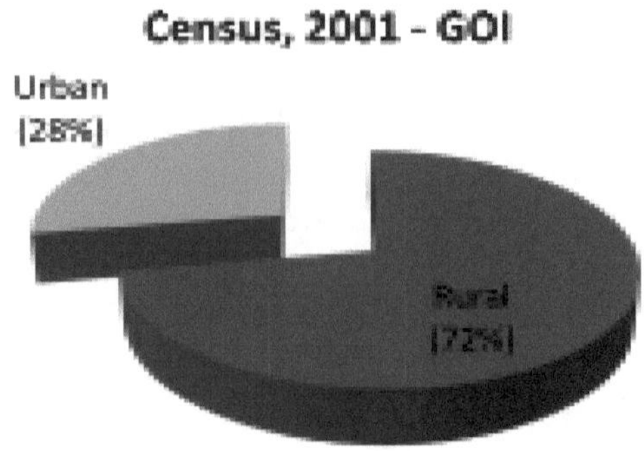

Fig 1: Distribution of the rural and urban population in India.
Source: Census, 2001 – GOI[6].

Multiple factors contribute to the migration of youth from the rural areas to urban areas (Fig. 2).Accordingto a survey conducted by the National Sample Survey Organization (NSSO) during July 2007 to June 2008, the migration of the population can be short term or long term. The causes of internal migration are variable and it is attributed to the migration for education or migration for employment[4].Maximum migration among the youth is observed in the age group of 25-29 years, the trend is similar for both males and females. Though the driving factors may differ, in case of male population the migration is derived by the need of employment opportunities or economic factors while in case of females the same is caused due to marriage or family movement from one location to the other[6]. Major reasons for migration are briefly discussed below:

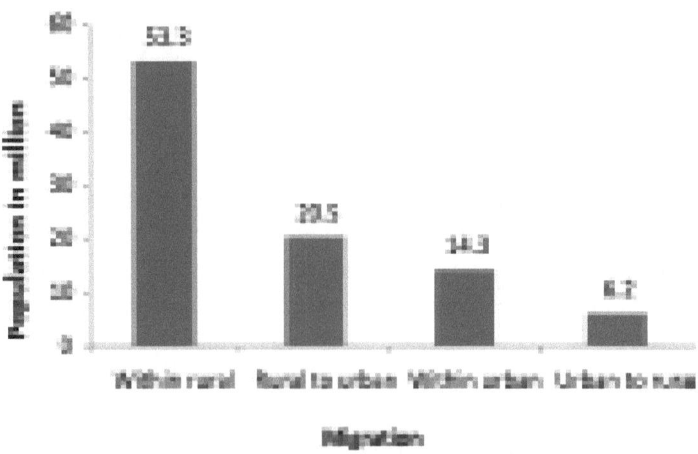

Fig 2: Trends of migration of the population and reasons for migration.
Source: Census of India 2001 - GOI[6].

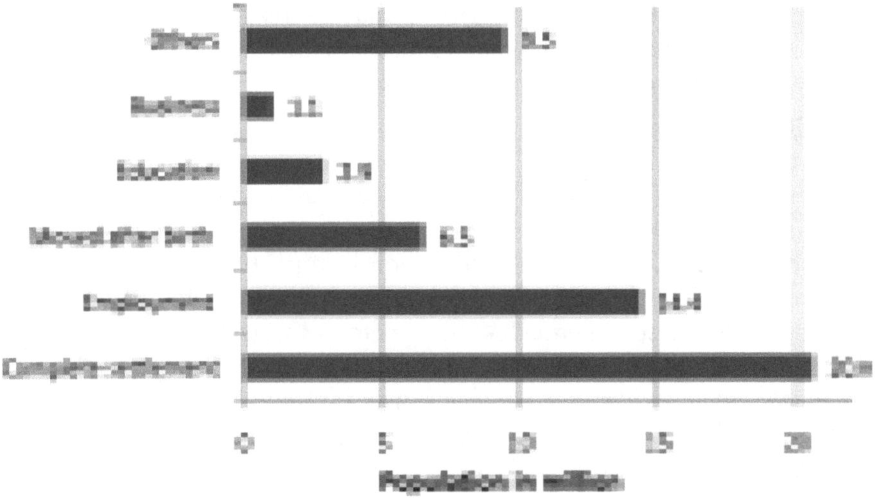

Fig 3: Reasons for migration for total population (excluding marriage).
Source: Census of India 2001 - GOI6.

1. **Migration for education:** India fall short-off achieving 100% literacy, though various policies and schemes are adopted by the government. The literacy rates are variable in different parts of country and some states perform better than the other on this index. Further, the literacy rates are not equal on the gender basis, and formal education of males is better than the female part. The regions of this country like Uttar Pradesh, Bihar, and Orissa lack the infrastructure for the formal primary and higher education. Therefore, maximum population migration for education is seen from these states[7]. This lead to outflow of the human capital from one region to the other, and thus lead to interstate migration for higher education. It should be highlighted here that this migration is very small in percentage (H"17%) as every state has cities where infrastructure for primary and higher education is available, and thus this migration is usually intrastate from the rural areas of a state to the urban areas of the same state (H"45%)[5]. Among the migrated population the highest no of both males and females are with education level of below matriculation or graduation, and population with high educational qualification have fewer tendencies for migration[6,7]. This trend suggests that migration is facilitated by the lack of educational infrastructure, which gives an idea for state to support and policy frameworkdevelopment for managing the problems arising due to migration.Rural female migrants have the largest share of illiterates in India, at 65.5%. Educational opportunities are yet to be made

available to women and girls in remote areas and villages. In other words, the educated youth in the rural areas migrate to urban areas for better education and employment opportunities.

Most important area for development of the education policies for rural areas is to understand the regional discrepancies occurring throughout the country, and consideration of regional differences is required for improving the situation of the youth in rural areas. Along with improving the main stream of (formal) education, overall development of education sector is required. Thus, at rural locations educational policy should focus on development of all forms of education including vocational education and skill development.Curriculum of the educational institutes should meet the demands of both rural and urban requirements, and should take into consideration the gender aspect (Table 1)[7]. Policies should be drafted such that young women's are encouraged for participating in income generating activities and are self-employed by easy access to funds and credit. At the same time parents should be encouraged for supporting their children in educational activities. Practices like child marriage should be discouraged as it will help in increasing the social status of the children and will improve the status of the women in society. But changing traditions, especially among parents who are not educated and who see no need to prioritize school, is a long process. When developing youth's opportunities, it is important to incorporate information campaigns that target parents in order to win their support.

Table 1: Reasons for migration in India.

Reason	Percentage of total migrants		
	Females	Males	Persons
Work/Employment	3.2	37.6	14.7
Business	0.3	2.9	1.2
Education	1.3	6.2	3.0
Marriage	64.9	2.1	43.8
Moved after birth	4.8	10.4	6.7
Moved with households	18.9	25.1	21.0
Other	6.7	15.7	9.7

Source: Census of India, 2001 - GOI[6].

2. **Migration for Employment:** Accordingto a survey conducted in 2009-10 distribution of the workers in different sectors is as follows: agriculture sector (53.2%), secondary sector (21.5%) and tertiary sector (25.3%). Unlike the case of migration for education which was primarily an intrastate phenomenon, 46% of individuals migrate to work in other states where as 54% work in the same state.

Employment has facilitated higher job opportunities to the youth as the share of non-workers is low in migrated population compared to the total population[6]. Further, it has been observed the male workers are available in the urban areas (60.9% of the total urban male migrants), while for female workers numbers are higher in the rural areas (24.8%).Indian agriculture is considered asthe provider of subsistence activity and employment generation by agriculture is unsatisfactory. Youth can be engaged in agricultural activities by offering high incentives and cash income.Thus, to uplift the condition of rural youth, policies related to agriculture should be promoted. Modern techniques and improved technologies for the crop production, livestock and agribusiness management should be adopted and disseminated among the rural youth. In addition, the linkages should be made between the markets and farmers to create the market access for the farmers.

It should be emphasized that lack of employment opportunities prompts the rural youth to migrate from rural areas to urban areas (Figs. 1,2&3); therefore, along with the agricultural activities other employment generation activities should be made available to them. These activities may include:

a. **Small livestock production.**Youth should be educated about the potential of small livestock production like poultry, pigs and goat production. Training programs should be initiated for making the youth well trained for production of livestock and small funding schemes should be initiated for making the credits available.

b. **Fishpond or aquaculture.** Youthrequire improved skills and knowledge in the construction and maintenance of fishponds, nutrition, health, breeding and fingerling management.

c. **Field and horticultural crop production.** Apart from rice production, a range of crops including fruits, vegetables and ornamental plants, have the potential for both food security and for selling. Suitable training will be required.

d. **Processing of NTFPs.** There is a wide range of items and commodities that could form very lucrative income-generating businesses and activities for rural youth. Adding value via processing or other marketing techniques (packaging, labeling, etc.) will expand the marketability of NTFPs.

e. **Mushroom production.**Mushroom production can be a viable alternative for the farmers having small land holdings or without land. Mushroom production is an indoor, labour intensive and high profit endeavor. Thus, this activity has ample scope of involvement of unemployed youth, women and small farmers. On the one hand mushroom production provides for the economic gains, on the other hand it supports the much desired nutrition security (much desired goal of Indian agriculture).

Migration and National Youth Policies

Migration results in unequal resources and workforces at the locations involved in the migration processes (Table 1). The youth faces many problems while migrating from one location to another, and these conditions are coupled with the social, economical and demographic situations of the area (Fig. 2). The policies framed by the government play an important role in migration of the youth from rural to urban areas. The negligence of the government policies towards the requirement of the rural needs and lack of infrastructure leads to the migration of the population from rural to urban areas. This phenomenon leads to the poor quality of life, risk, vulnerability and human trafficking of migrated population. These problems cause havoc in the host population and affect the acceptability of the migrated population in the new area. Thus, to ensure the social harmony effective policies for migration are necessary. Migration policy should include three aspects: admission, treatment and enforcement and it has been found that these issues are neglected in migration policies[8].

Young migrants have to face these multiple pillars of inequality, and the situation become chaotic for the government as well as migrants. In India, organisedlabour unions, employer organisations and civil society can play a significant role in lobbying for the effective formulation of public policies on migration. The enforcement of comprehensive migration policies will have a lasting impact on human development outcomes, not only for the individual migrants and their families but also for the communities at origin and destination[3].The National Youth Policy of the Government of India (NYP, 2003) points out that youth developments in India have been hampered because of a lack of adequate research on youth. In the NYP, 2003;India is recognized as the young nation for the large number of population between the age group of 15-35 years, constituting the workforce of the nation. But, at the same time, the NYP, 2003 failed to recognize youth migration and its different dimensions within the policy.

A Few Policies for Empowerment of Rural Youth

1. ***Integrated Rural Development Program (IRDP):*** IRDP was first introduced in 1978-79,to provide assistance to rural poor in the form of subsidy and bank credit for productive employment opportunities through successive plan periods. Various programs were launched in succession of IRDP to upliftthe rural youth:

a. **Training of Rural Youth for Self Employment (TRYSEM).** This scheme aims at providing basic technical and entrepreneurial skills to the rural youth/household in the age group of 18 to 35 years to enable them to take up income generating activities.

b. **Development of Women and Children in Rural Areas (DWCRA).** This scheme was initiated for strengthen the gender component of the rural population, the scheme started in 1982-83 to cover 50 districts initially. At present the scheme covers all the districts of the country. The scheme aims at improving the living conditions of the women in the country, thereby improving the status of the children and providing the rural women with the opportunities ofself employment by giving them training for skill up-gradation.

c. **Supply of Improved Tool Kits to Rural Artisans (SITRA).** This scheme was launched in 1992 in some selected districts, at present covers all districts of the country. Under the scheme improved hand tools were provided to the rural craft persons under the ceiling of Rs. 2000. Age bar of the TRYSEM scheme was removed under this scheme and people of all age groups are trained under this scheme.

d. **Ganga KalyanYojana (GKY).** The GKY started in 1996-97 to support small scale and marginal farmers, specifically belonging to backward classes (BCs), scheduled castes (SCs) and scheduled tribes (STs). The scheme provides for the subsidized loans to the farmers forborewells and pump sets. Thus, the scheme aims at providing the infrastructure to the farmers for better productivity.

2. ***Wage Employment Programs (WEP):*** Anti-poverty strategies (a few are exemplified in Fig. 3 as reason of migration), like assistance to the rural poor families to bring them above the poverty line by ensuring appreciable sustained level of income through the process of social mobilization, training and capacity building was included under wage employment programs. These programs provide for the employment opportunities during lean agricultural seasons and natural calamities like fold and drought by creating rural infrastructure which supports further economic activity. It includes the following programs:

a. **Swarnjayanti Gram SwarozgarYojana (SGSY).** The program aims at uplifting the rural households above poverty line by ensuring sustained level of income over a period of time. Self Help Groups (SHGs) are organized through process of social mobilization and capacity building.The SHG approach helps the rural household to build their self-confidence through community action. This process would ultimately lead to the strengthening and socio-economicempowerment of the rural poor as well as improve their collective bargaining power.

b. **National Rural Employment Guarantee Act (NREGA).** NREGA is an act of parliament. It aims at enhancing the livelihood security of the people in rural areas by guaranteeing hundred days of wage employment in a financial year, to a rural household whose members volunteer to do unskilled manual work. The objective of the Act is to create durable assets and strengthen the livelihood resource base of the rural poor.

c. **Employment Assurance Scheme (EAS).** EAS was launched in October, 1993 covering 1,778 blocks under drought-prone, desert, and tribal and hill area. The scheme was launched to provide employment in form of manual work in period when employment on agricultural activities is negligible. Food for work program was also started in 2000-01 as a component of EAS, and provide for the food through wage employment.

d. **Rural Housing.** Initiated in 1985-86, the IAY is the core program for providing free housing to families in rural areas. It targets scheduled castes (SCs)/scheduled tribes (STs), households and freed bonded laborers. The rural housing program has certainly enabled many BPL families to acquire permanent houses. Other schemes like SamagraAwasYojana (SAY) werealso initiated to support IAY.

Possible Attempts to Restrict Rural Youth Migration

To address the problem of youth migration it is inevitable to make the agriculture as the foundation stone for employment generation. This will facilitate increase in the workforce in the agriculture sector,and thereby, will help in increasing productivity of agriculture. This approach will open the opportunity to generate employment and improve the social architecture of rural India.In one hand, this will improve the rural condition in India and on the other hand growth in agriculture sector will stimulate the growth in the allied sectors as well, which will provide conditions for greater employment opportunities in other sectors as well. Thus,scenario and environment associated to the overall growth of the economywill develop. The need is for adoption of the policies with priority for education and up-gradation of skills of rural youth instead of providing them with menial work with meager salaries. The linkages should be strengthening between the agricultural research and the extension activity, and one approach could be establishment of village-based farm schools for giving training about agriculture. Credit flow for the agriculture should be increased and streamlined and crop loans and crop insurance should be made more accessible. Linkages should be made between the markets and farmers, these chains would therefore support the farmer-oriented policies.

CONCLUSION

Today, youth have unmitigated access to cheap and easy means of transport and are more likely than ever to migrate for reasons of distress or better employment, or for educational or economic opportunities (Fig. 3, Table 1). Despite their increasing mobility and labour market agility, Indian youth are largely absent from migration policies and labour discourse in the country. Further, the developmental policies have failed to meet the demands of masses as they are imposed vertically over the masses and lack of awareness in the people about these policies. Besides the reasons expressed in Fig. 3 and Table 1, the main causes of failure of developmental schemes for not delivering are: (1) no of schemes are troublesomely large, thus it is difficult to practically manage the schemes, (2) these schemes are launched by different authorities which have either no or negligible coordination among themselves, and hence they overlap on each other, and (3) implementation and effective reach of the programs to targeted population is slow[4]. Relative high rate of unemployment and low literacy rates show that potential of India's youth is underutilized and this untapped resource can form the driving force of Indian economy. Moreover, migration is an integral phenomenon in the rural population and results in imbalance in the workforce in rural and urban areas. It is important to understand the changing needs of the rural youth and address the same with targeted policies and schemes. Instead of subsidizing the demands of the rural people, emphasis should be given to skill development and self employment and youth should be educated about their rights. Training and counseling initiative should be undertaken for meeting this goal, public–private partnership can be viable alternative in reaching the goal. In conclusion, migration is a national phenomenon with deep rooted consequences; this should address with comprehensive policies for self employment, where agriculture sector can be of immense importance.

REFERENCES

http://www.thehindu.com/ (accessed on October25th, 2014).

http://india.gov.in/topics/agriculture (accessed on October25th, 2014).

http://www.fao.org/ (accessed on October26th, 2014).

http://planningcommission.gov.in/(accessed on October26th, 2014).

Chandrasekhar, S. and Sharma, A. (2014) Internal Migration for Education and Employment among Youth in India,Indira Gandhi Institute of Development Research (IGIDR), Mumbai, India.

Data highlights, Census of India, 2001.

Gangopadhyay, D., Mukhopadhyay, A.K., and Singh, P. (2008) Rural Development: A Strategy for Poverty Alleviation in India. Science and Technology, India.

Government of India (2010) Migration in India 2007-2008. Report No. 533, National Sample Survey Office, National Statistical Organisation, Ministry of Statistics and Programme Implementation.

Klugman, J. and Pereira, I.M. (2009) Assessment of National Migration Policies. Human Development Research Paper No. 48. New York: Human Development Report Office, UNDP.

NUEPA (2012) Secondary Education in India –Where Do We Stand? National University of Educational Planning and Administration, State Report Card 2010-11.

Rajan, S.I. (2013) Internal Migration and Youth in India: Main Features, Trends and Emerging Challenges.Discussion Paper,Centre for Development Studies (CDS), Kerala, India.

Registrar General and Census Commissioner of India. 2001. Census of India 2001. New Delhi: Ministry of Home Affairs, Government of India.

CHAPTER - 15

The Falling Farm Hands for Indian Agriculture: Understanding the Brain Drain of Rural Youth from Agriculture Farming to Non-Agricultural Sectors

Punit Kumar Agarwal[1], Pushpa[2], Smita Sirohi[2]
[1]Research Scholar, Dept. of Agriculture Economics, Banaras Hindu University Banaras, [2]Dairy Economics, Statistics & Management, National Dairy Research Institute, Karnal

ABSTRACT

Economic growth of a country depends upon the skills of its people and the application of knowledge, inherited or acquired by them. Optimum utilization of human resources undoubtedly leads to considerable degree of economic development of a country. There are over one billion youth (aged 15 – 24) in the world today. Eighty five (85) per cent of these youth live in the developing world, where 99 per cent of population growth is occurring. Around 50 per cent of the population in developing countries live in rural areas. Some 61.5 per cent of youth live in Asia. Many youth are unable to reach their potential because of poverty and associated hunger and poor health. Also, education and training is often of poor quality and is irrelevant and inappropriate for the needs of rural youth and of the labour market. So although many youth today are more educated than their parent's generation, they often do not possess employable skills. Even when good education and training are available and youth have been well trained, obstacles still exist, such as social customs (for example, caste and traditional values), lack of support mechanisms and lack of access to capital. Corruption is also a problem, as are structural impediments (institutional and governmental). Also, because youth numbers are growing faster than the rate of job creation, too few decent employment opportunities are available for the number of youth seeking them. The result is that a large number of youth are unemployed or underemployed. Not only do these youth lack income, they lack a means of gaining respect and a sense of belonging in their communities. Inability to find decent employment means youth often find employment in the informal sector, with poor working conditions and pay. Rural youth will often look to urban areas for employment and move to cities that adversely affect the household farming in India.

Key words: Migration, rural youth, farming, agri-clinic, agri-tourism

INTRODUCTION

Agriculture provides the principal means of livelihood for over 52 per cent of our population which lives in the rural areas. It is reported that 57 per cent of urban migrant households migrated from rural areas whereas 29 per cent of rural migrant households migrated from urban areas. Further, approximately 55 per cent of the households in rural areas and 67 per cent of the households in the urban areas had migrated for employment related reasons. Loss of job opportunities in agriculture is the primary factor of driving people away from agriculture. Agrarian crisis can be gauged from the fact that 240, 000 debt-ridden farmers committed suicide between 1995 and 2009. There is urgent need to slow the rate of migration from rural to urban areas to strengthen the agriculture. Thus, there is need for on-farm and off-farm rural employment to combat rural poverty and to secure adequate livelihood within the households of smallholders and land-less agricultural labourers. Higher growth in agriculture is needed for sustainable food security of the country and with the introduction of food security bill, the need for food grains will further increase. Thus, there is need to strengthen employment opportunities in the rural areas to sustain agriculture and agriculture based industries. There is need to arrest this trend. Rural India remain the focus of policymakers as 10.1 per cent of its labour force is unemployed compared with 7.3 per cent in urban areas. In rural areas, agriculture continues to be the dominant employer (52 per cent), followed by construction (7.2 per cent), manufacturing (6.7 per cent), and community services (6.3 per cent). Although there is a case for strengthening well-conceived schemes guaranteeing minimum employment, they could at best be bridge solutions. Upgrading the skills of the rural workforce to enable it to match the requirements of the manufacturing and services sectors would provide more lasting results. Agriculture is the most important component of development in the rural areas. Agriculture accounts for about 10 per cent of the total export earnings and provides raw material to a large number of industries. Agriculture in India uses 52 per cent of the work force, contributes 14 per cent of Gross Domestic Product of the country.

Interstate Migration in Rural India During 2007-2008: Few Evidences

It was estimated that almost a third of Indians (i.e. some 325 million people, out of a population of 1.14 billion in 2008) are migrants. The migration rate (proportion of migrants in the population) in the urban areas (35 per cent) was far higher than the migration rate in the rural areas (26 per cent); however, migration in India is largely confined to within the same state. 72 percent of migrant households in urban areas and 78 percent in rural areas have migrated within the same state.

Employment seems to be the most important reason for migration; in rural areas, 55 per cent of the households have migrated for employment related reasons. Migration rate in rural areas was lowest among the scheduled tribe (ST), nearly 24 per cent. For rural male, migration rate was lowest (nearly 4 per cent) among the 'not literates', and it was nearly 14 per cent among those with educational level 'graduate and above'. Among the migrants in the rural areas, nearly 91 per cent have migrated from the rural areas and 8 per cent have migrated from the urban areas, whereas among the migrants in the urban areas, nearly 59 per cent migrated from the rural areas and 40 per cent from urban areas. The reason for migration for male migrant was dominated by employment related reasons, in both rural and urban areas. Nearly 29 per cent of rural male migrants and 56 per cent of urban male migrants have migrated due to employment related reasons.

A higher percentage of the persons were found to be engaged in economic activities after migration: for males the percentage of workers have increased from 51 per cent before migration to 63 per cent after migration in rural areas while for females, it has increased from 20 per cent to 33 per cent in rural areas. For rural males, self-employment has emerged as main recourse to employment after migration. The share of self-employment in total migrants have increased from 16 per cent before migration to 27 per cent after migration, while the shares of regular employees and casual labours remained almost stable, in both before and after migration. (*Source: National Sample Survey 64th Round)*

Rural Job Scheme's are Good Substitute, but has its Problems

Economics has a strange way of looping back and whacking you on the back of your head. Governments take great pride in formulating grandiose schemes that are purportedly pro-poor but end up hurting the intended beneficiaries. Sometimes, the intentions are honest but the execution is poor, without much thought given to entire chain of consequences. With the election bugles having been sounded, speculation is rife about the contents of each party's manifesto. Some of it will be very predictable. For instance, the Congress will definitely try to display two large feathers — one for claiming to have sewn up the nuclear deal with the US and another on the spectacular success of its rural employment programme, the National Rural Employment Guarantee Scheme. Time for a quick look at the last one. The NREGS was implemented nationally through an Act in 2005 to provide a social safety net for the most vulnerable section of rural society. The scheme's operational guidelines define the NREGS goals: "Through the process of providing employment on works that address causes of chronic poverty such as drought, deforestation and soil erosion, the Act seeks to strengthen the natural resource base of rural livelihood and create durable assets in rural areas. Effectively implemented, NREGA (NREG Act) has the potential to transform the geography of poverty." The Act also stipulates that every worker be paid minimum wages fixed by the state for agricultural workers. This is where economics starts playing up — unseen and unheard. It works in two ways. One, where market wages are higher than the minimum wage rate (or, even the wage rate fixed under the particular works programme), the demand for labour for NREGA projects is never met fully. But — and this is the fun part of economics — the converse also holds true. In many cases, where the state has fixed a high minimum wage rate, which then acts as the floor in that localised market, other agricultural enterprises find it difficult to employ labourers at economical rates. One can argue that providing minimum wages isn't such a bad thing. For one, it provides an administered floor — beyond which wage rates can't possibly fall — to minimise economic exploitation. But, where political compulsions and one-upmanship lead to rising wage rates, it could even affect the cropping pattern of that state and lead to shortages of certain crops. A recent report by Swiss bank Credit Suisse predicts that many farmers in western Uttar Pradesh are likely to replace their sugar crop with a rice wheat crop. One of the reasons responsible for the crop-switch is the rural employment scheme. The report says: "The government's national rural employment guarantee scheme (NREGA) has resulted in shortage of agricultural labourers in this region. Migration of labour from eastern UP and Bihar has slowed considerably as work is easily available under NREGA. The average daily labour cost has increased from Rs 50-60 per day to Rs 80-100 per day (NREGA daily wage is Rs100 per day in UP).

Labour shortage and cost affect cane more than R-W (rice-wheat). For one, cane requires about 75-90 man days of labour per acre per year, while R-W requires about 60 man days per acre per year. Moreover, many jobs in R-W cultivation is readily mechanisable (harvester combines, threshers, etc, readily available for hire in the region), while cane is not amenable to mechanisation (due to lack of or high cost of machinery in the area or due to lack of technology)." There is another government policy — ostensibly to insulate poor farmers from the vagaries of an unpredictable marketplace — that is also boomeranging on the people it's supposed to protect. Many years ago, the government introduced the minimum support price (MSP) to assure farmers of a reasonable price for their produce. The government announces a price at the beginning of each crop cycle, thereby providing an artificial and administered floor price below which market prices can't fall. Sometimes, competitive politics forces the government to announce further price increases in mid-season. (Economic Times 16/3/09)

Now it's Time to Engage Agricultural Graduates and Youth in Household Farming

Government of India has launched various schemes like Farm Graduate Scheme, TRYSEM etc. to engage the unemployed agricultural graduates and rural youth in setting up of such ventures in agriculture sector which will not only accelerate the agricultural production with the dissemination of latest technical know-how to the farmers but also provide employment opportunities to lakhs of other unemployed youths of the country as well as the society. In such ventures, agri-clinics and agri-business centres, farmers'society are the most important. The agri-clinics and agribusiness centres are envisaged to provide expert services and advice to farmers on cropping practices, technology dissemination, crop protection from pests and diseases, market trends and prices of various crops in the markets and also clinic services for animal health etc. which would enhance productivity of crops and animals. Agricultural graduates may choose from a range of high potential areas given below to set up different ventures related to agriclinics and agri-business centres.Agricultural graduates who come from farming families and have an aptitude for agriculture can been couraged to undertake commercial farming on their own or leased land and/or to establish agroindustries and agri-services from which they can earn far higher income than in salaried employment. Special technical support can be provided to these agricultural and agri-business entrepreneurs to help them achieve financial self-sufficiency and high incomes. The Scheme of AgriClinic and Agri-Business Centres was launched in 2002 as a followup of Finance Minister's budget speech for 2001.The objective of the scheme is to provide fee based extension and other services to the farming community and also to create self-employment opportunities for agriculture graduates. The agriculture graduates are provided training in agribusiness development for two months in over 67 institutions in public/private sector located throughout the country and coordinated by National Institute of Agriculture Extension Management (MANAGE). These institutions also provide hand holding support to the trained graduates for aperiod of one year. The entire cost of training and handholding is being borne by the Government of India. Trained graduates are expected to setup Agri-Clinic and Agri-Business Centres with the help of bank finance. The scheme is being implemented with the help of Small Farmers Agri-Business Consortium (SFAC), MANAGE and

National Bank for Agriculture and Rural Development (NABARD). The training is open for graduates in agriculture and any subject allied to agriculture like horticulture, sericulture, veterinary sciences, forestry, dairy, poultry farming, fisheries, etc. Now the facilityhas been extended to the youth with agriculture diploma of one year after 10+2 examination. After the training, these graduates and diploma holders can start different agri-businesses and loan up to Rs. 20 lakh is provided by the NABARD and other nationalized banks out of which around 40 per cent is financial assistance. Agriculture graduates and diploma holders interested in starting such projects can take financial help from National Horticulture Mission, Nationalized Banks, NABARD, National Horticulture Board, State Cooperative Banks.

Why There is a Need to Engage Youth in Agriculture?

Young people are three times more likely to be unemployed than adults and more than 75 million youth worldwide are looking for work, according to the UN International Labour Organization (ILO). Due to their limited access to assets (in particular land), markets, finance and education and skills training, youth are often unemployed or work in formally – often in unpaid, very low-skilled, insecure and sometimes hazardous jobs (IFAD, 2012). Most of the educated youth find agriculture as an unattractive preposition; especially the way it is practiced traditionally by their parents. The society largely looks down upon farming, as also families of prospective brides do not prefer farming youth.

When specific youth policies do exist in developing countries, they often do not cater for poor rural youth but tendto be biased towards non-poor males living in urban areas. Consequently, there is growing disenchantment among rural youth towards agriculture vocation and they find it much lucrative to migrate to cities to do even menial jobs.

Large-scale migration of rural youth from farming to urban areas has caused concern among the agricultural policymakers, since, such a trend, if not checked, is likely to affect agricultural activities in the future. Thus, checking migration and retaining youth in agricultural sector is currently a big challenge.

On Going Programmes of Govt. to Stop Brain Drain of Rural Youth

Nehru Yuva Kendra Sangathan(NYKS): Established in 1987-88 as an autonomous organization under the Ministry of Youth Affairs and Sports, it has been channelizing the power of youth (13-35 yrs) on the principles of voluntarism, self-help and community participation. Over the years, NYKS has established a network of youth clubs in villages, where Nehru Yuva Kendras (501 at the moment) have been set up,harnessing youth power for development by forming Youth Clubs, which are village level voluntary action groups of youth at the grassroots level to involve them in nation building activities, working for community development and youth empowerment. However, only a few NYKS are involved in agricultural development. Much more could be achieved if NYKS could focus on skilling rural youth in agricultural activities.

Krishi Vigyan Kendra (KVKs): Imparting need based vocational training to farmers, farm women & rural youth to change their knowledge, skill & attitude as a result uplift the standard of living, is the main mandate of the KVKs. KVKs numbering over 631, exist practically in every district of India. A total of 4.86 lakh extension programmes/activities were organized by the KVKs during 2012-13, which attracted the participation of 170.16 lakh farmers and 2.61 lakh extension personnel. These KVKs organized 65,314 training programmes under capacity building with the participation of 18.8 lakh farmers/farm-women, rural youth and extension personnel during 2012-13.

NABARD Farmers' Clubs: NABARD's policy support for Farmers' Club Programme emphasizes on linking technologies with farmers' club members, while facilitating market access through capacity building of members of Farmers' Clubs including leadership training; linkage with technology/markets; Self Help Groups (SHGs)/Joint Liability Groups (JLGs) formation and forming Federations of Farmers' Clubs/Producers' Groups/Companies. Under this programme, the NABARD has so far (March 2013) assisted formation of 1.27 lakh farmers' clubs across the country. These clubs are organized by rural branches of banks with the support and financial assistance of NABARD for the mutual benefit of the banks concerned and the village farming community/rural people.

New Initiatives for Attractive Youth in Agriculture Business

ARYA (Attracting and Retaining Youth in Agriculture): The Indian Council of Agricultural Research (ICAR) has constituted a seven-member expert committee, to suggest ways of attracting youth to agriculture. The committee is working on issues such as providing training to farm youth on innovative and sustainable agricultural practices to help make agriculture a profitable venture. This would involve a series of activities, including exploring opportunities in secondary agricultural operations such as value addition of crops and hiring, and servicing of mechanized farm implements. This is going to be an important strategy of the ICAR focused on rural youth mobilization for agricultural transformation during 12th plan ICAR Reporter (October-December-2012).

National Rural Livelihood Mission (NRLM): NRLM aims at creation of opportunities for both wage employment and skill development for the rural youth, who lack skills in many areas of agricultural production and processing. National Skill Development Mission and the National Skill Qualification Framework are, thus, aggressively pushing the agenda of skill development to build the capacity of rural youth so that they are meaningfully employed in rural areas itself. Thus, need based experiential skill learning supported by public sector banks/organizations in rural areas is the key to strengthen the Rural Self Employment Training Institutes (RSETIs) being set up in all districts under NRLM to assist such youth (Likhi, 2013). NRLM requires suitable decentralized convergence of skill development programs run by multiple central ministries including the National Skill Development Corporation (NSDC).

Agricultural Skill Council of India (ASCI): Considering the need for skilling the work force in agricultural sector, the Agricultural Skills Council of India (ASCI) has been recently proposed by National Skill Development Corporation (NSDC), which could be one ideal institution to train rural youth. The ASCI proposes to train, certify and accredit 56.5 million workforce comprising of farmers, wage workers, entrepreneurs and extension workers, over 10 years through its training partners.

ASEAN-India Farmers Exchange: The Ministers of the ASEAN (The Association of South East Nations) concerned over the small number of young farmers' involvement in the agriculture sector, agreed on the importance of promoting innovation and entrepreneurship among young farmers to achieve more sustainable agriculture development in the region. This initiative is expected to create greater awareness among the young and innovative farmers on the promising career in the agriculture sector. The 1st Exchange Visit was conducted in Malaysia in conjunction with the 2012 ASEAN Farmers' Week and the Malaysian Agriculture, Horticulture and Agro-tourism (MAHA) International in November 2012, followed by the 2nd Exchange Visit conducted in India during December 19-30, 2012, wherein, farmers' delegation from nine ASEAN member states participated.

Agri-Tourism

In 2007, ATDC Launced Mahrashtra State Agri Tourism VistarYojana 2007 where in 52 farmers were selected to start Agri-tourism centers across the entire state of Maharashtra. Similary in 2008 – another 25 farmers were selcted, thus totally 72 agri tourism centers are operational currently in the entire state. Thus Maharashtra State Agri & Rural Tourism Co-operative Federation Ltd (MART) was registered on 12th Dec 2008, with main objective to coordinate the activity of planning, financing, marketing and liaison with various State and Central Agencies. The governing body of Federation has 12 representatives from agri& rural tourism co-operatives as well as 11 individual farmers who runs agri and rural tourism center along with women directors from the individual farmers and 2 government nominated directors. The Agri-tourism co-operative societies get financed from National Bank for Agriculture and Rural Development

(NABARD).The loan is utilized in creating infrastructure facilities in the village to support the tourism. Farmers main objective is aimed at utilizing the potential of agriculture farm by attracting tourists. Farmers carry-out normal agriculture practice during the week days and attend tourists in farm during the weekend days. This kind of arrangement did not interfere in the routine work of agriculture. Besides, the farmers generate additional income from the agriculture site using it for tourism purpose on the weekend. This has showed a source of additional income to farmers apart from allied agriculture activities like diary, poultry.

Agri-clinic

The Ministry of Agriculture, Government of India, in association with NABARD has launched a unique programme to take better methods of farming to each and every farmer across the country. This programme aims to tap the expertise available in the large pool of Agriculture Graduates. Irrespective of whether you are a fresh graduate or not, or whether you are currently employed or not, you can set up your own AgriClinic or AgriBusiness Centre and offer professional extension services to innumerable farmers.

Committed to this programme, the Government is now also providing start-up training to graduates in Agriculture, or any subject allied to Agriculture like Horticulture, Sericulture, Veterinary Sciences, Forestry, Dairy, Poultry Farming, and Fisheries, etc. Those completing the training can apply for special start-up loans for venture.

CONCLUSION

Due to high rate of migration and brain drain from more and more working hands and intellectual mind from rural areas to developed urban area in the search of job, affecting agricultural activities and agriculture production on the country. To attract agricultural graduates in household farming, central and state govt. has to take serious steps. Though it is true that attracting and retaining youth in agriculture is critical for Indian Agriculture. Most of the new innovations (both technical and institutional) require a skilled agricultural work force. For instance, promotion of high value agriculture, precision farming, organic cultivation, Hi-Tech horticulture, micro-propagation, Integrated Pest Disease & Nutrients Management, Post Harvest Management, development of backward and forward linkages etc, require well trained young farmers with enthusiasm and passion for farming and ability to take risks. The rural youth could be the ideal target for skill training in these new areas of agricultural growth and to do this effectively there is a need to mobilize young farmers. Organised groups of young farmers will be useful for introducing new production technologies and organizing effective input and output markets.India lives in rural areas and there is need to do every effort to make the lives comfortable and lucrative. There is need to develop some more rural centric schemes to enhance the rural infrastructure in terms of manpower and facilities so that the dream of Mahatma Gandhi of Gram Swaraj is realized in every village of the country.

REFERENCES

1. Bennell, P. (2007) Promoting livelihood opportunities for rural youth. Paper prepared for IFAD GoverningCouncil Roundtable: Generating Remunerative Livelihood Opportunities for Rural Youth. UK:Knowledge and Skills for Development.
2. DARE/ICAR Annual Report-2012-13, Page 93-99 Retrieved from http://www.icar.org.in/files/reports/icar-dare-annualreports/2012-13/technology-assessment-transfer-12-13.pdf
3. Gale F. (2002), The Graying farm sector: Legacy of Off-farm Migration, Rural America, Vol. 17, Issue 3.
4. Gardner B.L. (1992), Changing Economic Perspectives on Farm Problem, Journal of Economic Literature, Vol. 30, No. 1
5. Ghosh J, Chandrashekhar C P, Roychowdhury A, (2006), The 'Demographic dividend' and Young India's Economic Future, Economic and Political Weekly. Vol. 41, No. 49
6. GhoshJayati and C.P. Chandrasekhar (2003) Why is Agricultural Employment Generation Falling? www.macroscan.org, April 22.
7. Gladwell, M. (2000). The Tipping Point: How Little Things Can Make a Big Difference, BackBay Publishing.
8. ICAR Reporter (October-December-2012) from the DGs desk. http://www.icar.org.in/files/ICAR-Reporteroctober- december-2012.pdf
9. Glauben T, Tietje H, Weiss C, (2003) Agriculture on the Move:Exploring Regional Differencesin Farm Exit Rates, Working Paper EWP 0308, Department of Food Economics andConsumption Studies, University of Kiel.
10. Likhi, Abhilaksh.2013. Challenges for India's Livelihood: Youth Skill Development in Rural Areas.http://blogs.worldbank.org/publicsphere/challenges-india-s-livelihood-youth-skill-developmentrural-areas
11. Planning Commmission (2012) Working Group on Agricultural Extension for Agriculture and Allied Sectors, The 12th Five Year plan (2012-17), Planning Commission, Government ofIndia. http://planningcommission.gov.in/aboutus/committee/wrkgrp12/agri/wg_agriextn.pdf
12. NSSO, 55th round (2003). Migration in India, (July 1999 to June 2000), National Sample Survey Organisation, Ministry of Statistics & Programme Implementation, Government of India.

CHAPTER - 16

Migration Check Among the Rural Labourers through NREGA

A.S. Tigga
Asst. Prof.-cum-Jr. Scientist, Department of Extension Education, Bihar Agricultural College, Sabour, Bhagalpur

INTRODUCTION

The trend of rising unemployment is compounded by the existences of regional imbalances in development, within the country, which have collectively accelerated the phenomenon of migration. The history of migration is the history of people's struggle to survive and to prosper, to escape insecurity and poverty, and to move in response to opportunity. In developing countries like India, migration mostly takes place not due to the so called the full forces of the destination place, as usually happens in case of developed countries, but because of poverty, unemployment, natural calamities and under development of the origin place. Migration in developing countries like India is still viewed as a survival strategy. In India internal mobility is critical to the livelihoods of many peoples, especially for people from rural areas. In the Dindigul District of Tamil Nadu, NREGA works have stemmed migration from many villages. Ministry of Rural Development (2006). It has been conceptualized that, creating job out of NREGA programme will check the migration of the people from one place to another. In Pusa block, of Samastipur district the NREGA is in operation for about two and half years. It is with this background the present article has been presented with the specific objectives toexplore the extent ofmigration check among NREGA beneficiaries, running in this block.

METHODOLOGY

There are altogether 20 blocks in Samastipur district. Out of these Pusa block has purposively been selected for this study. Further, two panchayats from this Pusa block have been chosen by following the random sampling technique. Then again from each Panchayat, two villages selected following the same technique. In this way, the locales of the study were four villages selected from two Panchayat of Pusa block of Samastipurdistrict. For selection of respondents, simple random sampling techniques have been followed. From each selected village, a total of 30 respondents consisting of 15 male card holders and 15 female card holders selected from the list of NREGA card holder by following the random table. In this way the sample respondent of the study was consist of 60 male and 60 female NREGA beneficiaries/card holders. Karl Pearson's coefficient of correlation and multiple regression analysis, besides frequencies and percentage, were used to analyze the data. In order to measure the awareness of beneficiaries about theNREGA and its different provisions, which might affect the benefit of NREGA, to its beneficiaries, awareness test was developed and comprehensive score was obtained by the respondents on the question framed for the purpose.

RESULT AND DISCUSSION

Extent of check in migration of beneficiaries due to NREGA has been classified intocompletely checked, moderately checked, little checked and not checked. In this regards information have been gathered from beneficiaries as follows.

Table 1: Distribution of NREGA beneficiaries (Male & Female) according to the level of check in migration

Category	Male (n=60)		Female (n=60)		Total (n=120)		Rank
	F	%	F	%	f	%	
Completely checked	6	10	6	10	12	10	IV
Moderately checked	35	58.4	37	62	72	60	I
Little checked	11	18.3	12	20	23	19	II
Not checked	8	13.3	5	8	13	11	III

From the table 1, it is clear that, majority of the male beneficiaries reported about the moderately check in migration (58.4%), followed by little and no check (18.3%) and (13.3%). There is only 10 per cent of the male beneficiaries favored about the complete check in migration. In case of female NREGA beneficiaries, the situation is almost same. As the moderately check in migration had taken its share with (62%), followed by little completely and not checked category of migration check with the 20, 10 and 8 percent of the beneficiaries. In the case of totality it was found that migration check of the category of moderate check had again leaded with 60 percent of the beneficiaries.

Table 2: Benefits due to check in migration

S.N	Benefits	N=120		
		f	%	Rank
(i)	Enhanced livelihood security	108	90	I
(ii)	Enhanced quality of life	98	81.6	IV
(iii)	Potential for physical and emotional growth	92	76.6	V
(iv)	Increase in accessing to basic health	105	87.5	II
(v)	Increase in accessing to basic education	100	83.3	III
(vi)	Able to enjoining full civil and political rights	70	58.3	IX
(vii)	Able to participate fully and equally in public life	78	65	VIII
(viii)	Gender based violence decreased	80	66.6	VII
(ix)	Saving of money has increased	89	74.1	VI

From table 2, an attempt has been made to know the ranking for the benefits due to the check in migration. From table it is clear that, due to check in migration, the first ranking is for enhanced livelihood security (90%), followed by Increase in accessing to basic health (87.5%), increase in accessing to basic education took 3rd position (83.3 %), followed by enhanced Quality of life (81.6%). Potential for physical and emotional growth took 5th position in ranking with (76.6%), followed by enhanced saving of money (71.4%), and the rest benefits were decreased Gender based violence, able to participate fully and equally in public life (due to stability)and able to enjoining full civil and political rights (due to permanent resident of a particular place, village or town) with the 66.6, 65 and 58.3 per cent respectively.

CONCLUSION

As one of the significant objective of the NREGA is to arrest out migration of unskilled, landless labour force from the rural areas to urban area by ensuring up to 100 days of wage employment within their native jurisdiction so that the 100 days guaranteed wage employment can be judiciously and rationally utilized by the landless peasants during lean and distress seasons. This paper analyses the impact of this scheme in arresting out migration by taking the opinion of NREGA beneficiaries, who have enrolled under the NREGA scheme and who are in possession of job cards. Thus it could be concluded that although there is migration is going on, but there was a remarkable check in migration had taken place simultaneously along with theenhanced livelihood security drastically.

REFERENCES

Kumar, P.A. and Haorei, W. (2010). 'Impact assessment of MGNREGS on rural migration, Kurukshetra,59 (2) : 26-28

Ministry of Rural Development (2008).Performance Audit of Implementation of National Rural Employment Guarantee Act-Performance Audit Report No.11 of 2008 (New Delhi, Ministry of Rural Development, GOI).

CHAPTER - 17

Economic Opportunities for Rural Youth in Farming: Exploring The Alternatives for Migration to Urban Areas

Mahesh Chander[1], Hari R and Prakash Kumar Rathod
[1]Principal Scientist and Head, Division of Extension Education
Indian Veterinary Research Institute, Izatnagar-243122, Uttar Pradesh

INTRODUCTION

Farming acts as the primary occupation of majority of the population in India. According to National Sample Survey Estimate, about 49% of the Indian population is dependent on agriculture and allied industries (GOI, 2014, CIA, 2012) making it the largest source of employment to the people of India. Although, being a second largest food producing country in the world (Sarkar, 2009), the contribution of agriculture to the GDP has been declining. The contribution of agriculture to Indian GDP has declined from 30% in 1991 to 15% in 2011-2012 to 13.7% in 2013 (Economic Survey. 2013-2014). This decline is also accompanied by a decline in the participation of people in Indian farming sector. In fact between 2001 and 2011 about 9 million people have quit farming, which is about 2 million more than the decline noted in the previous decade (GOI, 2011). This figure equates to about 2035 farmers quitting farming per day. The recent census (GOI, 2011) also highlighted that for the first time in four decades an absolute decline (about 7%) in the cultivators has been noted.

The progress and prosperity of any sector depends on the quality of human resources in that sector. Among the working population, the highest quality of human resources lies with the youth as they are endowed with tremendous energy, imagination, dedication, creativity and enthusiasm (Gowda *et al*, 2013). Today, India has the highest number of youth in the world among which 60 percent lives in the rural areas (Ayappan, 2012). Most young people in developing countries such as India cannot afford to remain unemployed for long but, at the same time, struggle to find decent jobs in the rural economy (ILO,2012). Consequently, most youth are underemployed and seeking out a living in the farming sector resulting in large scale exodus to the cities, there by adversely affecting the rural development in general and agricultural development in particular. This situation is very much alarming and this has created a vacuum in the villages along with the deterioration of the quality of human resources in farming. Therefore there is an urgent need to prevent this urban exodus and attract rural youth to rural enterprise especially the farming sector.

Since huge expectations are placed on the farmer of today and on youth who represent the next generation of farmers, such stakeholders should be central to the debate together with representatives of agribusiness and investors in the agriculture sector. This paper raises some of the key emerging issues and identifies few of the opportunities in farming for youth as an alternative for migration to urban areas. Further, the paper also highlights the support of Government for youth in the form of various schemes and programmes to attract youth in agriculture.

Youth Defined

Youth is often understood to be the period of transition from childhood to adulthood, encompassing processes of sexual maturation and growing social and economic autonomy from parents and carers (Bennell, 2007).There is no universally accepted definition of youth, since the age ranges anywhere from 8 to 40 yrs. On the occasion of the International Youth Year in 1985, the United Nations General Assembly for the first time defined youth as people between the ages of 15 and 24 for its work on youth (with under 15s being classed as 'children'). The Government of India officially defines youth as persons between the ages of 13 and 35 years and it also varies depending on the programme. For instance, the National Youth Policy of India considers age group 10 to 34 years as youth. The United Nations (UN) and the International Labour Organisation (ILO), however, defined the youth as persons between 15 and 24 years of age for cross country comparison and analysis.

Population of World Youth (1950 and 2050)

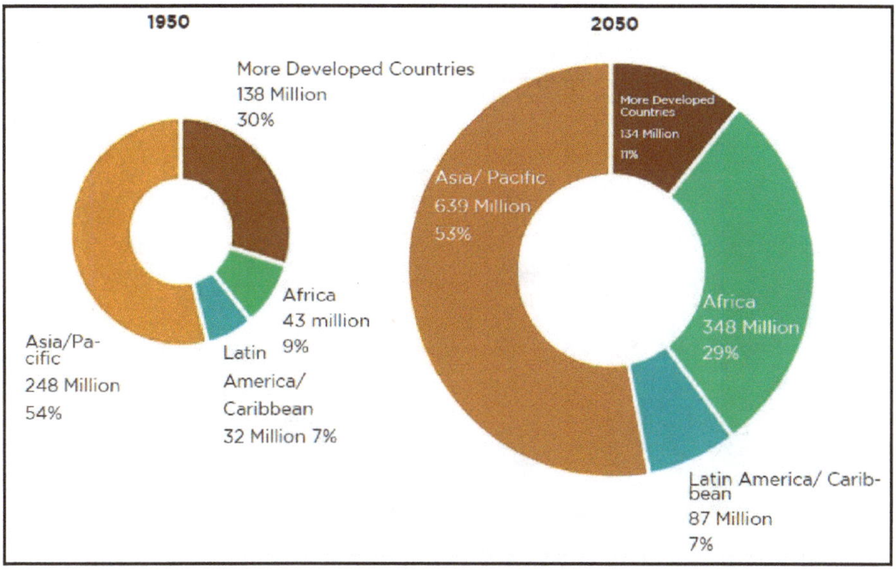

According to 2011 Census (GOI, 2011)., the youth population in the country including adolescents is around 550 million. In 2020, the average Indian will be only 29 year old, whereas in China and the United States of America the average age is estimated to be 37 years. We may utilize this demographic dividend for taking Indian agriculture to new heights by channelizing the creative energies of the youth through development of skills, knowledge and attitudes (Ayyappan, 2012).

Youth in Agriculture

Youth unemployment rate of India (9.9%) is higher than the total unemployment rate (3.7%) (GOI, 2014) and this is higher for rural youth (13.6%) compared to their urban counterpart (7.6%) (GOI, 2013). It is widely recognized that smallholder agriculture and non-farm production in rural areas are among the most promising sectors for youth employment in majority of developing countries (Bennell, 2010). Among, the non farm production sectors agro based rural industries also have immense scope for creating new employment opportunities for the youth (Economic survey, 2013-2014). India is losing more than 2,000 farmers every single day and that since 1991, the overall number of farmers has dropped by 15 million (Sainath, 2013). This has several implications for the future of Indian agriculture and India's food security. Young farmers can play an important role in ensuring food security if they are encouraged to involve in farming and the challenges they face are addressed. Over the past few years, rural youth have been shying away from agriculture and globally there is an increasing interest in finding ways of engaging youth in agriculture (IFAD, 2012; Paisley, 2013).

Agriculture population trends and projections to 2020 in key regions (millions)

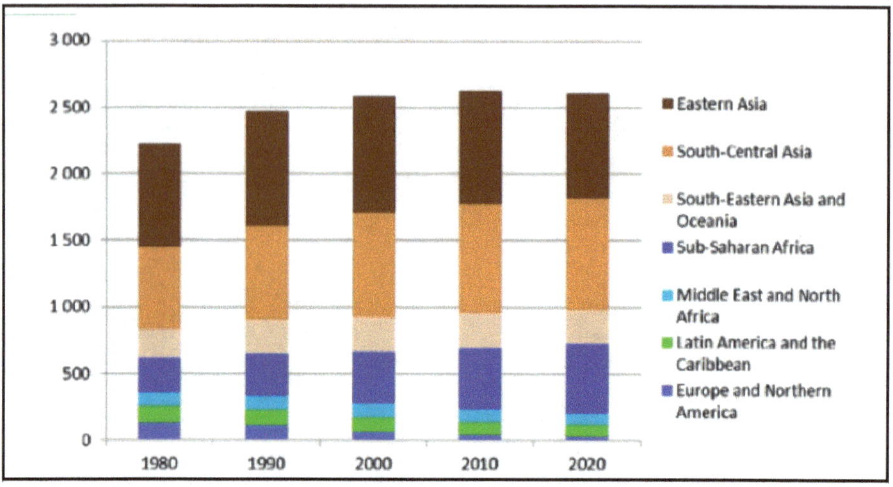

Source : Based on FAO (FAOSTAT)

Youth and Migration

Migration is a coping mechanism to escape poverty and an opportunity for rural young people to assist their family and their community. In addition, young people view migration as an avenue to improve their status, learn new skills, and transit into adulthood. As a consequence, migration continues to serve as the means to improve rural livelihoods (Harris, 2009). Youth are an integral component of the migrant population, both in terms of volume, and the effects they have on both their points of origin and destination (Buckman, 2013). Urban areas are becoming extremely overcrowded and overburdened, putting pressure on insufficient infrastructures, schools, health facilities, sanitation and water systems. This escalating urbanization has created a new context of poverty in which urban centres are overtaxed and unprepared to absorb increasing youth unemployment (Harris, 2009).

On the other hand, for the rural youth who wish to remain on the land, the obstacles include the lack of new arable land for cultivation, inadequate training and extension services, low prices for primary produce, poor markets and infrastructure, low levels of technology and limited credit. Land pressure often results in marginal and unsuitable land being used for cultivation leading to soil degradation and erosion, deforestation and other environmental problems of the region. For many, agriculture is thought of as a low status job and a last resort, which adds to the pull of urban life. Average earnings in most countries of the region are higher in urban than in rural areas (FAO,1991).

Although various census and surveys have pointed about the issue of migration, the exact depth of this disenchantment is not depicted by any census or survey as they are far beyond the measurement of any quantitative figure (Gupta, 2005). But these statistics point to the fact that Indian agriculture is moving from crisis to crisis. Not only has agriculture failed to provide gainful employment for rural youth, but also stagnating productivity levels and environmental degradation has made it the last resort for the rural youth (Tadele and Gella, 2012).

According to the planning commission report (2013) the poverty levels of rural India is 25.7%, which is much higher than the urban poverty level of 13.7%. This higher percentage of poverty in rural India is primarily due to agrarian distress, a chronic lack of employment, and farmer suicides. Migration statistics of India revealed that about 28.5% of the total population of India is migrants which account about 326 million people (NSSO, 2010). Of the total migrants 67.2% are from rural areas (Srivastava, 2011). As per the NSSO report, 2010 the major reasons for migration among the females are marriages and for males is employment. About 40% of the male migration is primarily for the purpose of employment and it accounted for the largest reason for migration among the males (NSSO, 2010). A detailed look at the nature of laborers (main and marginal) involved in farming and cultivators with very less land is the major group of people who move out to cities for jobs. Also, majority of these are seasonal migrants and there is a preponderance for illiterates to migrates (Rajan, 2013)

These data provides strong evidence to the fact that majority of the rural migrants to cities are young, marginal farmers and agricultural laborers who are disillusioned with the farming sectors. If drain of rural youth continues, India will be unable to exploit its demographic dividend in its farming sector. The International Labour Organization (ILO) report on the Global organization trends for youth states that "today's youth represent a group with serious vulnerabilities in the world of work. As a result, today's youth are faced with a growing deficit of decent work opportunities and high levels of economic and social uncertainty" (ILO, 2005). Lack of decent employment rather than open unemployment is the key problem in majority of the rural areas of world (Van der Geest, 2010) and this stands true for India also (Hari, 2014).

Reasons for Youth Migration

Agriculture is subject to risks of weather and price volatility that tend to reduce the overall demand for labour and to influence the contractual arrangements under which farm production is carried out. There can be major oscillations in the labour demand in the yearly agricultural cycle which causes seasonal employment-unemployment patterns, variations in the wages earned and results in dominance of casual labour over permanent employment (ILO, 2008). Waged agricultural workers typically find employment for only half the year. This has forced them to move out to cities as the income of agricultural labour gives them little income to sustain them between seasons. This non remunerative nature of the small scale farming is the primary reason for youth migration (Epistein , 2007; Sumberg *et al*., 2012; White 2012). Various studies have shown that youth migration is driven by bad socio economic conditions such as low wages, high rates of unemployment, underemployment in rural areas and poverty (UNICEF, 1995; Mukherjee and Das, 2011). Studies from southern India have pointed out that rural-urban youth migration was aggravated in village settings by reasons like limited land availability, the scarcity of water and a virtual absence of off-farm income-earning (Epstein, 2007). The increasing population pressure cause fragmentation of the land that land holding with each family is very less to continue farming in a feasible way.

While these may be some of the push factors that repel the rural youth from the country sides, there are also a set of pull factors that lure them towards the cities. These include the availability of educational opportunities and other services, and the allure of better paying and higher status jobs, most often in urban areas (Sumberg *et al*, 2012). Migration to urban areas seems attractive to rural youth who are taught urban values through the education systems which instill in them the idea that farming is a degrading occupation (Tadele and Gella, 2012). Infrastructure facilities and recreational facilities that cities offer add impetus to this migration including the fact that media too portray villages in poor lights. From the media, youth perceives the elasticity of jobs in cities as high compared to villages. The intentions to make "quick- large money" in short span propel the youth towards cities and these youth perceive agriculture as their last resort.

Further, few important issues and problems of youth to stay away from agriculture leading to migration are highlighted below as depicted by FAO (1991):

- For the rural youth who wish to remain on the land, the obstacles include the lack of new arable land for cultivation, inadequate training and extension services, low prices for primary produce, poor markets and infrastructure, low levels of technology and limited credit. Land pressure often results in marginal and unsuitable land being used for cultivation leading to soil degradation and erosion, deforestation and other environmental problems of the region.
- Fiscal and legal barriers complicate the farm inheritance as it must be fairly divided between all the heirs and also during the change of ownership.

- The women farmers also face various social and economic issues in India. They also have poor accessibility for information and Extension services.
- In some countries including India, agriculture is regarded as socially unviable, and associated with lack of self-esteem, living hand to mouth, and is not respected by society which offers no attraction to the youth. Young people see their parents struggling and in many instances, youth are discouraged by the family members themselves.
- Agriculture is no more considered as a profitable enterprise that provides income but is considered as a way of life to struggle and survive.
- Agriculture needs to develop a renewed profile – as a viable and innovative, modern sector that can offer decent work and can be attractive to youth.
- Education and the media must play a role in improving agriculture's image by exploring the success stories of youth.

Opportunities for Youth in Farming

Recent changes in the global food market, in science and technology, and in a range of institutions that affect competitiveness are creating new challenges to the competitiveness of smallholders, but are creating income opportunities, too. In order to create jobs, accelerated progress will be needed to increase agricultural productivity and to connect poor people to markets. Sustained growth that reduces rural poverty will require significant growth in agricultural value-added and multi-sector approaches that solve gaps as well as support agribusiness, and rural diversification. To create jobs that will increase rural income and welfare and, thereby, retain young people, it is necessary to increase investments in irrigation, water resources management, and research and extension; increase rural public services; and increase use of improved seeds, fertilizers and better agricultural practices(Zuehlke,2009). It is also necessary to address vulnerability to weather-related shocks and limited farmer capacity, distorted incentives (including Government policies) that keep farmers in subsistence farming, poorly functioning input/ output markets, and weak institutional capacity to manage the risk of food insecurity.

Youth though comprises a highly heterogeneous group are often referred as highly innovative and tech Savvy (Rustam, 2004). This tech savvy nature of youth could be exploited by the establishment of high end technology enterprise like precision farms, high tech dairies and farms for the production of designer agriculture products like designer milk and designer eggs. But it should be noted that most of these sectors are high investment sectors with high cost. So to start such enterprise an initial investment should be made initially by the government with low interest and prolonged recovery periods as these may take a few years to become successful. Another opportunity lies in formation of certain smart villages, with ICT employed tools enabled to maximize the benefits of good climate and minimize the effects of adverse weather. These ICT centres should also be linked to the markets so that demand, cost and other useful market information could be obtained so that maximization of the profit could take place. Educated youth from the villages could act as operators of these ICT centers. Certain Youth organizations in villages have become so innovative that they make direct farm to house delivery to clinch the advantage. The examples of the young women in the "nature Fresh" Dairy venture in Thrissur where hygienic milk is delivered at the door step of the customer within one hour of milking is a model that could be replicated elsewhere (Hari *et al*, 2012).

Another area where youth can lay their hand is the export oriented agriculture market. The globalization trend has created opportunities for an entrepreneur to consider world as his global market. Being more open to new and improved methods than adults, the youth of the country can produce quality products that could be exploited to markets outside the world if they follow the right standards. The educated agriculture and veterinary scholars themselves can start such ventures as they could have a better understanding of these global standards. Organic agriculture products are in high demand in the markets especially in the urban areas of India. The Economic boom in India especially has created a group of high income middle class with high spending power in the cities. Organic farming opens an arena where youth can try their skills to ensure better remuneration. The post harvesting technologies and value addition of agriculture and allied products is also growing very prominently. Production of value added products like Ice creams from milk, chips from potato etc. have been successfully demonstrated by various farmers and youth organizations.

Educated youth in rural areas can start Agri-clinics and Agri-business centres. It is ideal for a group of four to five graduates, who have specialized in agriculture, animal husbandry, fisheries, agribusiness and home science, could jointly launch an agri-clinic-cum-agri-business centre. Agri-clinics will provide the services needed during the production phase of farming, while the agri-business centre will cater to the needs of farm families during the post-harvest phase of agriculture. Thus, farm women and men can be assisted during the entire crop cycle, starting with sowing and extending upto value addition and marketing (Swaminathan, 2011). The multi-disciplinary expertise available within the group of young entrepreneurs will help them to serve farm families in a holistic manner. Educated rural youth have plethora of other opportunities like helping in establishing climate resilient agriculture practices like establishment of rain water harvesters aquafier recharge etc, establishment of farm field schools, preparation and selling of biological softwares like bio pesticides, vermin compost, testing the quality of soil and water, and operating climate risk management centers, to name a few. Youth could also be trained in integrated farming so that the profit earned could be increased and the uncertainties could be reduced. The B/C ratio of IFS is comparatively higher than single crop rearing farms in small holder systems (Manjula *et al.*, 2013).

Opportunities in livestock sector: Livestock rearing is the major subsidiary occupation in most parts of India. But keeping in view its potentiality for generating employment and providing food security to the country, it may be taken up as a major occupation. There are various livestock enterprises, which can be practiced on commercial scale and can be started with small investment on scientific lines by the rural youth.

- Development of commercial dairy farms
- Development of commercial sheep farms
- Development of commercial goat farms
- Prospects of meat export of goat and other animals
- Commercial poultry farms
- Development of piggery farm
- Buffalo calves rearing for meat purpose
- Livestock input industry
- Commercial feed and fodder production
    1. Establishment of area specific mineral mixture unit
    2. Commercial dry fodder enrichment unit
    3. Commercial hay/silage preparation
    4. Growing fodder crops commercially round the year
- Commercial breeding units
    1. Livestock breeding farms
    2. Providing private doorstep artificial insemination (A.I) services
    3. Establishing commercial livestock service centres
- Commercial livestock first aid clinics
- Organic livestock production units
- Livestock product post harvest processing units

- Processing units for important organs of dead animals
- Organic waste management and entrepreneurship generation

Government Initiatives to Promote Rural Youth

Acknowledging the importance of entrepreneurial ventures, various governmental agencies have started various schemes to assist in the development of skill required for starting small scale ventures. Many of these schemes are beneficial for the rural youth of the country and the youth can exploit the facilities of these schemes in starting small agro based rural industries. While many schemes focus on the skill development of the youth, a few support credit facilities and technological up gradation. A few of the entrepreneurial schemes and government initiatives are given below:

Rural Self-Employment Training Institutes (RESTI): Rural Self-Employment Training Institutes (RESTI) are promoted for the purpose of providing opportunities to the rural youth for skill up-gradation leading to self-employment. Rural Development and Self Employment Training Institute (RUDSETI) have also been successful. The Ministry of Rural Development has established RSETIs in all the rural districts of the country. These institutes are bank led i.e managed and run by the public/private sector banks with active support of the State and Central Governments. The RESTI's core offering includes its free, unique and intensive short-term residential training designed for rural youth. It is different from other vocational trainings by the fact that these are demand driven approaches, a priority for rural BPL youth, a wide choice of vocations and continued hand holding to ensure sustainability of micro-enterprise and credit linkage after training

Dairy Entrepreneurship Development Scheme (DEDS): A centrally sponsored scheme implemented through NABARD with an objective to promote setting up of modern dairy farms for production of clean milk, encourage heifer calf rearing, bring structural changes in the unorganized sector and generate self employment.

Trade Related Entrepreneurship Assistance and Development (TREAD): The scheme specially focuses on women. The young women from the villages can exploit the benefit of this scheme. Apart from making credit available to start any venture, the scheme also provides the necessary training required for starting small scale entrepreneurship ventures.

Credit Linked Capital Subsidy Scheme (CLCSS) for Technology Upgradation: The Credit Linked Capital Subsidy Scheme (CLCSS) by Ministry of Small Scale industries aims at facilitating technology up-gradation by providing upfront capital subsidy to SSI units, including agro & rural industry units (within the functional purview of the Ministry of Agro and Rural Industries) in addition to institutional finance (credit) availed of by them from banks, for modernisation of their production equipment (plant and machinery) and technology, both existing SSI units registered with the State Directorates of Industries, and new SSI units which have set up their facilities with the appropriate eligible and proven technology.

Dairy venture Capital Fund and Poultry venture Capital fund: To bring about structural changes in the unorganized sector, the measures like milk processing at village level, marketing of pasteurized milk in a cost effective manner, quality up-gradation of traditional technology to handle commercial scale using modern equipment and management skills, the Venture Capital Fund for Dairy and Poultry sector was started in X plan. However the dairy and poultry components have been separated into two new schemes during 11th Plan viz. Dairy venture Capital Fund and Poultry venture Capital fund. Financial assistance is provided as loan to the rural/urban beneficiaries under a schematic proposal through bankable projects.

Rajiv Gandhi Udyami Mitra Yojana: To provide handholding support and assistance to the potential first generation entrepreneurs, who have already successfully completed or undergoing Entrepreneurship Development Training Programme (EDP) / Skill Development Training Programme (SDP)/ Entrepreneurship cum Skill Development Training Programme (ESDP) /Vocation Training Programmes (VT), through the selected lead agencies i.e. 'Udyami Mitras' , in the establishment and management of the new enterprise, in dealing with various procedural and legal hurdles and in completion of various formalities required for setting up and running of the enterprise.

Aajeevika Skill Development Programme (ASDP): A sub scheme under NRLM, it addresses the issue of capacity building and training of the unemployed youth in the rural areas. The scheme is designed to equip the unemployed rural youth from BPL households with skills which would enable them to either secure placement in the industry or pursue sustainable self employment opportunities through Micro Enterprises.

There are certain other governmental schemes from Agricultural and Processed Food Products Export Development Authority (APEDA) (financial support for packaging development, export promotion and market development etc.), Ministry of Food Processing Industries (MoFPI) (Schemes relating to the Technology Upgradation/ Establishsment/ Modernisation of food processing industries), National Horticulture Board (NHB) (provides financial support for land development, cultivation expenses, poly house or shade nets, farm tools etc.), TEA, Coffee, Coconut and rubber board also provide financial support to young rural entrepreneurs.

Initiatives of Krishi Vigyan Kendra (KVKs): Imparting need based vocational training to farmers, farm women & rural youth to change their knowledge, skill & attitude as a result uplift the standard of living, is the main mandate of the KVKs. KVKs numbering over 631, exist practically in every district of India. A total of 4.86 lakh extension programmes/activities were organized by the KVKs during 2012-13, which attracted the participation of 170.16 lakh farmers and 2.61 lakh extension personnel. These KVKs organized 65,314 training programmes under capacity building with the participation of 18.8 lakh farmers/farm-women, rural youth and extension personnel during 2012-13.

Besides this, 8,486 skill-oriented training courses (both on-campus and off-campus) were organized exclusively for 1.91 rural youth, of which 69,163 (36.17%) were young women. These trainings were organized on various vocations viz. crop production and management, post-harvest technology and value addition, nursery management, livestock, fisheries, income generation activities, capacity building and group dynamics etc. These KVKs also conducted 5,730 capacity building programmes for 1.42 lakh extension personnel, which included 34,563 (24.35%) women extension personnel (ICAR, 2012-13). KVKs need more resources to organize more number of similar trainings.

Nehru Yuva Kendra Sangathan(NYKS): Established in 1987-88 as an autonomous organization under the Ministry of Youth Affairs and Sports, it has been channelizing the power of youth (13-35 yrs) on the principles of voluntarism, self-help and community participation. Over the years, NYKS has established a network of youth clubs in villages, where Nehru Yuva Kendras (501 at the moment) have been set up, harnessing youth power for development by forming Youth Clubs, which are village level voluntary action groups of youth at the grassroots level to involve them in nation building activities, working for community development and youth empowerment. However, only a few NYKS are involved in agricultural development. Much more could be achieved if NYKS could focus on skilling rural youth in agricultural activities.

NABARD Farmers' Clubs: NABARD's policy support for Farmers' Club Programme emphasizes on linking technologies with farmers' club members, while facilitating market access through capacity building of members of Farmers' Clubs including leadership training; linkage with technology/markets; Self Help Groups (SHGs)/Joint Liability Groups (JLGs) formation and forming Federations of Farmers' Clubs/Producers' Groups/Companies. Under this programme, the NABARD has so far (March 2013) assisted formation of 1.27 lakh farmers' clubs across the country. These clubs are organized by rural branches of banks with the support and financial assistance of NABARD for the mutual benefit of the banks concerned and the village farming community/rural people.

The broad objective of setting up Farmers' Clubs is to achieve prosperity for the farmers with overall agricultural development in its area of operation by facilitating credit counseling, technology counseling and market counseling. The NABARD provide a financial assistance of Rs. 10,000 to each club per annum for three years. The club members are expected to utilize this amount to meet routine expenses for formation, maintenance, and organizing awareness meets. Most of these farmers' clubs have a good representation of rural youth.

ARYA (Attracting and Retaining Youth in Agriculture): The Indian Council of Agricultural Research (ICAR) has constituted a seven-member expert committee, to suggest ways of attracting youth to agriculture. The committee is working on issues such as providing training to farm youth on innovative and sustainable agricultural practices to help make agriculture a profitable venture. This would involve a series of activities, including exploring opportunities in secondary agricultural operations such as value addition of crops and hiring, and servicing of mechanized farm implements. This is going to be an important strategy of the ICAR focused on rural youth mobilization for agricultural transformation during 12th plan (Ayyappan,2012). The initiative aims at analyzing the current policy environment and identifying supporting policies that can check the rate of migration of youth from rural areas. The ARYA will identify such mechanisms and models that would encourage the youth to avail the quantum of opportunities in allied sectors. It is expected that the youth educated in agriculture and allied enterprises will be able to earn a dignified livelihood from farming and other related pursuits. Educated youth in urban areas can also take up urban and peri-urban agriculture in which ample opportunities exist.

National Rural Livelihood Mission (NRLM): NRLM aims at creation of opportunities for both wage employment and skill development for the rural youth, who lack skills in many areas of agricultural production and processing. National Skill Development Mission and the National Skill Qualification Framework are, thus, aggressively pushing the agenda of skill development to build the capacity of rural youth so that they are meaningfully employed in rural areas itself. Thus, need based experiential skill learning supported by public sector banks/organizations in rural areas is the key to strengthen the Rural Self Employment Training Institutes (RSETIs) being set up in all districts under NRLM to assist such youth (Likhi, 2013). NRLM requires suitable decentralized convergence of skill development programs run by multiple central ministries including the National Skill Development Corporation (NSDC).

Agricultural Skill Council of India (ASCI): Considering the need for skilling the work force in agricultural sector, the Agricultural Skills Council of India (ASCI) has been recently proposed by National Skill Development Corporation (NSDC), which could be one ideal institution to train rural youth. The ASCI proposes to train, certify and accredit 56.5 million workforces comprising of farmers, wage workers, entrepreneurs and extension workers, over 10 years through its training partners.

ASEAN-India Farmers Exchange: The Ministers of the ASEAN (The Association of South East Nations) concerned over the small number of young farmers' involvement in the agriculture sector, agreed on the importance of promoting innovation and entrepreneurship among young farmers to achieve more sustainable agriculture development in the region. This initiative is expected to create greater awareness among the young and innovative farmers on the promising career in the agriculture sector. The 1st Exchange Visit was conducted in Malaysia in conjunction with the 2012 ASEAN Farmers' Week and the Malaysian Agriculture, Horticulture and Agro-tourism (MAHA) International in November 2012, followed by the 2nd Exchange Visit conducted in India during December 19-30, 2012, wherein, farmers' delegation from nine ASEAN member states participated.

Mobilizing Young Farmers: Most of the new innovations (both technical and institutional) require a skilled agricultural work force. For instance, promotion of high value agriculture, precision farming, organic cultivation, Hi-Tech horticulture, micro-propagation, Integrated Pest Disease & Nutrients Management, Post Harvest Management, development of backward and forward linkages etc, require well trained young farmers with enthusiasm and passion for farming and ability to take risks. The rural youth could be the ideal target for skill training in these new areas of agricultural growth and to do this effectively there is a need to mobilize young farmers. Organized groups of young farmers will be useful for introducing new production technologies and organizing effective input and output markets. In India, the recently initiated "The Young Farmers Association (YFA), Punjab" is involved in promoting and modernizing agriculture. The YFA, has started a programme called the Future Farmers Foundation (FFF) to encourage the youth to take up farming as a viable and lucrative occupation to increase farm income, while maintaining sustainability in agriculture. Some of these programmes are also reviewed by first author in a blog post for AESA (Chander, 2013).

CONCLUSION

High unemployment rates globally and the lack of interest in traditional agricultural pursuits, has been a major cause of concern. This can be solved by development of entrepreneurship which has the potential to contribute to the rejuvenation of farming by creating more employment opportunities and increasing profit. Although lack of youth participation in agriculture is well documented, plausible strategies must be framed to ensure that youth become the key stakeholders in farming. This paper has highlighted some of the key emerging issues about youth in farming & migration and identified few of the opportunities in farming for youth as an alternative for migration to urban areas. Further, the paper has also pointed out the support of Government for youth in the form of various schemes and programmes to attract youth in agriculture. The present day educated youth can tap the best potential based on the integrated application of science and social wisdom leading to technological and managerial up-gradation of farm operations. The research and development agencies on a participatory approach can support scientific technologies and development in agriculture and allied sectors. The support of public and private agencies in the form of finance and other relevant issues may be well explored and acted upon accordingly for promoting youth in agriculture.

REFERENCES

Ayyappan, S (2012). Message for ICAR Reporter. October-December, 2012.

Bennell, P. (2010). Investing in the future: Creating opportunities for young rural people, International Fund for Rural Development, Rome

Buckman, P (2013). Livelihood and Survival Strategies of Migrant Transport Apprentices: A Case Study of 'Tro-Tro' Mates In Accra. Dissertation Submitted to University of Ghana,

Central Investigation Agency (2012) The World Factbook (2013-2014) (51th Edition), Directorate of Intelligence Publication, Washington D.C. Assessed from: https://www.cia.gov/library/publications/the-world-factbook/index.html

Chander, Mahesh(2013). Youth: Potential Target for Agricultural Extension. AESA blog (http://aesagfras.net/blog. aspx?id = 59&title = Youth: %20Potential%20 Target% 20 For%20 Agricultural%20Extension).

Economic Survey (2013-2014). Ministry of Finance, Government of India. Accessed from: http://www.indiabudget.nic.in/ (19.09.2014)

Epstein, T. S. (2007) 'Poverty, Caste and Migration in South India' in Deepa Narayan & Patty Petesh Out of Poverty–Cross-Disciplinary Perspectives on Mobility, Mac Millan and the World Bank, Palgrave.

FAO (1991). Rural Youth Situation, Needs and Prospects - An overview with special emphasis on Africa. A paper presented at the sub-regional workshop on population and rural youth, Nairobi, Kenya.

Government of India (2001) Census 2001. Office of Registrar General and Census commissioner, ministry of Home Affairs, Government of India, New Delhi Accessed from : http://censusindia.gov.in/.

Government of India (2011).Census 2011. Office of Registrar General and Census commissioner, ministry of Home Affairs, Government of India, New Delhi Accessed from : http://censusindia.gov.in/.

Government of India (2014). *Employment and Unemployment Situation in India, NSS 68th Round (July 2011-June 2012)* (NSS report 554), National Sample Survey Office, National Statistical Organization, Ministry of Statistics and Program Implementation, New Delhi

Government of India. (2013). Report on Youth employment and Unemployment scenario 2012-2013, Ministry of Labour and Employment, Labour Bureau, Chandigarh.

Gowda, N.K., Nataraju, M.S and Veerabhadriah, V. (2013). Youth Agriculture and Rural Development, New India Publishing Agency, New Delhi

Gupta, D. (2005). Whither the Indian Village: Culture and agriculture in Rural India. Econ. And Pol. Weekly, 40 (8):751-757.

Hari, R. (2014). Determinants of rural youth participation in farming, Ph.D thesis (Veterinary Extension), Indian Veterinary Research Institute, Izatnagar

Hari, R., Sankar, S., Rajeev,T.S and Shyam,J. (2012). Empowerment of women through scientific dairy practices-A case study from Kerala. *In: Global Conference on Women in Agriculture* held on 13-15 March, 2010 at NASC Complex, New Delhi

Harris, C M (2009). Youth Migration and Poverty in Sub-Saharan Africa: Empowering the Rural Youth. Human Rights and Human Welfare. pp:159-186.

IFAD. (2012). Youth in agriculture: Special session of the Farmers' Forum Global Meeting, 18 February 2012, IFAD, Rome.

ILO (2008). Promotion of Rural unemployment for Poverty Reduction, 97th Session of International Labour Conference, Geneva.

ILO (2012). Decent work for Youth in India, International Labour Organisation Country office, New Delhi. Accessed from: http://www.ilo.org/newdelhi/info/WCMS_175936/lang—en/index.htm

International Labour Office. (2014). Global Unemployment Trend-2014, International Labour Organisation, Geneva.

International Labour Office. (2005). Youth Employment: From a National Challenge to Global Development Goal, Background paper contributed by the ILO to the G8 Labour and Employment Ministers Conference, London.

Lyocks, J.S., Lyocks S.W.J. and Kagbu J. H.(2013) Mobilizing Youth for Participation in Nigerian Agricultural Transformation Agenda: A Grassroots' Approach. Journal of Agricultural Extension. 17 (2):78-87.

Manjula, N., Dhahnunjaya, B. And Nagesha, G. (2013). Integrated Farming System –An Avenue for Rural Youth in the coastal zone of Karnataka. In : Youth in agriculture and rural Development (eds: Gowda, K.N., Nataraju, M.S and Veerabhadraiah, V.), New India Publishing Agency, New Delhi, pp:293-298

Mukherjee, S. and Das, K.C. (2011). Regional Disparity and Youth Migration in India, Indian Institute of Population Sciences, Mumbai.

National Sample Survey Organisation (2010) Migration in India 2007-2008.NSS 64 th Round (July 2007-June 2008), (NSS Report 533) National Sample Survey Office, National Statistical Organization, Ministry of Statistics and Program Implementation, New Delhi

Planning Commision (2013). Press note on Poverty Estimate 2011-2012, Planning Commission, Government of India, New Delhi

Paisley, Courtney (2013). Engaging youth in agriculture: Investing in our future. Global Food for thought. The official blog of the global Agricultural Development Initiative. http://globalfoodforthought.typepad.com/global-food-for-thought/2013/02/commentaryengaging-youth-in-agriculture-investing-in-our-future.html

Rajan, S. I. (2013). Internal migration and youth in India: Main features, trends and emerging challenges, UNESCO, New Delhi

Rustam, M. A. (2004). Youth and Globalism: A perspective, World Assembly of Youth, Malaysia.

Sainath, P. (2013). Over 2,000 fewer farmers every day. The Hindu, May 2. http://www.thehindu.com/opinion/columns/sainath/over-2000-fewer-farmers-everyday/article4674190.ece>

Sarkar, A.N. (2009). *Enhancing Global Competitiveness: Advantage India*, I K International Publishing House Private Limited, New Delhi, 416 p.

Srivastava, R. (2011). Internal migration in India : An overview of its Features, Trends and Policy Challenges, National workshop on Internal Migration and Human Development in India: Workshop compendium, Volume (2). UNESCO, New Delhi

Sumberg, J.; Anyidoho, N. A.; Leavy, J.; te Lintelo, D. J. H. and Wellard, K. (2012). Introduction: The Young People and the Agricultural Problem in Africa. IDS Bulletin, 46(6), 1-8.

Swaminathan, M.S. (2011). Youth as catalysts of accelerated agricultural and rural development, LEISA INDIA, 13(1): 6-7

Tadele, G. And Gella, A.A. (2012). 'A Last Resort and Often Not an Option at All': Farming and Young People in Ethiopia, IDS Bulletin, 46(6):33-43

Van der Geest, K. (2010). Rural Youth Development in Developing Countries: A global View-Gender Equity and Rural Employment. Rome. Food and Agricultural Organisation.

White, B. (2012). Agriculture and the Generation Problem: Rural Youth Employment and the Future of Farming, IDS Bulletin, 46(6): 11-19.

Zuehlke, E (2009). Youth Unemployment and Underemployment in Africa Brings Uncertainty and Opportunity. Population Reference Bureau.

CHAPTER - 18

Income Generating Activities for Youth in Farming Under Rural Development Programmes

Gopika M.H.[1], Preethi[2] and Kowsalya K.S.[3]
[1]*Assistant Technology Manager (ATM), ATMA Project, Dept. of Agriculture Hoskote Taluk, Bangalore Rural District- 562114*
[2&3] *P.G Students, Dept. of Agricultural Extension, College of Agriculture, UAS GKVK, Bangalore -560065*

ABSTRACT

Income generating activities are considered as those initiatives that affect the economic aspects of rural youth's lives through the use of economic tools such as credit and subsidy. If the youth in farming is not economically free and self-earning, he has to depend on others or migrate to the urban cities in search of jobs. Agriculture, self-employment and wage employment are the basic tools to fulfill the desire of economic independence. In the past, agriculture was placed on the top position as an income generation option and self-employment and wage employment to rural youth was the 2nd and 3rd priority respectively. Income generating activities such as dairy activities, goat and sheep rearing, poultry unit, bamboo products making, pickle making, leaf plate making, keeping powertiller, husking mill cum wheat grinding machine, tractor etc. Further, there are various self employment and wage employment opportunities through rural development programmes like Prime Minister Rojgar Yojana (PMRY), Swarnajayanthi Gram Swarojgar Yojana (SGSY), Pradhan Manthri Gram SadakYojana(PMGSY), Integrated Rural Development Programme (IRDP), Training of Rural Youth for Self -Employment (TRYSEM) etc. These are the best programmes especially for the upliftment of the rural youth to enhancing their income level. Some of these programmes are now helping rural youths to become more and more independent and free from economic dependence.

Key words: Income generating activities, self-employment, employment opportunities, wage employment, agriculture, rural youth

INTRODUCTION

Income generating activities are the prime determinants of level of income. Overall economic development strategies are incorporated in the Government's five-year development plans. Which would also reduce poverty among the rural masses, later on (perhaps in the 1960s) programmes specifically designed and addressed to the cause of poverty alleviation among the rural youth became indispensable.In the last four decades a number of such programmes have been and still are being implemented in the country.If the youth in farming is not economically free and self-earning, he has to depend on others or migrate to the urban cities in search of jobs. Programme for self-employment of the rural youth has been an important component of the anti-poverty programmes implemented through government initiatives in the rural areas in India.India's anti-poverty strategy comprises of a wide range of poverty alleviation and employment generating programmes, many of which have been in operation for past several years and have been strengthened to generate more employment opportunities, create productive assets, impart technical and entrepreneurial skills and aiming to increase productive assets among rural youth. Under these programmes, both wage employment and self-employment are provided to the rural youth below the poverty line.In 1998-99 government proposed to unify the various poverty alleviation and employment generation programmes under two broad categories such as self-employment schemes and wage employment schemes. Funding and organizational patterns will also be rationalized to achieve maximum beneficial impact of these programmes.

There is no doubt that one of the important aims of development in the developing countries of Asia and the Pacific has been the removal of poverty among the rural masses. Income generating programmes have a major role to play in alleviating poverty in these countries. Several governmental, non-governmental and voluntary organizations have taken upon themselves the task of training neo-literate as well as illiterate adults in income generating skills and activities. There is no central or even state level direction in regard to the type and content of these training programmes aimed at promoting income generation in rural areas. These organizations are locally based and build up their programmes and activities around local needs. Many of them are devoted to the empowerment of rural women so as to make them active partners, along with men in earning for their families.

Concept of Income Generating Activities

Income generating activities are considered as those initiatives that affect the economic aspects of rural youth's lives through the use of economic tools such as credit and subsidy. Freedom depends on economic condition even more than political. If rural youth in farmingis not economically free and self-earning, he has to depend on others and dependents are never free. Income generation confirmed that, overall there is poor understanding of the main principles of income generation schemes with routine service provision. Income Generation (wealth creation) scheme calls for loans rather than a grant, calls for a fair market rate of interest for the loans, provides management training to accompany the loans.Aims at participants, who are at the lowest levels of income in their communities, and provides and guides the organizing which results in effective groups (like co-operatives) that handle the credit and loans. Agriculture, self-employment and wage employment are the basic tools to fulfil the desire of economic independence. In the past, agriculture was placed on the top position as an income generation option and self-employment and wage employment had the 2nd and 3rd priority respectively.

Income Generating Activities

Income generating activities such as Dairy activities, Bakery unit, Goat and sheep rearing, Poultry unit, Bamboo products making, Pickle making, Leaf plate making, Keeping power tiller,Husking mill cum wheat grinding machine, Tractor etc.

Dairy Activity

Central and State Governments are giving considerable financial assistance for creating infrastructure facilities for milk production.The ninth Five year plan outlay on Animal Husbandry and Dairying was Rs. 2345 crores.The dairy industry plays an important role in the socio-economic development of India. It generates huge employment opportunities for people in rural places. Dairying is an important source of subsidiary income to small/ marginal farmers and agricultural labourers. The manure from animal provides a good source of organic matter for improving soil fertility and crop yields. The surplus fodder and agricultural by-products are gaining utilized for feeding the animals. Since agriculture is mostly seasonal, dairy can be considered for subsidiary income source.

Bakery Unit

Bakery enterprise is on the market which enables rural youth to obtain as a lot as you may hold the opportunity to by selling bakery products. Youth may hold the opportunity to be able of run a bakery at your own and can get a great amount of money in return of this all. So this is how rural youth will hold the opportunity to be successful of earn money from running this company.

Sheep and Goat Rearing

The sheep unit represents a very reasonable livelihood option for agricultural labourers as it require less resources and not demand very specialized skills. Training institutions should organize training for rural youth and women in small ruminant production technologies and its importance for income generation. Efforts should be made by both research and development institutions to link up rural youth and women with funding agencies. Rearing sheep requires less resource and is appropriate where capital is scarce.Sheep rearing can provide part time self-employment without affecting the main occupation for small and marginal farmers.

Leaf Plate Making

Now a day's people are using leaf plates for all kinds of occasions. There is a lot of demand for leaf plates in market.SHG women members have been trained in leaf plate making. They are able to meet the basic needs and other needs like education of children, creation of assets etc. by this activity.

Self-Employment and Wage Employment Opportunities Through Rural Development Programmes

There are various self-employment and wage employment opportunities through rural development programmes like Prime Minister Rojgar Yojana (PMRY), Swarnajayanthi Gram SwarojgarYojana (SGSY), Pradhan Manthri Gram Sadak Yojana (PMGSY), Integrated Rural Development Programme (IRDP) and Training of Rural Youth for Self-Employment (TRYSEM) etc.

Analyzing the Rural Development Programmes

1. Prime Minister Rozgar Yojana
2. Swarna Jayanthi Swarozgar Yojana

1. Prime Minister Rozgar Yojana (PMRY)

PMRYwas launched on 2nd October 1993 to assist educated unemployed youth to set up self-employment ventures. The target for the year 2004-05 and 2005-06 under the yojana has been enhanced from 2.20 lakh beneficiaries to 2.50 lakh beneficiaries per annum.

Objective of PMRY

The PMRY aimed to provide employment to more than a million persons by setting up of 7 lakh enterprises by the educated unemployed youth. It relates to the setting of self-employment ventures through industry, service and business routes.

Salient Features of PMRY Scheme

1. This is centrally sponsored scheme.
2. The development commissioner (small scale industries) under ministry of small scale, rural and agro industries government of India is the apex body for this scheme.
3. The respective commissioners or directors of industries implement the scheme at the state level.
4. The implementation agencies at the grass root level are district industries center (DIC) who would be the instrument for the grounding of the units.
5. The DCSSI has setup a special PMRY division in the headquarters under the able guidance of IAS officers. The DCSSI formulates the rules, regulations and guideline instructions and provides clarifications on all the matters pertaining to the PMRY scheme.
6. It also devised complete feedback information by the means of getting monthly, quarterly, and annual progress report.
7. Similarly at the state level, state level PMRY committee meetings monitors the progress of the scheme every quarterly.
8. The yearly targets for number of beneficiaries for each state are fixed by DCSSI.

Parameters of PMRY

- **Age**: for all educated unemployed between the age group of 18-40 years, in general with a 10 years relaxation for SC/STs ex-service man, physically handicapped and women.
- **Educational qualification:** 8th passed. Preference will be given to those who have been trained for any trade in government recognized or approved institutions for duration of least six months.
- **Family income:**Neither the income of the beneficiary along with the spour nor the income of parents of the beneficiaries shall exceed Rs 40,000 per annum.

- **Residence:** Permanent residence of the area of at least 3 years.
- **Defaulter:** Should not be a defaulter to any nationalized bank or financial institutions or co-operative bank. Further, a person already assisted under other subsidy linked government schemes would not be eligible under this schrme.
- **Activities covered:** All economically viable activities including agriculture and allied activities but excluding direct agricultural operations like raising crop, purchase of manure etc.
- **Project cost:** Rs 1.00 lakh for business sector. Rs 2.00 lakhs for other activities. Loan to be of composite nature. If two or more eligible persons join together in a partnership, project up to 10.00 lakhs covered.
- **Subsidy and margin money:** Subsidy will be limited to 15% of the project cost subject to ceiling of Es 7,500/- per entrepreneur.
- **Collateral:** No collateral for project up to RS 1 lakh. Exemption from collateral in case of parternership project will also be limited to an amount Rs 1 lakh per person participating in the project.
- **Rate of interest and repayment schedule**: Subsidy will be limited to 15% of project cost subject to ceiling of Rs 7,500 per entrepreneur. Bankswill be allowed to take margin mony from the entrepreneur varying from 5% to 16.25% of the project costs as to make the total of the subsidy and the margin money equal to 20% of the project cost.
- **Training and other assistance:** The training expenses and operational expenditure to be covered with in the ceiling of Rs 2000 per case. The existing system of revising the scale of expenditure in consultation with the finance for various activities to the implementing agencies of the state and central subject to remain within the ceiling of Rs 2000 per case sanctioned.
- **Implementing agency**: The district industry centers and the director of industries shall mainly be responsible for schemes implementation along with banks.
- **Linkages of targets with recovery:** Basic minimum targets based on the population and the number of educated unemployed, additional targets would be linked to the recovery of loans sanctioned past performance of sanctions or special circumstances prevailing the state.
- **Reservation:** Preference should be given to weather sections including women. The scheme envisages 22.55 % reservation for SC/ST and 27% for other backward class. In case of SC/ST/OBC

candidates is not available, state government will be competent to consider other categories of candidates under PMRY.

How the Scheme is Implemented?

1) Fixing of targets
2) Allocation of targets by states
3) Allocation of financial targets to banks
4) Calling for PMRY applications
5) Constituting the force committee
6) Providing training to the selected candidates
7) Monitoring and reviewing of PMRY

Income generating activities under this programmes

Si No	Name of the Activity	Financial Support
1	Power tiller (2 no)	2,26,000
2	Milch cow (cross breed)10 number	2,55,000
3	Bamboo products	65,000
4	Brick making	87,000
5	Clay modeling	70,000
6	Ice and ice candy manufacturing	4,00,000
7	Husking mill cum wheat grinding	3,50,000
8	Bakery unit	3,50,000
9	Piggery	2,50,000

Income generating agriculture and other activities

Si No	Name of the Activity	Financial Support
1	Papaya cultivation	20000
2	Leman cultivation	20000
3	Vegetable cultivation	30000
4	Pumpset(up to 5H.P.)	25000
5	Milch cow	42000
6	Piggery	30000
7	Ginger cultivation	30000
8	Turmaric cultivation	30000

2. Swarn Jayanti Gram Swarozgar Yojana (SGSY)

SGSY has been launched from April 1999 to eradicate unemployment problems and give economic assistance to the rural poor and downtrodden peoples by restructuring various self employmentprogrammes.

Objective of SGSY

To bring the assisted poor families (Swarozgaries) above the poverty line by providing them income –generating assets a mix of bank credit and government subsidy .

Salient features of Swaranjayanti Gram SwarozgarYojana (SGSY)

1. It is a holistic programme for promoting self-employment of rural poor.
2. Covers all aspects of self-employment of rural poor organizations of rural poor into Self Help Groups (SHGs)
3. It ensures poor families to have monthly income of at least Rs2000 .
4. It is particularly focus on the most deprived groups among the rural poor.
5. It is coordinated and consolidated efforts of all the concerned agencies, block, panchayath, bank, NGO,etc
6. The scheme provides for the creation of Risk Fund with 1%, consumption of loan not exceeding Rs2000 per Swarozgari.
7. The bankers play Avery critical role in the implementation of SGSY programme right from the identification of activities, formation and evacuation.

Income generating activities adopted by Swa- Shakti SHG

Sl. No .	Income generating activities
1	Sheep and goat rearing
2	Dairy
3	Weaving
4	Bakery
5	Leaf plate making
6	Vegetable vending
7	Pickle making

Key activities for Self Help Group under SGSY:

S. No	Name of the Activity	Tentative Project Cost
1.	Power tiller (2 no)	2,26,000
2.	Milch cow (cross breed)10 number	2,55,000
3.	Bamboo products	65,000
4.	Patishilpa	87,000
5.	Clay modeling	70,000
6.	Ice and ice candy manufacturing	4,00,000
7.	Husking mill cum wheat grinding	3,50,000
8.	Bakery unit	3,50,000
9.	Piggery	2,50,000

CONCLUSION

Rural Development programmes like PMRY, SGSY, IRDP and TRYSEM are the best programmes especially for the upliftment of the rural youth to enhancing their income level. Some of these programmes are now helping rural youths to become more and more independent and free from economic dependence.

REFERENCES

Arau Prakash, R. (2004). Analysis of Swarnajayanthi Gram SwarojgarYojana in Salem and Thiruvallur district of Tamil Nadu.M.Sc. Thesis unpub.University of Agricultural Sciences, Dharwad.

Arunakumara, B. (2005). A Critical Analysis of Swashaktiprogramme in Karnataka. M.Sc. Thesis Unpub., University of Agricultural Sciences, Dharwad

http://www.ministryofrural development.com

CHAPTER - 19

Engaging Youth in Agriculture - A Key to Food Security

Niketha L., Bagish Kumar, Minu Singh and Sanjeev Kumar
Ph.D Scholars, Dairy Extension Division, National Dairy Research Institute Karnal, Haryana

ABSTRACT

Engaging youth in agriculture has been a prominent topic recently and has risen up the development agenda, as there is growing concern worldwide and also in India that young people have become disenchanted with agriculture. Over 60 per cent of the world's rural population is made of youth, with half of them being young women and girls. Due to their limited access to assets (in particular land), markets, finance and education and skills training, rural youth are often unemployed or work informally – often in unpaid, very low-skilled, insecure and sometimes hazardous jobs. If young people living in rural areas do not find enough incentives, profitable economic opportunities and attractive environments in which to live and work, they will continue to migrate to cities. This trend would not only contribute to the mega urbanization and growing urban unemployment that is already under way, but is expected to affect global food production: who will then feed the global population that is projected to reach 9.2 billion people by 2050. With most young people around 85% living in developing countries, where agriculture is likely to provide the main source of income it is vital that young people are connected with farming. Currently around the world we're living in an era where rapid urbanisation has led to a decline in rural populations and for the first time ever the majority of the world's population lives in a city. The UN World Health Organization predicts that "by 2030, 6 out of every 10 people will live in a city, and by 2050, this proportion will increase to 7 out of 10 people" meaning that more young people than ever before are moving to cities and towns to find work, leaving few behind to work in rural areas. With this predicted concentration of the global population in urban areas it is easier to understand why the number of young farmers is in decline. Government should emphasize the need to invest in "the rural youth of today, the farmers of tomorrow". Investing in young people living in rural areas is key to enhancing agricultural productivity and food security, boosting rural economies, and reducing rural-to-urban migration. Young people have enormous potential for the innovation and risk-taking that is often at the core of growth and development in rural areas, particularly in smallholder agriculture. Young farmers and producers often have greater capacity for innovation and entrepreneurship than older adults. This capacity may better equip them to address the emerging requirements of agriculture and the rural non-farm economy. Farming offers the young generation a chance to make a difference by growing enough food to feed the world. Those who become farmers now have the opportunity to be the generation that end world hunger and alleviate malnutrition, as well as helping the sector adapt to climate change. There are many challenges ahead for the sector but if young people are offered education in agriculture, a voice at policy level, and in the media, and are engaged with innovations then the agriculture industry can attract youth again. As we look to find solutions to feeding a world of nine billion people by 2050, it is this new generation that – working together – can help to achieve global development. The present paper is conceptual in nature and deals with present conditions of youth their further growth, hurdles and how their energy can be mobilised to pump the fresh blood in the age old Indian agriculture.

Key words: Youth, health, food security, farming, media

INTRODUCTION

Recent trends suggest that India might very well be at the crucial point of the transition in its agriculture dependent population. A large proportion of youth in the countryside is on their way out of agriculture. Rising disenchantment with the profession pushes them out of agriculture while opportunities in other sectors of the booming economy pull them out of agriculture. This is suggested by this study which attempts to identify the drivers of this process of withdrawal and assess the odds of an average farmers' move out of agriculture. Youth who possess non-farm skills are 1.4 times more likely to move out of agriculture. Age has a significant role as the odds ratio in favour of shifting out of agriculture is high among farmers below 30 years of age.

The phenomenon deserves attention since it challenges the myth of persistence of overpopulated workforce in agriculture. With the current 8 percent growth rate of the Indian economy the fall in agricultural population is likely to only get steeper. This will have far-reaching implications for the way agriculture is likely to be organized in future. We foresee aging of the farmer population, increased feminization of agricultural workforce and mechanization of farms. This will put great demands on the extension agencies, technology providers and the agricultural institutions that will have to wake up and recognize the requirements of the new-age farmer. The demands will be greater on the irrigation institutions which have by far ignored the role of women as farmers and as irrigators.

It's probably too early and too ambiguous to make such assertions– more than 50 per cent of the population is currently engaged in farming; the rate of growth of non-farm jobs is abysmal, the service sector has too less job opportunities to offer and the manufacturing sector has only been experiencing jobless growth. The transition, even if it were to happen, will take a long time. However, if one took cognizance of the surroundings and the developments therein, there are a sufficient number of reasons for us to think the contrary. First, the growth performance of the Indian economy gives us sufficient grounds to expect a significant change in the employment scenario 3. The 6th economic census which reports that the growth in labor opportunities in the non-farm sector is outpacing the growth in labor force gives further hope. Second, the current farm crisis in the country seems to be another dominant force causing many to quit out of desperation. The mounting number of farmer suicides and the rising disenchantment with the profession among farmers are some subtle but disturbing indicators. Further, if we look back in time we find that in most cases withdrawal from farming took place in a very short concentrated period of time either by way of growth in the non-farm sector or farm crisis as experienced by US in 1980s. Most of the East Asian economies such as Japan, South Korea, and Taiwan which are similar to India in being land scarce and labor surplus underwent steep fall in agricultural population within a short span of less than 20 years. To support the argument furthermost of them were experiencing similar economic growth rates (that India is experiencing now).

The National Commission on Farmers stressed the need for attracting and retaining educated youth in farming. The National Policy for Farmers, placed in Parliament in November 2007, includes the following goal- "to introduce measures which can help to attract and retain youth in farming and processing of farm products for higher value addition, by making farming intellectually stimulating and economically rewarding". At present, we are deriving very little demographic dividend in agriculture. On the other hand, the pressure of population on land is increasing and the average size of a farm holding is going down to below one hectare. Farmers are getting indebted and the temptation to sell prime farmland for non-farm purposes is growing. Over 45 per cent of farmers interviewed by the National Sample Survey Organisation wanted to quit farming. Under these conditions, how are we going to persuade educated youth, including farm graduates, to stay in villages and take to agriculture as a profession? How can youth earn a decent living in villages and help shape the future of our agriculture?

There is a need to tap the potential of youth to revive agriculture which is the backbone of rural economy in the country. Large-scale migration of rural youth from farming to urban areas has caused concern among the country's agricultural policy makers as such a trend, if not checked, is likely to affect agricultural activities in the future.

Possession of skill seems to be an important factor in determining out-migration from agriculture. The odds of a farmer moving out of farming increase with skill attainment. Education too lends a positive push to migration but is not significant at 5 per cent level. Most of the out-migration is visible in the lower age group making age another critical variable. Among the farm level factors, farm size has an impact but the relationship manifests itself differently in smallholder group and among the large farmers. Both appear to be withdrawing but for different reasons. Irrigation has no significant impact on the withdrawal behaviour. Finally, proximity to towns is found to be a critical determinant fuelling out-migration decisions of farmers.

For young people to take to agriculture farming must be both intellectually satisfying and economically rewarding. This will call for a technological and managerial up gradation of farm operations. We have to harness the best in frontier science and marry it with the best in traditional knowledge and ecological prudence. Such a blend leads to the science of ecotechnology. In addition to ecotechnology, Universities have to develop excellence in biotechnology, space technology, nuclear technology, nanotechnology, renewable energy and management technology. The Universities in India should enable every scholar to become an entrepreneur.

During his recent visit to India, US President Barack Obama pointed out that India is fortunate to have over half of its total population of 1.2 billion under the age of 30. Out of the 600 million young persons, over 60 per cent live in villages. Most of them are educated. Mahatma Gandhi considered the migration of educated youth from villages to towns and cities as the most serious form of brain drain adversely affecting rural India's development. He, therefore, stressed that we should take steps to end the divorce between intellect and labour in rural professions.

A three-pronged strategy as a panacea as given by M S Swaminathan:

- Through appropriate land use policies, technologies and market linkages; develop for this purpose a "4C approach", i.e., Conservation, Cultivation, Consumption and Commerce.
- Enlarge the scope for the growth of agro-processing, agroindustries and agribusiness and establish a "Farm to Home" chain in production,

processing and marketing.

- Promote opportunities for the services sector to expand in a manner that will trigger the technological and economic up gradation of farm operations.

In the short term we need to recognize that the current withdrawal from agriculture by youth is not only inevitable it is, to some extent, good for the economy. It would reduce the burden on agriculture and raise effective income for the residual population. But the situation as of now is troubling. while a large mass of youth is trying to make its way out of farming few have the necessary skills to be able to move out of farming profitably. The result is poor quality migration, creating problems for urban habitations while not necessarily reducing the burden on rural areas. In the short term, skill building of rural youth could be treated as a priority area. This would not only increase the pay-off to migration but facilitate withdrawal from farming as well.

Engaging youth in agriculture has been a prominent topic recently and has risen up the development agenda, as there is growing concern worldwide that young people have become disenchanted with agriculture. With most young people – around 85% – living in developing countries, where agriculture is likely to provide the main source of income it is vital that young people are connected with farming. Youth lacked access to training and education on farming and therefore were not being encouraged to perceive agriculture as a future career.

Despite the decline in interest for agriculture as a career there are still young farmers working all over the world. To encourage others to join the sector it is vital that they are offered a voice, and that we take note of what they have to say. Particularly this includes giving young farmers at policy level a chance to their opinion and experiences. In this way, they can show other young people that farming can be a rewarding career as well as highlighting the important role of agriculture on a global scale. Establish a Youth Council that includes a farmer and livestock keeper as representatives. Demonstrating the recognition of the Youth Council to include representation from agriculture at government level. Another, rather different, way of offering young farmers a voice is to use media. At the beginning of 2013 a British television programme followed the stories of a selection of young farmers across the country, entitled First Time Farmers the programme aimed to defy stereotypes and demonstrate that there is a role for young people in the agriculture industry. A younger generation can help introduce new technologies whilst also learning from traditional methods, holding the potential to offer the perfect fusion of new and traditional solutions to some of the world's biggest challenges. ICTs could provide new opportunity for making agriculture more interesting for young people. One of the ways to do that is to train them and given them opportunities to access ICTs so that they can engage in value chains. The increased use of mobile phones in farming can also help deter young people away from stereotypes. Of traditional farming and help change their perceptions on agriculture, helping them to view it as an exciting and innovative industry.

CONCLUSION

Farming offers the young generation a chance to make a difference by growing enough food to feed the world. Those who become farmers now have the opportunity to be the generation that end world hunger and alleviate malnutrition, as well as helping the sector adapt to climate change. There are many challenges ahead for the sector but if young people are offered education in agriculture, a voice at policy level, and in the media, and are engaged with innovations then the agriculture industry can attract youth again. As we look to end solutions to feeding a world of nine billion people by 2050, it is this new generation that – working together – can help to achieve global development.

CHAPTER - 20

The Ways to Make Family Farming Look Charming with the Involvement of Youth

S.Jegan Karuppiah and Arokia Marry Eddolin
Scientist-cum-Research Scholar, M.S. Swaminathan Research Foundation Chennai

ABSTRACT

Agriculture is backbone of our country, India is lives in its villages, these are all vibrant statements about our nation and our farming system but now the situation has been turned out in to urbanization in terms of migration.UN also announces this year (2014) as an international year of family farming in this context, the real status of FF in India is a biggest question mark. This means FF is not happening as expected even though Tamil Nadu is one of the familiar state across India for family farming but due to lack of youth involvement in agriculture, Family Farming is not happening in a full-fledged and unfortunate thing is it is even more decreased.If this situation persists over a period of time, the above said vibrant statements will become spiritless, hence the solutions for this problem is Indiais reverse migration and every home should have farmer. These are all possible, if the youth is attracted towards farming and adopt Family Farming System (FFS). The statistics says in Tamil Nadu, 13 percent of youth only involved in farming and the population according to the census 2011 is about 7.2 crore, which means meager youth were involved in farming. Although Tamilnadu is one of the fastest growing urban state yet more than half of the population lives in villages and noticeable fact is more than half of them are under the age of 30. So that youth isthe only and powerful source to make agriculture through Family Farming. We believe that if and only if agriculture becomes intellectually, economically viable and bounty which fetches rewards, the younger generation of Tamilnadu will be interested to take forth farming as profession and make Family farming as an attractive one.This paper will explain possible areas to attract youth in Farming and adopt Family Farming System.

Key words: Youth, Family Farming System, Tamil Nadu, Farming Agriculture

INTRODUCTION

The United Nations declared 2014 as the International Year of Family Farming (IYFF) in order to understand the significance of family farming in alleviating poverty and increase in global food security. According to the UN, the IYFF has planned to encourage new growth in policies particularly at the countrywide, but also regional levels that will help farm families eradicate hunger, through small scale sustainable agricultural production. Family farming involves about 500 million families consisting of over two billion people. This paper will clearly explain how youth can be attracted and retained in Agriculture especially in family farming to reduce poverty and achieve zero hunger in Tamilnadu, India.

Geographical Area of Tamil Nadu

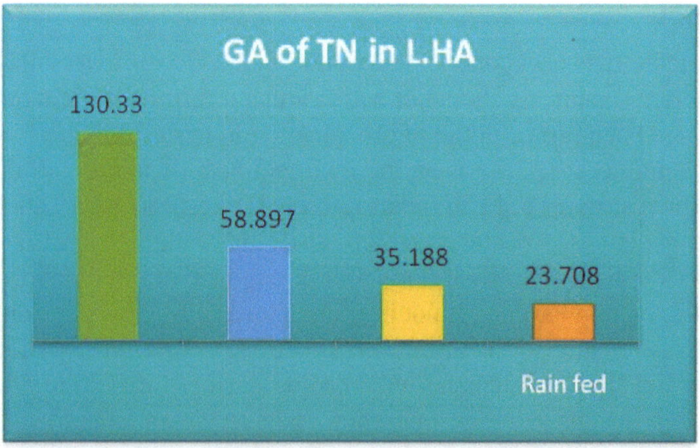

Geographical area - 130.33 L.Ha, Gross area sown - 58.897 L.Ha. (44.1%), Gross Area Irrigated- 35.188 L.Ha. (60%), Rain fed-23.708 L.Ha. (40%), Net area had sown - 49.858 L.Ha.Net Area Irrigated- 29.64 L.Ha. Landholdings: No. of land holdings - 81.18 L.Nos, Average size of holding - 0.80 Ha, Small farmers/Marginal Farmers - 91% which covers 60% of the total cultivable area in Tamil Nadu according to the said statistic of Tamil Nadu geographical area and its land holding pattern.Also from the figure it is understood that out of the total cultivable area only 45% are being cultivated, but remaining 55% which is more than half is still left untouched as per the data taken in the year 2013.

Tamil Nadu is one of the states with maximal area for agriculture but as a contradiction the number of farmers involved in agriculture is very minimal particularly involvement in farming as family is very meager.As per the recent researches, there is a 76% fall in the availability of food and the population is growing at the rate of 1.3% every year. Hence by 2050, we have to at least double our productivity to achieve zero hunger.It is predicted that India will be the one with higher number of youths by 2050, so the only powerful and effervescent community at that time is Youth whom we consider as the only hope to overcome such crisis situation.

Factors Hindering Youth's Interest

Even though many factors influence, there are two main factors that affect youth from involving in agriculture is social and environmental factors. In view of social, today's youth readiness to involve in agriculture is considered to be 'idiotic' and societal respect for the farmers is questionable when compared to other professionals. Many young farmers are facing harassment like denial of bride when searching marriage proposals. So they find it difficult to get married at the right age. In substantiating with this kind of disgrace, many experienced/ traditional farmers were also repeatedly voicing that agriculture is not profitable anymore. And so the youth were in a state to decide that farming is not their cup of Tea.

In terms of environmental factors, Ground water depletion, failure of monsoon or inadequate rainfall is a major threat for irrigation to carry out farming. Apart from this issues, unavailability of agricultural inputs like quality seeds, fertilizers and pesticides, unawareness and lack of access about government schemes on time and no minimum support price for agriculture products.. With all these issues, climate change also possesses a major threat in the form of rainfall, arise of new and uncontrollable pest attackand change in seasonal cropping and so on. Amidst of these tough environment, farming is considered as one of the most challengeable occupation.

Captivating Ways

Organization of farmers as groups or federations helps them in availing various benefits like government schemes and cross learning's. Also mass number of farmers should be made aware about the export opportunities of agricultural products along with the proper procedures to get export licenses. In addition to that, infrastructure facilities should be enhanced like cold storage rooms and depository need to be established in each block or at least district level. The Government has to set up a processing plant for grain processing and such preparations. Despite of all these, in order to make farming economically viable, strong linkage with marketing opportunities so as to reduce the middle men intervention need to be strengthened.

Agricultural students after finishing their studies should contribute for farming to bring some good change rather than tediously preparing for government jobs and getting placed. Policies need to be made stringent in buying and selling agriculture lands for non agricultural purposes.

Exploitation of ground water is upsetting the water table. We are exploiting more than 37% of the total water reserves without substituting them. Simultaneously salt content was increasing in the agriculture lands. In this case depletion of ground water should be altered by rain water harvesting methods and it should be enforced regularly by the government.

Rain water harvesting is one of the best and sustainable solutions to address the irrigation challenges in farming. Most of the villages have at least one pond, regular maintenance operations like de-silting can be adopted through the MNREGA. Actually Farm pond cleaning which is defined as one of the task in MNREGA should be changed because people can't do properly. This activity should be given responsibility to the agriculture engineering department to ensure its objective of water storage to its fullest capacity.

In this context NGO and other international organizations those working with water management techniques should propagate on micro irrigation methods and other new technologies like AWD (Alternate Wetting and Drying), Blue green manure, drip and sprinkler irrigation, mulching, irrigation scheduling, Rain gun methods etc should be proactively introduced.

There were also concerns that youth were being excluded from farming because of external risk like non-security of job or low income. To cope with this, youth has to be provided with field trainings and exposures along special schemes which encourage farming youth has to be developed in order to encourage and make them involved in agriculture. Mobile technology have greatest role in reaching mass farmers so SMS services must be provided for agriculture timely appropriately and accurately on government schemes, subsidies, availability of inputs regarding.

Suggestion and Recommendations

- Capacity Building of youth on agricultureand latest technologies
- Engage youth activelyin to farming with respect
- Include youth to planning and policy efforts and permit youth to identify their individual interests.
- Create platform for Facilitation, Youth Communication, Advocacy and Networking.
- Institutionalizing Youth Policy Engagement at Primary, Secondary and Academic institutions level.

CONCLUSION

Youth will definitelyattract to agriculture and get involved in family farming if and only if farming becomes economically rewarding and intellectually stimulating not only in Tamil Nadu but also all over India. Subsequently appropriate policies need to be developed to make agriculture more attractive, profitable and a credible career to the youth through the application of appropriate and modern technologies.

Hence focusing specifically on Family farming will help in paving way to retain youth enthusiasm towards farming inorder to address the serious issue of food insecurity which we are expected to face in future. Hence this kind of special emphasis like commemoration of the "International Year of Family Farming" and materializing the suggested recommendations will definitely not only bring motivation among youth to achieve zero hunger but also Farming will be considered as one of the high paid and highest seeking profession in the near future.

ACKNOWLEDGEMENT

Sincere thanks to Prof.M.S. Swaminathan, Dr.Ajay.K.Parida, Executive Director of MSSRF, Dr.Parasuraman, Principal Scientist of Youth Development wing, Ms.Nancy J.Anabel Director of IEC, MSSRF and research supervisor Dr.N.Seeninatarajan, Professor, Department of Life-long learning, Gandhigram Rural Institute for their great guidance and inputs.

REFERENCES

1. Alston, J., C. Chan-Kang, M. Marra, P. Pardey, and T. J. Wyatt (2000) A Meta-Analysis of Rates of Return to Agricultural R&D: Expede Herculem? IFPRI Research Report No. 113. Washington, D.C.: International Food Policy Research Institute.
2. Jamison, D., and L. Lau (1982) Farmer Education and Farm Efficiency. Baltimore: Johns Hopkins University Press.
3. Jayne, T., T. Yamano, M. Weber, D. Tschirley, R. Benfica, A. Chapoto, and B. Zulu. (2003) Smallholder Income and Land Distribution in Africa: Implications for Poverty Reduction Strategies. Food Policy 28: 253–275.
4. United Nations (2013) Human Development Report 2003. New York: Oxford University Press.
5. World Bank (2014a) The CGIAR at 31: An Independent Meta-Evaluation of the Consultative Group for International Agricultural Research. Washington, D.C.: World Bank Operations Evaluation Division.York: McGraw Hill.

CHAPTER - 21

Role of KVKs in Preparing Next Generation Farmers

S.K. Dhanda[1], R.S. Sheoran, P.K. Chahal and Narender Kumar
[1]Sr. E.S (Agro-forestry)-cum-Programme Coordinator
CCS Haryana Agricultural University, Hisar, Krishi Vigyan Kendra, Hisar Haryana - 125004

ABSTRACT

The future of any country depends on how the youth are moulded. Employment is the key to economic and social development and providing youths with job opportunities is a major issue for all countries irrespective of their level of development. Villages have become a source of cheap labour for growing cities. Meanwhile, rural communities are deprived off the services of rural youth. Increasing pressure on land is compelling youth to get out of farming. In India, there are 639 Krishi Vigyan Kendras (KVKs) located in every district and also known as the AKC (Agriculture Knowledge Centre). One of the mandate of KVKs is to impart vocational trainings to rural unemployed youth, landless laborer, school dropout boys and girls.The Hisar Districthas been divided into nine blocks and three sub divisions covering 274 villages, with average annual rainfall about 400 mm and the soil type issandy to sandy loam. Sh. Joginder Singh s/o Sh. Mahender Singh, 10+2 passed, 26 yrs. old youth was engaged in traditional cultivation of agricultural crops on his farm at village Saharwa (Hisar). In 2010, he came in contact with KVK, Hisar and got trainings from this centre. He was motivated to diversify his traditional system of agriculture with integrated farming system of vegetable production+ dairy farming+ horticulture+ boundary plantation of agro-forestry multipurpose tree species. With this diversified farming system, his income increased many folds i.e. Rs. 35000/acre to Rs. 4.40 lac/acre/year.

Key words: Labour, Employment, Skill, Rainfall, Bio-gas, Strawberry Dairy Farming

INTRODUCTION

The future of any country depends on how the youth are molded. Employment is the key to economic and social development and providing youths with job opportunities is a major issue for all countries irrespective of their level of development. For the unemployed youths, the problem gets worse. The huge numbers of frustrated and unemployed youths are easy target for terrorists, criminals, drug dealers etc.The migration of rural youths is tremendously increasing to urban areas in search of government; private, industrial jobs etc. and most of them are not taken up low paid jobs under unhealthy working condition. Villages have become a source of cheap labour for growing cities. Meanwhile, rural communities are deprived off the services of rural youth. Increasing pressure on land is compelling youth to get out of farming. Not only farming, there are hardly any opportunity for such rural youth to be gainfully employed in the rural areas, thus migratingto urban areas.Unemployment being the most visible poverty system, many developmental programmesare now aiming at promoting self employment by equipping the rural youth with the necessary skill & motivation.

In India, there are 639 Krishi Vigyan Kendras (KVKs) located in every district and also known as the AKC (Agriculture Knowledge Centre). The KVKs are preparing the next generation farmers through imparting vocational training/ skill in different discipline of agriculture and its allied fields. Because, agriculture is the main source of livelihood of the people in rural areas.KrishiVigyanKendra, Hisar (Haryana) has done a commendable job, converting rural youth in innovative farmers in different fields of agriculture through imparting vocational trainings and skill developmental programmes. These young innovative farmersnot only improved their socio-economic status by diversifying the traditional agriculture, natural resource management, adopting new and innovative ideas, eco-friendly techniques through integrated pest & nutrient management and subsequently reduced the cost of cultivation. These trained youths are motivated to adopt latest agriculture technology and act as master trainers/resource person for KVK, Hisar.

One of the mandate of KVKs is to impart vocational trainings to rural unemployed youth, landless laborer, school dropout boys and girls. The KVK Hisar has trained several unemployed rural youth in the field of horticulture, bee-keeping, dairy farming, seed production, fruits & vegetable preservation and cutting & tailoring etc.

Brief Description of Agro-Ecology of Hisar District

The Hisar District lies in the south- west of Haryana. It has divided into nine blocks and three sub divisions covering 274 villages. The average annual rainfall of the district is about 400 mm. and the soil of the area aresandy to sandy loam. The total area of the district is 4467 sq. kms. out of which cultivable area is 357418 ha. The major cropping patterns are Paddy-wheat, Cotton-wheat, Guar-wheat, Guar-raya, Cotton-barley etc. Irrigated area is 87% of the total cultivable area in the district. The following case study illustrated how vocational training can inspire youth to derive livelihood from integrated farming system. The idea of vocational skill is helpful for rural youth in their daily life as well as choosing career in a particular trade.

Case Study

Sh. Joginder Singh S/O Sh. Mahender Singh, 10+2 passed, 26 yrs. old youth started traditional cultivation of agricultural crops on his farm at village Saharwa (Hisar). In 2010, he came in contact with KVK, Hisar & got trainings from this centre. He learned a lot of new techniques and ideas through KVK activities like vocational trainings, demonstrations, field days etc. He was motivated to diversify his traditional system of agriculture. After getting soil & water tested, the experts of KVK suggested him for strawberry cultivation in his sandy fields. As irrigation water is a limiting factor, he was motivated for micro, sprinkler and drip irrigation system. Boundary plantation of agro-forestry multipurpose tree species was also done at the farm.

Different Component of Integrated Farming System Adopted by Mr. Joginder Singh

Horticulture+ Vegetable+ Dairy Farming+Agroforestry

He started strawberry cultivation integrated with vegetable production and dairy farming. Strawberry was grown on raised beds with micro irrigation system. Mulching with polythene sheets during winter to avoid frost injury, moisture losses and weed infestation. He cultivated strawberry crop from August to March followed by vegetable production on the same beds like chillies, bottle gourd, bitter gourd, pumpkin, cucurbits etc. from April to July. He has also installed a bio gas plant on his dairy farm. Besides energy, the slurry from this bio-gas plant is being utilized in these crops as organic manure. Experts of KrishiVigyan Kendra suggested forming an association of strawberry growers and getting it registered. After registration, this association was linked with Strawberry markets in Delhi, Jaipur, Amritsar and Ludhiana to fetch the remunerative price of their produce to avoid exploitation of the farmers by the middle man.

Economics of the Farm

Farm Component	Number/Area (Acre)	Input cost (Rs./yr.)	Net Income (Rs./yr.)
Straw berry	2	3 lacs / acre	2.8 lacs/ acre
Vegetable as Relay crop on strawberry beds	2	20,000/ acre	70,000/acre
Livestock	10 Buffaloes	50000/yr.	90000/yr.

Economics of Conventional Farm

Farm Component	Number/Area (Acre)	Input cost (Rs./ Acre)	Net Income (Rs./Acre)
Cotton- Wheat	2	17000	35000

A) Strawberry cultivation

Input cost- It includes the cost of all operations i.e. field preparation, bed formation, seedling preparation or purchase, cost of fertilizer, irrigation, drip irrigation, mulching, low tunnel, cost of packing kits, transportation and overall labor – Rs. 3 lacs / acre.

Gross income- 5.8 lacs/ acre

Net income- 2.8 lacs/ acre. It could be increased uptoRs. 3.5 lacs/ acre but an unknown disease affected the strawberry crop very badly during the current season.

B) Vegetable production

On the same beds which were used for Strawberry cultivation, different vegetables are to be sown in the month of April and inputs used for vegetable production costs around Rs. 20,000/ acre (cost of all inputs including labor) and the gross income is around Rs. 90,000/acre and the net income is Rs. 70,000/ acre.

C) Dairy farming

Joginder Singh is also rearing six Murrah buffaloes and four cross bred cows. For rearing these animals, the input cost is Rs. 50,000/ yr. and he is getting net income of Rs. 90,000/ yr. after selling the milk of these animals.

D) Economics of conventional farm

The conventional farm in HisarDistt.includes the existing traditional crop rotation i.e. Cotton- wheat. The input cost for growing these two crops for both the seasons i.e. kharif&rabi is as under:

Gross income from both the crops	:	Rs. 52000/ acre
Total input cost	:	Rs. 17000/ acre
Net income	:	Rs. 35000/ acre

Net income from integrated farming system

Strawberry	:	Rs. 2.8 lacs/ acre
Vegetables	:	Rs.70,000/ acre
Dairy farming	:	Rs. 90000/ year
Total income	**:**	**Rs. 4.40 lacs/ year**

CONCLUSION

As an occupation, agriculture seems to be considered flow esteem. Thus, unemployed youths are forced to choose another career. There are not enough initiatives from the government to provide opportunities to learn vocational skills. This skill can be used in their daily life but also helps in choosing alternative careers. Half hearted support to policies that encourage rural youth to take up agriculture farming will not yield expected results. The focus on formal education alienated these youth from their traditional occupations forcing them to migrate to urban areas. The discussion with young farmers/ rural youth revealed that they are innovative and can be helpful if vocational trainings/ skills are imparted to them. This case study indicated that these rural youths can see their future in agriculture related entrepreneurs if proper guidance, skills and opportunities are provided at young age.

CHAPTER - 22

Good Agricultural Practices (GAP) and Good Dairy Farming Practices (GDFP) - A Way Forward to Promote the Family Farming and Attract Rural Youth

S.P. Lal[1], S.R.K. Singh[2], Jayant Goyal[3], Mradula Tiwari[4]
[1]Ph.D. Scholar, Dairy Extension Division, NDRI, Karnal
[2]Sr. Scientist, Zonal Project Directorate, Zone VII, Jabalpur, JNKVV (MP)
[3]Ph.D Scholar, Division of Extension Education, IVRI, Izzatnagar
[4]Veterinary Officer, Department of Animal Husbandry, Uttarakhand

ABSTRACT

Good Agricultural Practices (GAPs) and Good Dairy Farming Practices (GDFPs) is the need of hours because of the several burgeoning issues like-climate change, low food quality etc. GDFPs concept was introduced in 2004 by International Dairy Federation/Food and Agriculture Organization (IDF/FAO) and it was later updated in 2011. GAPs ensure quality food produce that ultimately leads to better heath. Good Agricultural Practices for dairy farmers is about implementing sound practices on dairy farms – collectively called Good Dairy Farming Practices. These practices need to ensure that the milk and milk products produced are safe and suitable for their intended use, and also that the dairy farm enterprise is viable into the future, from the economic, social and environmental perspectives. Since, dairy farmers are in the practice of producing food for human consumption, they must have confidence in the production of safety and quality of the milk. Since the introduction of the concept in 2004,the importance is growing in various countries, and agencies including private, individual farm house or farmers etc. GAPs and GDFPs underpin the production of food and milk that satisfies the highest expectations of the food industry and consumers. Both of these practices are capable of bring good prices to the farmers at their door steps. But the real issues are that both the practices are labour intensive and requires extra care — so, here the concept of 'family farming' may come handy. Today, in Punjab and Haryana due to urbanization the land has become too costly and that's way these days the whole attention of the farmers including their sons' are on real estate and that is threat to our sustainable development. So, if we want to attract and retain youth in Agriculture then GAPs and GDFPs are panacea for that- as both have greater scope to fetch fair and remunerative prices of the agricultural produce and product.

Key words: GAPs, dairy, youth, development, consumption

INTRODUCTION

If mountain doesn't go to the Muhammad then,
Muhammad must go to the mountain.

(Swami Vivekananda)

Half of the children in India are chronically malnourished, 23 crores of the countrymen are suffering from chronic hunger and some 2 billion people in the world are suffering dietary deficiencies that have lethal impact on their health. A large number of farmers in India are compelled to commit suicide and are at the verge to quit farming than to live in a miserable livelihood conditions. Today, India is world's suicide capital with a suicide in every 2 minutes. Moreover, Suicides as a whole rose nationally in the 1997-2005 period, but the rate of increase in farm suicides was far higher than the rate of increase in suicides by non-farmers. So, what can be the possible solution to these entire monstering problems? Imagine if there was a better way for marginal and small farmers to cultivate paddy (the staple food of half the world and paddy straw is the important fodder for the dairy animals) — one that required less costly inputs, less water, and considerably increased yields? Or wonder the potential benefits if there were a simple way to increase the milk production from cows by substantial amount? Wouldn't these things be developmental miracles? Well, they are not miracles; they are realistic possibilities. The problem is that, most farmers in the developing world (majority of the farmers in India) don't have proper knowledge about them — or don't know how to implement them successfully due to poor extension delivery system in India.A cultivation practice known as 'Method of Rice Intensification' (SRI) which requires 1/10th of the seed than conventional methods and involves transplanting rice saplings, spacing them in a grid, keeping the soil drier, and carefully weeding plots, can increase rice productivity (SRI comes under GAP). SRI has been called one of the most important agricultural innovations of the past 50 years, yet it is employed by a fraction of farmers. An aquatic fern— azolla, which can be readily cultivated and added to dairy animal feed, can enhance production of cow's milk by 15 to 20 percent (Azolla feeding comes under GDFP). And there are countless other opportunities and avenues! So, what can be the possible solutions for these types of problems? The answers lie somewhere in theGAP andGDFP, and for this we require strong extension approaches!

What is Good Agricultural Practices (GAP)?

A multiplicity of Good Agricultural Practices (GAP) codes, standards and regulations have been developed in recent years by the food industry and producers organizations but also governments and NGOs, aiming to codify agricultural practices at farm level for a range of commodities. Their purpose varies from fulfillment of trade and government regulatory requirements (in particular with regard to food safety and quality), to more specific requirements of specialty or niche markets. The objective of these GAP codes, standards and regulations include, to a varying degree:

- ensuring safety and quality of produce in the food chain
- capturing new market advantages by modifying supply chain governance
- improving natural resources use, workers health and working conditions, and/or
- creating new market opportunities for farmers and exporters in developing countries.

Good Agricultural Practices are "practices that address environmental, economic and social sustainability for on-farm processes, and result in safe and quality food and non-food agricultural products" (FAO COAG 2003 GAP paper). These 'four pillars' of GAP(economic viability, environmental sustainability, social acceptability and food safety and quality) are included in most private and public sector standards, but the scope which they actually cover varies widely.The concept of Good Agricultural Practices may serve as a reference tool for deciding, at each step in the production process, on practices and/or outcomes that are environmentally sustainable and socially acceptable. The implementation of GAP should therefore contribute to Sustainable Agriculture and Rural Development (SARD).

What are Good Dairy Farming Practices (GDFP)?

Good Dairy Farming Practices (GDFP) concept was introduced in 2004 by International Dairy Federation/Food and Agriculture Organization (IDF/FAO) and it was later updated in 2011 (FAO and IDF, 2011). Good Agricultural Practice for dairy farmers is about implementing sound practices on dairy farms – collectively called Good Dairy Farming Practice. These practices need to ensure that the milk and milk products produced are safe and suitable for their intended use, and also that the dairy farm enterprise is viable into the future, from the economic, social and environmental perspectives. Since, dairy farmers are in the practice of producing food for human consumption, they must have confidence in the production of safety and quality of the milk. Since the introduction of the concept in 2004, the importance is growing in various countries, and agencies including private, individual farm house or farmers etc. GDFP underpins the production of milk that satisfies the highest expectations of the food industry and consumers. The international framework to ensure the safety and suitability of milk and milk products is contained in the Codex Recommended International Code of Practice – General Principles of Food Hygiene (CAC/RCP 1- 1969, Rev. 4, 2003)1 together with the Codex Code of Hygienic Practice for Milk and Milk Products (CAC/RCP 57-2004)2. GDFPs recognizes that dairy farmers are an integral part of a larger dairy food production and processing chain and that all participants in the chain - dairy farmers, suppliers to dairy farmers, milk carriers and haulers, dairy product and food manufacturers, distributors, retailers and consumers - should be part of an integrated food safety and quality assurance management system. Dairy farmers can play their part by ensuring that good dairy farming practices are implemented at the farm level. Good dairy farming practice also ensures that the milk is produced by healthy animals in a manner that is sustainable and responsible from the animal welfare, social, economic and environmental perspectives. So implementing good dairy farming practice is good risk management for the short and long term future of the dairy farming enterprise. GDFP encourages dairy farmers to adopt 'proactive' preventative practices rather than waiting for problems to occur. GDFP underpins the production of safe, quality-assured dairy products in a sustainable manner which would leverage the future of dairy farming on a local, national and international scale.

Potential Benefits, Challenges and Solutions Related to GAP and GDFP

Potential benefits of GAP: Appropriate adoption and monitoring of GAP helps improve the safety and quality of food and other agricultural products.It may help reduce the risk of non-compliance with national and international regulations, standards and guidelines (in particular of the Codex Alimentarius Commission, World Organization for Animal Health (OIE) and the International Plant Protection Convention IPPC regarding permitted pesticides, maximum levels of contaminants (including pesticides, veterinary drugs, radionuclide and mycotoxins) in food and non-food agricultural products, as well as other chemical, microbiological and physical contamination hazards.Adoption of GAP helps promotes sustainable agriculture and contributes to meeting national and international environment and social development objectives.

Challenges and solutions related to GAP: In some cases GAP implementation and especially record keeping and certification will increase production costs. In this respect, lack of harmonization between existing GAP-related schemes and availability of affordable certification systems has often led to increased confusion and certification costs for farmers and exporters.Standards of GAP can be used to serve competing interests of specific stakeholders in agri-food supply chains by modifying supplier-buyer relations.There is a high risk that small scale farmers will not be able to seize export market opportunities unless they are adequately informed, technically prepared and organized to meet this new challenge with governments and public agencies playing a facilitating role.Compliance with GAP standards does not always foster all the environmental and social benefits which are claimed.Awareness, sensitizationand motivation is needed for 'win-win' practices which lead to improvements in terms of yield and production efficiencies as well as environment and health and safety of workers. One such approach is Integrated Production and Pest Management (IPPM).Both of these practices are capable of bring good prices to the farmers at their door steps. Both the practices (GAP and GDFP) are labor intensive and require extra care — so; here the concept of 'family farming' may come handy.

Why Paradigm Shift In 'Extension Approaches' Is Required To Promote GAP, GDFP, Family Farming And Attract Rural Youth

Now let's talk about; what is the role of Extension (in the 21st century)? The role is not only to inform and train rural youth, instead—to empower farmers and youth, induce innovation and that is endorsed by the definition of few social scientists. Extension is a series of embedded communicative interventions that are meant to induce innovations which supposedly help to resolve (usually multi-actor) problematic situations. (Leeuwis, C. and van den Ban, A.).

What is extension approaches: An extension approach is the basic planning philosophy that is being adopted by an agricultural extension organization. This helps extensionists to understand the fundamentals, concepts and functional methods of extension adopted to fulfill its aims, especially in the planning phase (Leeuwis, 2004). Extension approach refers to the doctrine for an organization, which informs, stimulates and guides farmers including rural youth such aspects of the organization as its structure, mission, vision, leadership, its programs, strategies, its resources and linkages beneficiaries.So, to conclude we can say that extension approaches are different guiding principles, doctrine or basic planning philosophy.

Categorization of Extension Approaches

Axinn (1988)	Gêmo et al. (2005)
1) General agriculture	1) Public
2) Commodity	2) Commodity
3) T&V (training & visit programme)	3) T&V
4) Agricultural participatory approach	4) NGOs
5) Project approach	5) Private sector
6) Farming systems research and extension	6) Farmer field schools (FFS)
7) Cost-sharing	
8) Educational institute approach	

Convergence of various 'extension approaches' and linkage of agricultural input delivery system is required to promote GAP, GDFP, family farming and attract rural youth:

Over the years, manpower and paucity of funding have adversely affected the performance of public sector agricultural extension services. In India there are 638,596 villages and have about 120 million farm holdings. If it is proposed to provide one village extension worker for every 800-1000 farm families then the requirement of field level extension workers is estimated to be about 13-15 lakhs, against which the present availability is only about one lakh workers and whose job is to translate research into useable information for the 50 percent of Indians who earn a livelihood from farming, many of whom cannot read and lack electricity. Historically, extension efforts have tended to bypass the poorest, however "Only 40 percent of Indian farmers receive any kind of agricultural information and only 6 percent are reached by the public sector extension agents.

The irony is that only about 5% of the farm households in India do access information on livestock.The genuine question is how to scale up all the farmers of India. UK once prided itself that the sun never set on the British Empire, but today it is far behind the USA, Germany and Japan. This is due to the fact that it never declared any long-term vision, planning or project (The UK has recently started a Technology Foresight programme). Similar is the case with our agricultural sector, We rarely declare any long term plan and if it is somehow declared we rarely implement them truly. Let's take the case of genetic upgradation of our indigenous breed; we have hardly achieved anything in it, since independence. But some distant nations are smartly using our local breed, let take the e.g. of Brazil. The majestic cow from Brazil belonging to the Gir breed of Gujarat, this cow — named She-ra — clocked 62.033 litres of milk/day in a 3-day milk competition at the 40thExpaja in Brazil, beating her own record of 59.947 litres. Today, Brazil is the biggest exporter of Indian breeds of cows. Gir cow now records over 62 *l/day* in Brazil. Gujarat due to their long term plan like— micro irrigation (to conserve water from irrigation), silage making (to ensure year round availability of fodder for the dairy animals), good AI and breeding programme is now achieving agricultural growth rate of more than 10% from the past 10 years. Similar is the case with Madhya Pradesh, "The state has outsmarted even Haryana and now stands third after Punjab and Uttar Pradesh in per hectare yield of food grains," .When asked about this achievement—a government spokesperson said, "it was only due to intensive agriculture extension programmes, motivational programmes, training to farmers and long term planning". Motivational programmes! Yes, the role of extension personnel are not only to give training to the farmers or completing their organizational/institutional mandate but also to motivate them and empower them with current knowledge, this is prime essential to maintain rapport for a long run. For example if you are an extension personnel then tell them the importance and procedure of making the AADHAAR card aka UID (because in coming year subsidies/money for social welfare schemes/direct benefit transfer "DBT" would be given only through UID), if you are in Maharashtra then tell them that the state government gives Rs.65 per day per animal as subsidy for fodder allocation to fully grown cows and buffaloes in their fodder camps, if you are in Rajasthan then inform them that the state's livestock development board is providing Rs. 20,000 subsidy on purchase of each Murrah bull .

Today, the situation is that very few farmers know about silage making, AI (the real benefit of using AI is that it gives all farmers the possibility of gaining from genetic improvements created elsewhere, privately or collectively), Embryo transfer technology (could be utilized for the faster improvement of livestock),Azolla feeding (it is a highly nutritive and cheap organic feed substitute for dairy cattle, poultry, pig and fish. It is also used as a bio-fertilizer for wetland paddy. A part from milk productivity enhancement tool the farmers also have very less knowledge about the optimum use of cattle-dung and urine. Cattle-dung can be use for the preparation of vermicompost. There is a common myth among the farmer that we need permanent concrete pit for large scale production of vermicompost. Now, high quality fabric tarpaulins bags (can be compared with simple polythene bag) are available in the market in various range and these beds are also cost effective and portable in nature. These beds are UV stabilized and can resist with sunlight and heavy wind.

Panchgavya is a low cost and environment friendly bio-pesticide prepared from the product and by-product of plants and animals, ICAR has declared Panchgavya as one of the best farm innovation in 2010. Farmers have all the right to know that there fodder crop like Guar or cluster bean (Cyamopsistetragonoloba) has now converted in cash crop, for the high demand of guar gum in the world market (so that they can harness new economic benefit and that would ultimately lead to more Guar production due to competition in the open market and more importantly faith in agriculture).So, convergence of different extension approaches are quite essential.

Using information methods to promote family farming and attract rural youth

Victor Hugo once said, “There is nothing more powerful than an idea, whose time has come”. So, it’s time to divide all our information method (IM) into 3 categories:-

First line of IM Second line of IM Third line of IM 1) Mobile SMS based IM 2) Video based IM 3) Internet based IM 1) Information imparted through training or live-Demonstration by the local body/institute. 1) Information imparted through training or live-Demonstration by far-flung body/institute.

Note: The cost per information/adoption increases from left to right

A) First line of IM: - In this knowledge and attitude towards family farming and rural youth are changed. But, there are certain limitations of it. For e.g.) Mobile SMS based IM is suitable for information like—"After every milking dip teats in *Mastrite* (a chemical to control mastitis) and keep cattle standing for 30 minutes. Do this by feeding them after milking”. But, if we want to teach the family farmers about complete package of clean milk production or Azolla production cum feeding, then video based IM is more feasible.

B) Second line of IM: - In this skill is more pronounced by imparted training or live-Demonstration. For e.g. if we want to train farmer/ rural youth about milk fat testing or how to perform AI, then Second line of IM is far more practicable. It would be worth mentioning that before imparting training or live-Demonstration the outline of the subject matter explained by video based IM is far more successful.

C) Third line of IM: - It is given when requisite expertise is either not available at all or not available up to the required degree. In this IM there is much more chances for the exposure of farm and dairy innovation to the farmers. It can be broadly classified into 3 categories namely:-

1) Training within nation by exchange or without exchange of farmers (Inter-district or Inter-state):- e.g. Farmers from all across India coming to NDRI (KVK, Karnal to acquaint them with the latest dairy farming progress and innovation).

2) Training Overseas: - In 2012 more than 2000 farmers (majority of them were youth) fly to Israel to learn GAP and GDFP technology in

18th International Agricultural exhibition in Tel Aviv, Israel (biggest Agricultural fair in the world).

3) Training by Foreign Experts: Self explanatory

As an estimated 70 million rural milch animal households are engaged in milk production and only about 5% of the farm households in India do access information on livestock. So, it's time to scale up all the milch animal households of our country and it's very easy to do that by Mobile SMS based IM & Video based IM(and even at very low expenditure). For this we have to create the database of all the dairy farmers and this would be done by all 637 KVKs (various SAUs, CAUs, and NGOs etc may extend their help to the KVKs.). Today majority of the rural youth/farmers have Mobile/cell phone and this could be harnessed by sending them agricultural tips through KVKs. But, There are two problems in this IM: First —All farmers don't have mobile and second— not all type information sharing are viable through SMS (as discussed earlier). And, Here Video based IM comes handy as this approach is 7 times more cheaper than the traditional approach and according to Edgar Dale's 'Cone of Experience farmers' learning ability are up to 70% as farmers watch the video and also participate in discussion. In 2011 GOI declared 60,000 villages, as pulse village in order to promote pulse production. In the similar format GOI should declare the least milk producing villages as 'dairy village' and at least one youth in each village as resource person (who will have sound knowledge of farming and will guide other framers to adopt suitable practices and declaring him/her as resource person will boost higher moral). So, if our policy makers will take 'first line of IM' seriously, they can easily provide information to all the farmers up to the end of 12th 5 year plan. Note information/knowledge is like the nuclear chain reaction and once it starts, it would diffuse to other farmers automatically.

Action Research: a component of GAP and GDFP is required to promote family farming and attract rural youth

The government should promote extension personnel to solve the problem of young and budding farmers through Action Research (AR) instead of just addressing the problem. The Action Research is broadly classified into 3 types:-

TYPE 1: Technical/Technical-Collaborative/Scientific-Technical/Positivist AR

TYPE 2: Mutual-Collaborative/Practical-Deliberative-Interpretivist Perspective AR

TYPE 3: Enhancement approach/Critical-Emancipatory Action research

There are four basic themes of AR viz. empowerment of participants; collaboration through participation; acquisition of knowledge; and social change. By extension personnel mostly TYPE 2 AR is used. Where, the researcher and the practitioners come together to identify potential problems, and a mutual understanding is reached about the anticipated solution. For example, if extension personnel want to solve the problem of drought hit village(In 2009, the country faced a severe drought in 338 districts out of total 640 districts in India and so 2009 was declared as 'drought-year'), then s/he has to take care of soil topography, rainfall received per year and a series of other paraphernalia. In that region— watershed can be the possible solution for water harvesting and for that we have to dig the pond. Now, where the soil type is supportive then average size pond is okay—but where silting is problem, we have to dug large size pond with higher dimension and capacity. Let's take the example of a village—'Hiware Bazar' in Ahmednagar district of Maharashtra, where the villagers solved the problem of water crisis through Action Research. This village hardly receives 400 mm of rain and lies in rain-shadow area. So, they decided to build watershed in their village in order to harvest the rain water (a series of other actions/initiatives/measures were also taken to prevent indiscriminate use of stored rain water through collective community participation).Today, due to the application of Action Research they are now able to take 4 crops in a year (previously they could hardly take one crop in a year), in 1992 land under irrigation was 40 ha and in 2009 it was 550 ha, in 2000 Fodder production was 100 metric tonnes but in 2009 it was more than 8000 metric tonnes, in 1992 milk production was 150 *l/day*but in 2009 it was 4000 *l/day*. Today, the village has more than 112 biogas plants for providing fuel, electricity and manure. In 2008 the village received 'National Water Award' and that's way the village is popularly known as "The miracle water village". We can see the magic of AR, family farming, community and rural youth participation— once there was no sufficient water for the cattle to drink and today 70% of the villages are engaged in dairy. In another case, Bihar and Punjab/Haryana were facing the problem related with Rice production. Bihar was facing the problem of low productivity and they solved their problem by SRI method. As a result of that— the production of rice in Bihar increased to 8.2 million tonnes in 2011-2012 compared to 3.1 tonnes in the previous year.

The problem of Punjab/Haryana is 'the perennial shortage of labour' and the state governments of Punjab & Haryana came up with paddy transplanter as the solution (even with heavy subsidy). But, the programme failed due to several reasons—one of the prominent reasons was lack of training and poor demonstration. Now, see the difference—in Bihar, SRI was implemented through a lot of public consultation and AR, and the programme is successful but in Punjab/Haryana the government didn't do AR and homework and as a result the programme suffered despite providing a huge subsidy.

Role of Governmental Organizations, SHGs, NGOs, Civil Society, Corporate World to Facilitate GAP, GDFP, Family Farming and Attract Rural Youth

Remember, India is a vast nation—so; they all have to be active participant. For e.g. Oil India Ltd. (OIL) plans to tie up with the Assam govt. to start a dairy co-operative in Assam and later extend it to entire north-east as a part of their CSR, Corporate Social Responsibility(Remember whole north east is facing acute shortage of milk, the highest per capita availability is in Nagaland 93 gms/ day much short of ICMR recommendation of 310 gms/day). In another case—KVK, Kannur (won the National Best KVK Award for 2009) has taken many initiatives to empower and motivate the rural youth for e.g. it has open a mall on the campus to display and market the branded value added food products produced by the farmers. It also set up a loan window with a tie-up with State Bank of India in its premises, in order to provide hassle-free loans to willing young entrepreneurs in agriculture. Thus, it is providing 3 in 1 window i.e. training, credit and marketing. These types of initiative must be taken by other KVKs for the uplift family farming. Remember varieties of agricultural products have potential to capture the food market through GAP and GDFP, but it has unfortunately spread its wing in some pockets only. So, if we will follow the trend of KVK, Kannur then Indian farming could reap huge dividend and that would ultimately attract rural youth towards agriculture.

Today, 80% of the farmers own less than a hectare of land. So, it's time to reinvent the importance of GAP and GDFP for sustainable farming. Indian farmers have been raising crops and livestock together for centuries and for them these components have been an integral part of agricultural farming systems. The results of more than 15 years of investigation on the development of suitable farming systems for semi-arid tropical situations in Haryana, India, indicate that mixed farming systems of crops and animals are more efficient and remunerative, and generate more employment for rural youth, than arable farming systems under small land holdings for irrigated and dry land conditions.

The country has already witnessed the success story of Ramesh Chandra Dagar, a farmer in Sonipat district, Haryana for his success in integrated GAP and GDFP,consequently reaping more than 10 lakhs rupees in a hectare (his success story is already being taught in NCERT Biology,Class XII text book and in 21 different nations where, Central Board of Secondary Education (CBSE) branch exists). But, these success are scattered, unorganized and few. So, to meet the need of GAP and GDFP for sustainable agriculture, a radical rethinking for reshaping agricultural education, research and extension is urgently required. Agricultural professionals including farmers will need specific types of reorientation and training with emphasis on holistic concepts, institutional behaviour, cooperative attitude, respect for nature, local farming systems and indigenous knowledge.

After that a clear understanding of three factors—the market demand, the natural resources available and maintaining the product quality – helps them succeed. We must understand that Extension is not "Cinderella magic wand" where we can get the solution of everything. Other parameter must be improved if we want to see the real effect of extension. In 1971 GaribiHatao (abolish poverty) slogan was very popular— but, now the situation is that our country has more poor people today, than at the time of independence.

So, keep in mind that extension is just one part (other sine qua non to be the part of GAP and GDFP for the holistic approach to promote family farming and attract rural youth are)

Physical Need	Human Need	Social Need	Financial Need	Digital Need
Buildings And Infrastructure	Functional literacy	Better equipped Institutions	Initial and operational cost	Hardware Software
Road, seawaysair waystransport	Computer literacy	Norms, taboos	Maintenance Cost	Connectivity and content
Training institute	Motivation and awareness	Political and administrative support	Training cost	Latest gadget and apps

Then, only synergy of policy with extension can be seen. Finally, GAP and GDFP is not about, what can be done by the extensionist. But it is all about what can be done by both farmers and extensionist together. For this convergence of rural youth to the agricultural system is must.

Rural Youth is Applying "Adam's Equity Theory" of Motivation in Agriculture

Suppose youthA gets a 20% raise and youth B gets a 10% raise in their income for the same effort and labour. Will both be motivation level for both the case is same? In fact, here Awill be twice motivated and Bwill be negatively motivated? Adam's equity theory says that it is not the actual reward that motivates, but the perception, and the perception is based not on the reward in isolation, but in comparison with the efforts that went into getting it, and the rewards and efforts of others. Today, in Punjab and Haryana due to urbanization the land has become too costly and that's way these days the whole attention of the farmers including their sons' are on real estate and that is threat to our sustainable development. Rural youth see other fellow farmer that by selling his land and starting his own non-agricultural business is fetching more money than agriculture then he gets disenchanted with agriculture.

So, if we want to attract and retain youth in Agriculture then GAP and GDFP are panacea for that- as both have greater scope to fetch fair and remunerative prices of the agricultural produce and product. But, the effort should be from both side- farm family and government. As GAP and GDFP are labor intensive which can be met by family farming and it also requires better market linkage to sell the produce that can be provided by government. So, GAP and GDFP can be a win-win-win situation for farmers, government, costomers and it has a huge potential for creating employments for rural youth.

REFERENCES

Axinn, G. H. (1988). Guide on Alternative Extension Approaches. Rome: FAO.

FAO and IDF.(2011). Guide to Good Dairy Farming Practice. Animal Production and Health Guidelines.No. 8. Rome.

FAO (2008). What are Good Agricultural Practices?

Gêmo, H. C.K. Eicher,and S. Teclemarian (2005). Mozambique's Experience in Building: A National System. East Lansing, MI, US: Michigan State University Press.

CHAPTER - 23

Involvement of Tribal Women in Agriculture

Pradeep Kumar[1], Sundaravadivu[2],O.P. Mishra[3] and Kalyan Ghadei[4]

[1&2]Research Scholar, Department of Extension Education, Institute of Agricultural Sciences, Banaras Hindu University, Varanasi-221005,Uttar Pradesh [3]Associate Professor and [4]Assistant Professor, Department of Extension Education Institute of Agricultural Sciences, Banaras Hindu University, Varanasi-221005,Uttar Pradesh

ABSTRACT

Agriculture can act as a catalyst in accelerating growth and reducing poverty. Agriculture can be an important engine of growth and poverty reduction. But in spite of its futuristic capabilities, the sector is underperforming because the productivity is reduced because of the constraints faced by the women in agriculture and allied sector. The women are the backbone of agricultural workforce but worldwide their hard works have mostly been unpaid. They have to perform most tedious and back breaking part of work in agriculture, animal husbandry and at homes. The tribal women play a significant and crucial role in agricultural development and allied fields including in the main crop production, livestock production, horticulture, post harvest operations, agro / social forestry, fisheries, etc. Tribal women have to manage, often, complex and multiple livelihood strategies. Their activities typically include producing agricultural crops, tending animals, processing and preparing food, working for wages in agricultural or other rural enterprises, collecting fuel and water, engaging in trade and marketing, caring for family members and maintaining their homes. Women's participation in rural labour markets varies considerably across regions, but invariably women are over represented in unpaid, seasonal and part-time work, and the available evidence suggests that women are often paid less than men, for the same work.

Key words: Involvement, Agriculture, Tribal Women

INTRODUCTION

Tribal women plays a major role in the co-management of their natural, social, economic resources and agricultural development including crop production, livestock production but they remain backward due to traditional values, illiteracy, superstition and many other social and cultural factors. The participatory role of tribal's in improving their living conditions by fully exploring natural endowments and alternative uses must find an appropriate place in the strategic approach (Sanjay Kanti Das, 2012).The women play a significant and crucial role in agricultural development and allied fields including in the main crop production, livestock production, horticulture, post harvest operations, agro/ social forestry, fisheries, etc. (Prasad & Singh 1992). Improvement of women's community participation for the livelihood is not only an important goal of socio-economic development of the country but it is as well as a necessary input to the development process. Rural Indian women's are extensively involved in agricultural activities. However the nature and extent of their involvement differs with the variations in agroproduction systems. The mode of female participation in agricultural production varies with the landholding status of the family. Their roles range from managers to landless labourers. In over all farm production, women average contribution is estimated at 55% to 66% of the total labour. In the Indian Himalayas a pair of bullocks works 1064 hours, a man 1212 hours and a woman 3485 hours in a year on a once hectare farm, a figure that illustrates women.s significant contribution to agricultural production. (Shiva FAO,1992). The National Research Centre for Women in Agriculture (NRCWA) has been functioning at Bhubaneshwar, Orissa, for developing methodologies for identification of gender implications in farming systems approach and to develop women specific technologies under different production systems.

Scope of Tribal Women in Agriculture

Alam, S.M. (2006) has pointed out in his article entitled "Production Hazards, Marketing Risks" mentioned that majority of tribal women is self-employed and work in dangerous environments. Their daily tasks includes keeping and caring for the livestock at farms. They grow grains, cotton, fibers, fruit, and vegetables. The farmer plants, tills, fertilize, sprays, harvests, packs and stores the product. The livestock farmer feeds and cares for animals, while the horticulture farmers produce ornamental plants and nursery products.

Tribal women performs numerous labor intensive jobs such as weeding, hoeing, grass cutting, picking, cotton stick collections, separation of seeds from fiber. Tribal Women are also expected to collect wood from fields. This wood is being used as a major fuel source for cooking. Because of the increasing population pressure, over grazing and desertification, women face difficulties in searching of firewood. Livestock is the primary subsistent activity used to meet household food needs as well as supplement farm incomes. The majority of farms own some livestock. The pattern of livestock strength is mainly influenced by various factors such as farm size, cropping pattern, availability of range-lands including fodder and pasture. It is common practice in the rural areas to give an animal as part of a women's dowry. Studies have revealed rural women earn extra income from the sale of milk and animals. Mostly women are engaged in cleaning of animal, sheds, watering and milking the animals. (Lal and Khurana, 2011).

Since time immemorial women have significant contribution in almost every aspects of our society. Besides having an anchoring role in the management of their families, women equally participate in different economic activities like crop production, vegetable cultivation, fruit raising, livestock management, pisciculture and other miscellaneous activities such as mushroom cultivation, leaf plate making, bamboo works, badi and papad making etc. price shops, selling of cloth, sarees and bangle, embroidery unit,, stationery production and sale etc.

- Tribal Women Contributions as Food Producers
- Tribal Women as Gatekeepers
- Multi-Dimensional Role of Tribal Women
- Tribal Women's contribution to decision making
- In Preservation

Tribal Women's

- Tribal Women's contribution to animal husbandry
- Tribal Women in Poultry Production
- Tribal Women in Drudgery Reduction
- Maintenance of Biodiversity
- Tribal Women's contribution to agricultural Marketing
- Policy and Strategies for Tribal Women
- Tribal Women in Selection of Enterprise

Tribal Women Contributions as Food Producers

The sustainable production of food is the first pillar of food security. Millions of women work as farmers, farm workers, and natural resource managers. In doing, so they contribute to national agricultural output, maintenance of the environment, and family food security. Women labour is the most important input in crop cultivation, mainly comes from households itself and on some occasions from hired manpower (Gautam et al. 2012). Maximum agriculture activities are performed by women except some heavier tasks of land preparation (clearing, felling trees and burning vegetable, ploughing and land preparations). Rest activity performed by women such manuring and carrying; owing and weeding. Threshing and winnowing; transportation and storing activities are mainly done by women.Tribal Women's participation in crop production, livestock production, horticulture, post harvest operations, agro / social forestry, fisheries, etc. Tribal women have to manage, often, complex and multiple livelihood strategies.Tribal Women are active in all aspects of crop production, expect in ploughing in which Tribal women are not actively involved. TribalWomen are involved in sowing, weeding, plant protection, maintenance, harvesting, food preparation, storage, preservation, treatment and marketing of crops.

Tribal Women as Gatekeepers

The second pillar of food security is making sure that household members, particularly the children, receive an adequate share of the food that is potentially available. Tribal Women act as the "gatekeepers" of their households' food security through the allocation of their own time and income.

Multi-Dimensional Role of Tribal Women

Agriculture: Sowing, transplanting, weeding, irrigation, fertilizer application, plant protection, harvesting, winnowing, storing etc.

Domestic: Cooking, child rearing, water collection, fuel wood gathering, household maintenance etc.

Allied activities: Cattle management, fodder collection, milking, Forest product collection, herbal drugs and medicine collection etc.

Tribal Women's Contribution to Decision Making

Women's role in agricultural operations, animal husbandry and other economically productive activities is very significant. They contribute about 60-70 percent of the labour required for these activities thus playing a pivotal role in sustaining economy. The decision making process is an important segment of every household because it makes implementation of a plan or programme quite easy.

In Preservation

Indigenous crops and seeds storage

Agro diversity plays multiple roles in rural economy and has a strong human dimension as manifested through socio-cultural link and involvement of women. Earlier work has brought into focus the important role of women in the agro diversity management and use of environmental resources including soils, crops, water, forest and the indigenous plants on which local communities depends. Storage of Indigenous crops and seeds are critical to achieving food security on both a national and local level by easing price variations and assuring more stable supplies. In India, storage facilities are inadequate and ineffective. It is estimated that one-fourth of all food produced is lost due to spoilage, insects and rodents. Tribal Women play an important role in Indigenous crops and seeds storage and handling, and as such could benefit from improved methods and facilities. Improvements in traditional systems of storage and food handling could make a major contribution to food security in terms of stabilizing supplies and helping to maintain the nutritional value of stored and processed foods. Seeds are the most important component of agriculture productivity and a striking symbol of agricultural research as it contains embodied technology. In India, both private and public sector plays commendable role in seed production and its distribution.

Food Preservation

Traditional method of management such as mulching with cleared weeds and stamps of crops after harvesting to conserve soil moisture, fallow period ranges from 3-6 month after a complete crop cycle. Deliberate leaving weed and tree present in agricultural field check soil erosion, are practiced as well. Tribal Women perform majority of the work in food processing in developing countries. Tribal Women areresponsible for supplying their families with food and care; they often have special knowledge of the value and diverse use of plants for nutrition, health and income. Consequently, they are frequently the preservers of traditional knowledge of indigenous plants. Tribal Women process and preserve the fruit and vegetable produce from their home gardens and from the forests. Women are universally responsible for preparing food for their households and thus for the nutritional well-being of its members.

Tribal Women's contribution to animal husbandry

Agarwal (1987) reported that the involvement and participation of rural women were more in animal husbandry than in agriculture. It is true that a nucleus of dairy unit more or less exists as a traditional component in every rural household as a source of draught power or nutritional substance for the family, or a secondary source of income. Venkatachalam (1993) and Sethi (1991) expressed that almost 80-90 per cent of the total work related to cattle care is performed by women. In many places the entire management of livestock, starting from cutting, collection, carrying and chaffing of fodder to feeding and milking, preparation of milk products, cleaning of cattle shed, collection of cowdung to the manure pits, preparationof cowdung cakes and their storage was done by women **(Achanta, 1982).**

Tribal Women in Poultry Production

Poultry production is an occupation that supports the livelihoods of many women, especially the tribal and rural poor in India. Poultry act as financial reserves in crisis situations, and thus reduce vulnerability. The majority of the rural poor still depend on agricultural labour, which does not offer a year round earning opportunity, and livestock rearing supplements their income. Poultry is a source of low cost, high quality food for the family and small cash for women through sale of birds and eggs. Most of the procurement and sale of birds and eggs are done within the village and in local weekly markets.

Tribal Women in Drudgery Reduction

Tribal Women in agricultural communities work for longer hours compared to men and due to their heavy work load, Tribal women often find it difficult to attend training and other extension programmes. Releasing time from these heavy workloads is essential in order to get meaningful participation of women in agricultural technologies. There is a need to understand the daily and seasonal workloads of women farmers and accordingly suggest her for drudgery reducing improved tools and techniques. Promote sowing tools, paddy threshers, winnowers,sprayers, weeder, harvesting tools, cleaner, grader, parboiling units, maize shellers, dal making machines etc.

Maintenance of Biodiversity

Nature has given us abundance; Tribal women's indigenous knowledge of biodiversity, agriculture and nutrition has built on that abundance to create more from less, to create growth through sharing. Tribal Women farmers share seeds freely and with sharing as a base, there can never be scarcity. The giving and sharing of food in abundance has been the basis of ensuring food security. The giving and sharing of knowledge has been the basis of the growth and evolution of knowledge. An economy of sharing is an economy of abundance. This worldview of abundance is based on sharing and on a deep awareness of human as members of the earth family. This awareness that in impoverishing other beings, we nourish ourselves is part of our present ecological knowledge and ancient wisdom. Without giving and sharing there can be no sustainability; with our sustainability, there can be no space. Diverse Women for Diversity (DWD) movement celebrates sharing for sustainability and peace through food festivals, exposure tours, training programmes in biodiversity conservation, sustainable agriculture, indigenous healing systems, water conservation, capacity building for leadership and good governance.

Herbal Drugs, Plants and Medicine

A strong local management and a large biodiversity base provide the best insurance for both food security and political stability. Agro-biodiversity can only be assured if the people's vast and varied traditional knowledge base is guaranteed an enabling environment to survive and thrive. It is critical that agricultural policy objectives, such as food sovereignty and seeds are put first and trade rules subject to them and not vice-versa.There is an urgent need to promote/revive indigenous crop varieties and reverse the loss of agro-biodiversity caused due to market drivers. Indigenous crops are more resilient to climate variations, farmers have better knowledge of handling them, and traditional crops generally meet the food preferences of communities, making it all the more important to create measures to promote and revive them.

Small Millets

Tribal Women farmers have a high level of awareness in environmental management. Their direct contact with the natural environment as providers of water, fuel, food and fodder,small millets and their high level of awareness of environmental practices. Need to be more widely utilized for sustainable management in farming systems.Tribal Women's central role in food preparation means that they have direct responsibility for fuel wood supply and are therefore worst affected by fuel wood scarcity.

ITKs

Indigenous technical knowledge is one of the most important resource in which indigenous communities should have control over and access to various prospects of nature. It can be defined as the synergy and interaction among organisms, land, technology and social organisms that serve to fulfill production goals and sustain livelihood systems. The rate of biodiversity loss has been hastened by the combination of several economic, social, environmental and political factors in many marginal hilly village ecosystems. Such ecosystems are characterized by diversity in both space and time dimensions. The conservation of agrodiversity in village ecosystem the context of global food security assumes greater importance with specific reference to women.

Tribal Women's Contribution to Agricultural Marketing

Tribal Women's participation in rural labour markets varies considerably across regions, but invariably women are over represented in unpaid, seasonal and part-time work, and the available evidence suggests that women are often paid less than men, for the same work.The participation of rural women in processing activities is found higher than in marketing related activities such as purchase of animals, selling of milk, disposal of animals etc. Women are involved in a big way in the marketing of agricultural crops (whether produced by them, purchased from others or produced by their spouses) - most times travelling from rural areas to the urban municipal markets, spending between one to three days at these locations, before returning to their communities. There are also those tribal women who travel from urban areas to purchase directly from farmers and resell to various markets, including overseas. Tribal Women are also involved in the marketing of value-added products manufactured from our primary produce.

Policy and Strategies for Tribal Women

Tribal Women's participation in agriculture varies widely from different ecological sub-zones, farming systems, castes, classes etc in India. The involvement of tribal women in crop production varies according to the type of crop grown and the cropping systems and the socio-economic status of the family. There is also a need to make distinctions between the involvement of women as agricultural laborers' and involvement of women in agricultural operations on their own farm. In relatively prosperous areas where land holdings are large and most of the agricultural operations are mechanized, women play only a marginal role. The women from poorer families work as agricultural laborers' irrespective of the community to which they belong. Female labour participation rates are noticeably higher among scheduled caste and scheduled tribal communities (Prameela, 1990). The problems of tribal women are different from other rural women and they need a totally different approach to overcome some of these constraints. Tribal Women must be empowered by enhancing their awareness, knowledge, skills and technology. Planners, social welfare agencies and women's development organization should provide a rightful place to tribal women. There are some points developedof Policy and Strategies for Tribal Women:

1. These professionals include policymakers, extension officers, agricultural teachers in schools, researchers, food technologists and a host of others. In these areas, too, Tribal women occupy prominent positions and therefore play an important role in agricultural development.
2. Tribal women devote much of their time to daily domestic tasks, including family care and household chores. They also devote a considerable amount of time to crop and livestock production on the farm and to marketing these products. They maintain household gardens, and prepare a number of other food and non-food items for sale or home consumption. These activities contribute directly to family income.

3. The tribal women will have farm management skills and capable of taking even complex decisions like shifting from subsistence farming to diversified agriculture, organic farming with stand competition from market forces improvement in work or farm efficiency etc.
4. Suitable agricultural interventions will enhance production of quality food grains, vegetables and fruits. This will help in improving per capita availability of food, which will, in turn, impart food and nutritional security.
5. Improved access to credit, establishment of milk cooperative and marketing will enhance the managerial abilities of farm women in dairying. Women will play a significant role in livestock management and production and will be the member of dairy cooperatives.
6. Greater awareness will bring in reduction in malnutrition in rural areas through fortification of conventional foods with nutrient rich produce such as soybean.
7. Suitable gender-specific technologies will empower the tribal farm women.
8. Reduce losses of produce by improved storage structures and technologies will raise income of the tribal farm women.

Tribal Women in Selection of Enterprise

Many of the tribal women trained under different programmes have initiated new enterprises as individuals and as groups and could increase their income substantially from these enterprises. But it is the selection of the right kind of enterprises that is critical. The choice of enterprise depends primarily on the time available with women, their risk taking capacity, cost of inputs, cash flow, markets etc. The capacity of the women groups to understand the likely outcomes of different enterprises and to help them take a decision on what enterprise would be suitable for them, could be developed through appropriate training by KVKs, NGOs, SAUs and others organizations. The tribal woman farmers require activities that are not too time consuming, not drudgeries and will allow them to generate income. With this in mind, the most common income generation activities i.e. - Purchase/leasing of agricultural land, Improved seeds, bullocks, implements and other inputs for crop production, Purchase of dairy cows, goats and chicks, Tailoring, Paddy processing, Rope making, Fish-net production and repair, Mat weaving, Production of processed milk products, Pottery, Vermi-composting, Nursery raising, Bee-keeping, Food processing (pickle and papad production and sale, curry and spice powder units), Leaf cup production, Mushroom production and cultivation of crops, Garment production, Pasture development,

Soap powder, Agarbatti making, Operation of flour mill, Production of greeting cards, Consumer store and fair price shops, selling of cloth, sarees and bangle, embroidery unit,, stationery production and sale etc.

CONCLUSION

Tribal Women play a crucial role in all farm-related activities from land preparation to marketing. They contribute a higher proportion of labor in agricultural sector than men. However, they are not active in decision making. It stresses the need for a new agricultural research and extension agenda which integrates gender analysis into the process of technology generation and dissemination. It also comes up with future strategies to make tribal women a more active part of important farm decisions both at the household and legislature level. This research note emphasizes the importance of balancing agricultural research systems, biodiversity, indigenous technical knowledge, integrated farming system, extension education, and policy-making bodies to attain tribal women empowerment in agriculture. The tribal women play a significant and crucial role in agricultural development and allied fields including in the main crop production, livestock production, horticulture, post harvest operations, agro / social forestry, fisheries, etc. Tribal women have to manage, often, complex and multiple livelihood strategies. Their activities typically include producing agricultural crops, tending animals, processing and preparing food, working for wages in agricultural or other rural enterprises, collecting fuel and water, engaging in trade and marketing, caring for family members and maintaining their homes. Women's participation in rural labour markets varies considerably across regions, but invariably women are over represented in unpaid, seasonal and part-time work, and the available evidence suggests that women are often paid less than men, for the same work.

REFERENCES

Achanta, I.D., (1982). 'Role of Women in Agricultural Developments, Kurukshetra, vol.31(2), pp. 15-17.

Agarwal, Hema, (1987). 'A Study on Contribution of Farm Women in Animal Husbandry Enterprise'. An unpublished M.Sc. thesis submitted to IVRI, Izatnagar.

Aggarwal, M. (2003). "Economic Participation of Rural Women in Agriculture" in Economic Empowerment of Rural Women in India, edited by Gopal Singh, 2003, RBSA Publications, Jaipur, Rajasthan.

Gautam, U.S., Singh, A. and Singh, S.R.K. (2012). Participatory Approach of Women in Agriculture: Vision 2025. Indian Research Journal of Extension Education, Special Issue (Volume I), January, 2012:38-42

http://www.jamaicaobserver.com/magazines/allwoman/The-role-of-rural-women-in-agricultural-development_9503459

Lal, R. and Khurana, A. (2011) Gender issues: The Role of Women in Agriculture Sector. ZENITH International Journal of Business Economics & Management Research, 1(1):29-39

Prameela, V. (1990). A Study on Knowledge and Participation of Rural Women in Agricultural Operations with Respect to Paddy Crop in Kurnool District of Andhra Pradesh, M.Sc.Ag.Thesis (unpublished).

Prasad, C. and Singh, R.P. (1992). Farm Women: "A Precious Resources" in Women in Agriculture, Vol.2, Education,Training and Development edited by R.K. Punia, 1992, Northern Book Centre, Ansari Road, New Delhi.

Sanjay Kanti Das (2012). An Analysis of Constraints in Women Empowerment in Tribal Area: Evidences From Assam. Asian Journal of Research in Social Science & Humanities,2 (4):61-74.

Sethi, R.H. (1991) Women in Agriculture. Rawat Publications, New Delhi, pp. 55-90.

Shiva Vandana, (1992) Staying Alive: Women, Ecology and Survival in India. Published by Kali for Women, A-36, Gulmohar Park, New Delhi.

Venkatachalam, S. (1983). Training of rural-Must for Sound Dairy Extension Programme', Tamil Nadu Journal of Cooperation, vol : 74 (8-9), p.475

CHAPTER - 24

Analysis of Constraints Faced by Farm Women in Agriculture

Kumari Pooja[1] and Kumari Arunima[2]

[1]*M.Sc. Scholar, Department of Extension and Communication Management,College of Home Science,R.A.U., Pusa,Samastipur-848125*

[2]*Associate Professor, Department of Extension and Communication Management College of Home Science, R.A.U., Pusa,Samastipur-848125*

ABSTRACT

Women are vital human resources and play a significant role in agricultural and allied sectors undertaking both on-farm and off- farm activities. Yet their contribution has not been recognized and they are considered as invisible workers. Considering this, the present study has been conducted to analysis the constraints faced by farm women in agriculture and allied sectors. A total of 100 farm women were selected as respondents through proportionate random sampling. The selected respondents were interviewed personally using pre-tested well structured interview schedule. The finding showed that the major constraints that farm women faced were limited access to productive resources like land, credit, input,extension services,storage and technical assistance, their absence from decision-making process , less wages and their illiteracy also hindered their progress. Women perform all un-mechanized agricultural tasks and perform multiple tasks, which add more burden to them.

Key words: Farm women, constraints, agriculture

INTRODUCTION

According to Food and Agriculture Organization (2011), women farmers account for more than quarter of the world's population. Women comprise, on an average 43 percent of the agricultural work force in developing countries, ranging from 20 percent in Latin America to 50 percent in Eastern Asia and

Sub-Saharan Africa. As farmers, agricultural workers and entrepreneurs, women constitute the backbone of India's agricultural and rural economy. They constitute approximately 50% of agricultural and livestock workers. In rural India, the percentage is as high as 84%.Rural Indian women are extensively involved in agricultural activities.The mode of female participation in agricultural production varies with the landowning status of farm households. Their roles range from managers to landless laborers. In over all farm production, women's average contribution is estimated at 55% to 66% of the total labour with percentages, much higher in certain regions. In the Indian Himalayas a pair of bullocks works 1064 hours, a man 1212 hours and a woman 3485 hours in a year on a once hectare farm, a figure that illustrates women's significant contribution to agricultural production. (Shiva FAO, 1991) Yet women have less access than men to agricultural related assets, inputs and services. The constraints and opportunities that women face in agriculture vary across the agro-ecological and geographical regions of the country depending upon, among others, the socio-economic-cultural context. Their hard work has not only been unpaid but also remained mostly unrecognized. They perform on a daily basis the most tedious and back-breaking tasks in agriculture, animal husbandry and homes. They are invariably paid lower wages than men for the same agricultural work. Land ownership titles are most often in a man's name. Men often either take or dictate the decisions concerning farming and women have to compulsorily carry out. Farm produce is marketed commonly by men and that gives them complete control over household finance. Only 11 percent women have access to land holdings, that too, mostly as small and marginal farmers. With grossly inadequate access to education and technology, a host of other socio-economic factors have an adverse impact on the lives of women farmers in recent years. Despite their eagerness they have often not been able to take advantage of opportunities from new technologies, innovations and markets.

If they have easy, timely and reliable access to productive resources as men, women can increase yield by 20 percent to 30 percent, raising the overall agricultural output in developing countries by 2.5 percent to 4 percent. This gain in production can reduce the number of hungry people in the world by about 12 percent to 17 percent, besides increasing women's income.

The primary objective of this study is to analyze the constraints faced by farm women in agriculture.

METHODOLOGY

The study was conducted in the Samastipur district of Bihar. The selection of the district was made purposively with the consideration of agriculture as the main source of livelihood of rural people. A total 100 farm women as respondents were selected. The data were collected with the help of a pre-tested interview scheduled.The selected respondents were interviewed personally with the help of a well structured and pre-tested interview schedule in order to get relevant information and to draw the conclusion.

RESULT AND DISCUSSION

Socio-economic characteristics of the respondents: The findings of the research revealed that majority of the women who participate in home farm and allied activities were belonged to middle age group (52%). This age was regarded as the most active and potential for different farm and allied activities. This was supported with the similar study conducted by Padma(1987) and Puneetha (1989). Table 1. also depicted that respondents of forward caste showed least participation (15%) while the respondents of backward caste (49%) followed by SC/ST (36%). This was supported by the study conducted by Swarnjet (1981). On the basis of observed trends in terms of education it can be concluded that still the importance of education is being under mind in our rural masses specially the girl's education. The obtained results supported the fact which were also explained by the findings of Kaur et. al. (1998) that women's participation in agricultural activities tend to decline with increase in literacy. Table 1. Revealed that members of small and medium (32%) (58%) were mostly engaged in agricultural activities which was also supported by studies of munjal et.al. (1985). The respondents belonging to low income group were mostly labourers who were basically dependent on the wage earnings for their livelihood. Similar observations were borne out of the findings of Patel and Mehta (1988) and Vidhale et.al. (1993) who reported that among rural women, the income level has been influential factors for their participation in agricultural work. Land is regarded as an income generating assets in the rural areas. the findings about land distribution depicted in table 1. Were supported by the findings of Prasad and Singh (1992) who found that rural women belonging to small, marginal or landless agricultural labourers household bore a disproportionately large share of the family's work plus there proportion in agricultural activities was also higher.

Constraints faced by respondents : The major constraints faced by respondents are depicted in Table 2. The perusal of table indicated that playing dual role of a housewife and working women was perceived as a very important problem by 81% of the selected respondents. Illiteracy and lack of knowledge and skill was the other factors that hindered their growth as agricultural workers. Since women are considered as secondary workers in the area of agriculture hence they were not found to be actively involved in the dissemination of knowledge about new technology given by extension workers. As a result they were found to remain confined in their traditional package of practices. The negative attitude of society (37%), restricted mobility (22%) were observed as another problem as perceived by rural women. In our male dominated (73%) society, still the majority of the decision are being taken by the male members in which women were not allowed to play any important role.

Table 1: Socio-Economic indicators

Indicators	Frequency	Percentage
Age		
Upton 25 years	31	31
25-55 years	52	52
Above 55 years	17	17
Caste Composition		
Forward caste	15	15
Backward Caste	49	49
SC/ST	36	36
Educational Status		
Illiterate	42	42
Primary education	26	26
Middle education	18	18
High school	12	12
Inter and Above	2	2
Family Size		
1-4 members	32	32
5-7 members	58	58
Above 7 members	10	10
Annual Family Income		
Low up to Rs 15,000	21	21
Average up to Rs 20,000	57	57
High above Rs 25,000	22	22
Land holdings		
Upton 1 acre	48	48
1-5 acre	30	30
Above 5 acre	22	22

Table 2: Distribution of respondents according to the constraints faced by respondents

Constraints	Frequency	Percentage
Cultural norms	70	70
Male dominance	73	73
Dual role on house and farm	81	81
Lack of fund	62	62
Lack of knowledge and skill	57	57
Illiteracy	42	42
Restricted mobility	31	31
Lack of economic resources	52	52
Negative social attitude	37	37
Lack of command in theimplementation of a decision.	54	54
Unequal distribution of household work	61	61
Health problems	41	41
Lack of training	48	48
Lack of financial inclusion	80	80

CONCLUSION

The Indian economy is dependent on agriculture and more than 70% of our population living in rural areas is engaged in agriculture. Among the rural population, women are associated directly or indirectly with agriculture. Hence farm women must be recognized as an important unit contributing to the economic growth of the country. The result of the study showed that the rural women seldom have a control over the production resources like land, credit, finance, and marketing facility. The constraints were related with their social, economical, financial and technological conditions. Hence based on the results and discussions some recommendations can be suggested for them:

- Literacy rate of women in this area must be improved.
- Equal pay for equal work to both men and women.
- Sufficient loans should be made conveniently available to women at lower interest for agriculture.
- The cost of agricultural inputs must be reasonable . Agricultural inputs must be made available at the right time by government seed/fertilizers stores for women farmers.

REFERENCES

Butt. et.al. (2010) Role of Rural Women in Agricultural Development and Their Constraints. J.Agric.Soc.Sci.,6:53-36.

Kaur,J. et.al. (1998) Role of Education and Involvement of Rural Women in Decision-Making. Indian Economic Panorma,3(3):51-54.

Makkena,P. (1987).An assessment of the nature and type of participation of women in ricecultivation practices in agriculture with special reference to training & visit system inAndhra Pradesh,M.Sc.Thesis,Andhra Pradesh Agricultural University,Hyderabad.

Munjal,S.,Punia,P.K, and Sangwana,V.(1985).Women's economic contribution in farm households in Haryana (summary). India J.Agricul.Econo.40(3);279

Nimbalkar ,M.D et al,(2013) Role performance and constraints face by farm women engaged in agriculture. Indian Journal of applied research .Vol 3,Issue 7.

Pallavi (2001) Study of time utilization pattern of rural women with respect to their involved in farm, and allied activites.M.Sc Thesis ,Dept of Ext.Edu.R.A.U,Pusa

Patel,R.K., and Mehta,R.K. (1988).Appropriate livestock production technologies for farm women. In :Abstract conference on farm women ,ICAR ,New Delhi,P.69.

Puneeta Sagar Pala Kurthi (1989).Role of women in rice cultivation .M.SC Thesis,Andhra Pradesh Hyderabad.

Swarujet,K. (1981).Role of farm women in selected agricultural operations in five villages of Ludhiana district:M.Sc(Ag) Thesis,P.A.U,Ludhaina.

Tiwari.N.(2010).Economic and technological constraints facing farm women.International Journal of rural studies.Article 7,Vol 17,P(1-5)

Vdhale, A.Deshpode.,P.V,Magendar,S.D. (1993).Participation of tribal women in agriculture operation.Maharastra J.Ext.Edu.Vol.9

CHAPTER - 25

Information Needs of Farm Women Related to Kharif Paddy: A Study in Bolpur Sub- Division of Birbhum District (W.B.)

Suparna Bose[1] and Sarthak Chowdhury[2]
[1]*Department of Human Resource Development, Jharkhand*
[2]*Department of EES, Palli Siksha Bhavana, Visva-Bharati, West Bengal*

ABSTRACT

Information is the data that have been put into a meaningful and useful context and communicated to a recipient who uses it to make decisions. Information is needed because of it affects individuals living activities. The present age has rightly termed as an 'information age'. Information cannot provide finance, inputs or infrastructure. But, it can help in creating awareness about technologies and mobilise people to use them. It also helps in imparting training to people, organising community and ultimately the development of the whole nation.

The present study was undertaken to assess the information needs of young farm women (18-35 age group) with respect to kharif paddy cultivation. The study was conducted in Bolpur sub-division of Birbhum district, West Bengal. Results of this empirical study indicated that farm women wanted to know more about selling of product, pest control,nursery raising and disease management. The study also revealed that most preferred sources of information for the young farm women personal localite sources.

Key words: Young farm women, information needs, personal localite, personal cosmopolite, kharif paddy

INTRODUCTION

Women have been active participants in every stage of agricultural production and in every period of agricultural history. In a society that has recognised farming as a male occupation, however, many of women's contributions to agricultural production have been ignored or misunderstood. In India, 86% of rural women workers are in agriculture compared with 74% of rural male workers, and this gender gap is growing. In addition, a large percentage of households are female-headed. Women's economic contributions to the farm and farm household have historically been undervalued, that rarely portrays women as significant contributors to the farm income. Farm women were represented as spouses or farm helpers, but not as producers or decision-makers (Walter and Wilson, 1996). It is observed that farm family labour can be differentiated by gender such that the work of men is accorded greater value while the work of women is reduced or ignored (Alston, 1994). The World Women's Report,1985 paid Women in agriculture faced problems like lack of economic resources under their command, lack of contact with information sources and lack of pertinent facts or knowledge about agriculture. They also revealed that other problems faced by rural women were lack of support from extension agencies, lack of command in implementation, low educational status and low mobility (Talukdar and Karmakar, 1997)

In this paper an attempt has been made to assess the information needs of the young farm women (18-35 age group) with respect to kharif paddy cultivation. Information is knowledge that was unknown to the receiver prior to its receipt. Information is an essential input apart from land, labour, capital and organisation in the present day agriculture. Hence assessment of information needs of the farmers is an essential prerequisite of successful farming. Research on information needs and information seeking concurs that information is tailored to individual's job and to their tasks within those jobs (Ingwerson, 1996 and Zeffane and Gul, 1993).

The following objectives provided for the study:

- To determine socio-economic traits of young farm women.
- To identify the information needs of farm women related to paddy cultivation
- To ascertain the preferred communication sources of farm women for seeking information.

RESEARCH METHODOLOGY

The present study was undertaken in randomly selected two blocks; i.e. Bolpur-Sriniketan Block and Illumbazar Block of Bolpur subdivision of Birbhum District in West Bengal. For the present study 120 farm women were selected by the use of simple random sampling technique. The age group of present study was 18 to 35. The data were collected with the help of a pre-tested structured interview schedule by personal interview method.

RESULT AND DISCUSSION

Socio-Economic Profile of the Farm Women

From the table 1, it is evident that majority of young farm women falls to marginal farmer category (78 percent).Most of the young farm women (58 percent) belonged to small family and rest of them belonged to the family with more than 5 members. Family income of the respondents indicates that major part of the farm women belonged to APL (40 percent), second part of the respondents belonged to BPL (32 percent).Most of the respondentsAbout 57 percent of the respondents had obtained formal education, among them 22.5 percent of the respondents had education up to primary level, 19.2 percent of the respondent had education up to middle school, 14.2 percent of the respondents had attained high school education and only 0.8 percent of the respondents were graduates. As far as media exposure is concerned, most of them (88 percent) had mobiles, while about 75 percent had radio and about 63.3 had television.

Table 1: Socio –Economic Profile of the Respondents

Characteristics	Categories	Frequency	Percentage
1. Land Holding	Marginal (upto1.0 ha)	94	78.3
Small (1.01-2.0 ha)	22	18.3	
Medium (2.01-4.00 ha)	4	03.3	
Total	120	100	
2. Type of Family	Upto 5 Members	69	57.5
More than 5 Members	51	42.5	
Total	120	100	
3. Family Income	BPL(Rs.12000PA)	39	32.5
APL(Rs.12000-25000PA)	48	40.0	
Medium Income (Rs.25000-50000PA)	25	20.8	
High Income (more than Rs.500000PA)	08	06.7	

Contd...

4.	Education	Illiterate	52	43.3
	Primary Level	27	22.5	
	Middle School	23	19.2	
	Matriculate/High School	17	14.2	
	Graduate and Above	01	0.83	
	Total	120	100	
5.	Media Exposure	Radio	91	75.8
	Television	76	63.3	
	Mobile Phone	106	88.3	

Information Needs Related to Kharif Paddy Cultivation as Perceived by the Respondents

Leckie (1996) reported that occupational information often was not passed on to female farmers, most of the female farmers in this study came to farming with an inadequate knowledge of farming practice.It was clear from the results (table 2) that farm women wanted to know more about sellingofproduct,pest control, nursery raising,disease management and seed variety which were assigned first, second, third,fourth and fifth rank. Transportation, threshing, storage, pre-harvest desiccation and winnowing occupied the next five positions, while fertilizing,transplanting and seed treatment were also sought after operations for information at eleventh, twelfth and thirteenth rank. Respondents comparatively showed less interest in information on land preparation,water management, puddling, harvesting, and weed control.

Table 2: Distribution of respondents according to relative position of items of information need related to kharif paddy cultivation as perceived by the respondents

n=120

SlNo.	Information regarding Kharif Paddy Cultivation	Most (3)	Moderate(2)	Somewhat(1)	Total Score	Rank
1.	Selling of Product	96	11	13	358	I
2.	Pest Control	109	05	06	343	II
3.	Nursery Raising	108	02	10	338	III
4.	Disease Management	93	19	08	325	IV
5.	Seed Variety	98	07	15	323	V
6.	Transportation	83	32	05	318	VI
7.	Threshing	72	46	02	310	VII
8.	Storage	91	07	22	309	VIII

Contd...

9.	Pre-Harvest Desiccation	93	6	11	302	IX
10.	Winnowing	78	23	19	299	X
11.	Fertilizing	77	12	25	280	XI
12.	Transplanting	78	02	40	278	XII
13.	Seed Treatment	67	31	08	271	XIII
14.	Land Preparation	70	15	06	246	XIV
15.	Water Management	61	08	26	225	XV
16.	Harvesting	55	17	12	211	XVI
17.	Puddling	40	33	18	204	XVII
18.	Weed Control	21	17	13	110	XVIII

Preferred Communication Sources for Seeking Information

The table 3 shows the various information sources preferred by the respondents as per their need. For paddy cultivation mostly chosen sources were personal localite sources such as husbands, friends, relatives and neighbours as these were ranked at first, third, fifth and sixth positions respectively. Rest of the personal localite sources experienced farmers and local leaders were ranked at eighth and ninth positions respectively. Most desired personal cosmopolite sources were input supplier which is placed at second rank. KPS personnel, ADO and KVK/PSB staff and which were placed at eleventh, twelth and fiftinth rank. Sadaf *et al.* (2006) also found in their study that that male head/husband, neighbours and fellow farmers were perceived to be the better existing information sources by the respondents as compared to others. They further noticed in their study that rural women were not benefited from television and radio in spite of delivery of different agricultural programmes by these sources of information.

Table 3: Distribution of respondents by their preferred communication sources for seeking information

n=120

Sl.No.	Information Sources regarding Kharif Paddy Cultivation	Extent of Preference				
		Most (3)	Moderate(2)	Somewhat(1)	TotalScore	Rank
1.	Husband	105	12	03	342	I
2.	Input Supplier	100	20	00	340	II
3.	Friends	98	18	04	334	III
4.	SHGs[X]	102	09	09	333	IV

Contd...

5.	Relatives	92	15	13	319	V
6.	Neighbours	81	12	27	294	VI
7.	Radio	55	41	24	271	VII
8.	Experienced Farmer	51	13	10	189	VIII
9.	Village Leader	44	16	07	171	IX
10.	TV	17	14	04	83	X
11.	KPS Personnel*	06	17	11	62	XI
12.	ADO	07	03	12	36	XII
13.	Contact Farmer	06	02	08	28	XIII
14.	Farmer's Organisation	02	06	02	20	XIV
15.	KVK/PSB Staffß	03	04	01	18	XV
16.	Newspaper	02	01	05	13	XVI
17.	Farm Magazine/ Farm Literature	0	0	0	0	XVII

[X]Self-help groups,*KrishiPrayuktiSahayak, ßAgri. College Staff

CONCLUSION AND RECOMMENDATIONS

From the above study it can be concluded that young farm women were eager to get information regarding various operation of kharif paddy cultivation. The study indicated that farm women wanted to know more about selling of product, pest control and nursery raising, The study also found that young farm women preferred localite sources of information such as husbands, friends, relatives and neighbours over the various personal cosmopolite sources.Most desired personal cosmopolite sources were input supplier.

The results of the study will help to understand flows of farm information and information sources that farm women use. The results of this study will help to bridge the information gap regarding farm technologies.The findings of the study can also be used in guiding policy makers and development planners who are concerned about gender issues while designing agricultural projects within the region and elsewhere in the country.

REFERENCES

Alston, Margaret (1994). Meeting the Needs of Rural Women. In: Issues Affecting Rural Committees. Proceedings of an International Conference held by the Rural Education Reseach and Development Centre (Townville, Queensland, Australia).

Ingwerson, P. (1996). Cognitive Perspectives of Information Retrieval Interaction: Elements of a Cognitive IR Theory. Journal of Documentation,Volume 52(1) pp. 33-50.

Leckie, Gloria J. (1996). Female Farmers and the Social Construction of Access to Agricultural Information. Library & Information Science Research, Vol. 18, pp. 297-321.

Sadaf, Saima; Javed, Asif and Luqman, Muhammad (2006). Preferences of Rural Women for Agricultural Information Sources: A Case Study of District Faislabad, Pakistan. Journal of Agriculture and Social Sciences, Vol. 2(3), pp. 145-149.

Talukdar, R.K. and Karmakar, Anjali (1997).Problems in Decision Making in Agriculture- A Study in Golaghat District of Assam. Journal of the Agricultural Science Society of the North East, Vol. 10(2), pp. 281-282.

Walter, G. and Wilson, S. (1996). Silent Partners: Women in Farm Magazine Success Stories, 1934-1991. Rural Sociology (USA), Vol. 61(2), pp. 227-248.

Zeffane, R. M. and Gul, F. A. (1993).The Effects of Task Characteristics and Sub-Anti Structure on Dimensions of Information Processing.Information Processing and Management, Vol. 29, pp.703-719.

CHAPTER - 26

Empowerment of Women Members of Self-Help Group through Rural Pisciculture in Birbhum District of West Bengal

Rajyosree Roy[1], Mihir Mondal[2] and Debabrata Roy Gupta[3]

[1]Department of Zoology, Abhedananda Mahavidyalaya, Sainthia, Birbhum West Bengal, India

Email. royrajyosree@gmail.com

[2]State Poultry Farm, Kakdwip, South 24 Parganas. West Bengal, India

[3]Livelihood Development Officer, District Rural Development Cell, Birbhum West Bengal

ABSTRACT

A detailed survey was carried out in all the 19 rural blocks of Birbhum district of West Bengal. Pisciculture is a powerful income generating activity (IGA) by the women Self Help Group (SHG) members related to poverty reductionand employment generation for a vast section of the rural society. Self Help Groups are non-formal cooperatives that are causing a silent revolution in rural India through micro credit and various other economic activities involving women are the major change agents. Nearly, thirty thousand SHGs have been formed in Birbhum district of West Bengal that had made a notable success in micro finance. This study has focused on the functioning of 38 women SHGs of Birbhum district comprising 644 members mainly belonging to Below Poverty Line family.Totally 38 SHGs have been selected for the study. Each group had it unique name and consists of about 10 to 18women members. Information was elicited from all the members of 38 women SHGs, which together formed 644 members, constituting the study sample. The data were collected with the help of semi-structured interview survey schedule. The survey was carried out during the year 2009. Different social class of these fisher women members of the

SHGs were assisted under Swarnjayanti Gram SwarojgarYojana. These SHGs were found to save a portion of their meager income and pool that money for their inter-lending among them at an interest decided by the SHG. The survey revealed that cash credit and project loan was used for better sustainability and enhancement of income from pisciculture. Most of the cultural water bodies i.e. ponds were under lease condition. Self Help group members particularly belonged to scheduled caste community actively take part in retail marketing of fish in rural areas.Empowerment of women was undertaken through training and other assistance given by local Panchayat, NGOs and Govt. Officials.The SHG women members increase in self-confidence, leadership qualities, skill to undertake pisciculture and marketing, knowledge of book keeping, enhanced income, linkage with bankand protest against social injustice.

Key words: Self Help Groups, pisciculture, loan, women empowerment

INTRODUCTION

Pisciculture is a powerful income generating activity (IGA) of the women Self Help Group (SHG) members related to poverty alleviation and employment generation for a vast section of the rural society. The SHGs are mostly non-formal groups and have a common perception of need and an impulse towards collective action (Batiwala,1994 and Satish, 2001).Fishery sector contributes to nutritional security and in earning foreign exchange. Women Self Help Group members play a crucial role in various piscicultural activity and fish processing work. Fish farming requires less space, time, money and has a higher feed conversion rate (Olagunju *et al.*, 2007). Aquaculture can benefit the livelihoods of the poor either through an improved food supply and/ orthrough employment and increased income (Samantary *et al.,* 2009). Micro finance is making headway in the effort for eradicating poverty and empowering women (Jayalakshmi, 1998). Empowering women needs a holistic approach to participate in decision-making in the household, community and local democratic sector and prepare women to take up leadership position in agricultural activities (Pinto, 1995 and Sabri, 1998). Self Help Groups in rural areas are causing a silent revolution through micro credit and various other activities to make the pisciculture sustainable. In this backdrop, the present study is an attempt to understand the various activities undertaken by SHGs to make pisciculture as a livelihood activity by the rural women members.

The specific objectives of the study are:

1. To assess the performance of the women fishery groups and analyse constraints.
2. To measure the impact of women Self Help Groups in fishery on household levels.
3. To understand the social and economic empowerment of the women members of SHG.

MATERIAL AND METHODS

The above objective of the study was achieved through quantitative as well as qualitative assessments.The area of focus in this study was 19 rural blocks of Birbhum district of West Bengal. Totally 38 SHGs have been selected for the study. Each group had it unique name and consists of about 10 to 18 women members. Information was elicited from all the members of 38 women SHGs, which together formed 644 members, constituting the study sample. The data were collected with the help of semi-structured interview survey schedule. The scheduleincluded questions relating to the area of culturalwater bodies,owner of the pond/tank, leasing period and rent, availability of inputs, yearly fish production, profit distribution among members, advantages, disadvantages and constraints of pisciculture, savings &loan accounts, training thereof etc. available to the SHG members. Focus group discussions were conducted to cross-check the collected data. Secondary information was collected from books, ledgers and registers maintained by SHGs.The survey was carried out during the year 2009. The data was analysed as per the method described by Snedecor and Cochran, 1967 using MS Excel software package.

RESULT AND DISCUSSION

Performance of the fishery groups and their constraints

Social class of the SHG women members is presented Table 1. It was evident from table that majority (50 %) of the members belonged to Scheduled Caste formed the largest class of fishery activity SHGs. The minority community constitutes the second largest (33 %) fishery activity SHGs followed by other community (11 %) and Scheduled Tribe (6 %). So, the Scheduled Caste community was dominant over the other community for carrying out fishery as an economic activity. It was evident that fisher women community who belong to Scheduled Caste was properly represented and there was no major adverse selection.

Table 1 : Distribution of membership by social class of fisherwomen

Sl.No.	Social class	No.of members	% of total
1.	Scheduled Caste	322	50
2.	Minority	212	33
3.	Others	71	11
4	Scheduled Tribe	39	6
	Total	644	100

Distribution of SHGs by production of fish in fresh water bodies is shown in Table 2. The majority (53 %) of the SHGs produced fish from 901 Kg to 1000 Kg followed by 24 % of the SHGs produced 701 Kg to 900 Kg fish. More than 1000 Kg fish was produced by only 18 % of the SHGs because they followed improved i.e. scientific method of pisciculture and up to 700 Kg fish was produced by 5 % of the SHGs. Survey indicated that the scale of technology adopted was not adequate to optimise yield potential.

Table 2 : Distribution of SHGs by production of fish in freshwater bodies

Sl. No.	Fish production (kg/acre/year)	No. of SHGs	% of total
1.	Up to700 kg	2	5
2.	701 to 900 Kg	9	24
3.	901 to 1000 Kg	20	53
4.	More than 1000 Kg.	7	18
	Total	38	100

It may be mentioned here that pisciculture is the part time workof the SHG members. They were involved in other income generating activities (IGA) like paddy to rice processing, house dairy, goat/sheep/pig rearing, agricultural/ horticultural activities etc.Distribution of SHGs by constraints encountered in fishery project is presented in Table 3. It is observed from table that 92 % of the SHGs faced financial hardship for pisciculture. Majority (66 %) of SHGs suffered pilferage of fish stock because all the dwellers by the pond side were fish catcher. Problems like getting lease of pond and pond excavation/ renovation were envisaged by 47 % and 24 % of the SHGs respectively. Shortage of water in summer was faced by 18 % of the SHGs. It was found that the SHGs had to face many operational problems especially availability of inputs like quality of fish seeds,feed etc.

Table 3 : Distribution of SHGs by constraints encountered in fishery project

Sl. No.	Operational problems	No. of SHGs	% of total
1.	Financial difficulties.	35	92
2.	Pilferage of fish stock.	25	66
3.	Getting lease of pond	18	47
4.	Pond excavation	9	24
5.	Water shortage during summer	7	18

Impact of Self Help Groups In Fishery on Household Levels

Distribution of SHGs by increase in income per month through fishery project is presented in Table 4. From table, it is clear that 5 % of the SHGs increased monthly income up to Rs. 1000/-because they followed traditional method of fish culture. 24 % of the SHGs increased income between Rs.1001 to 1500/- per month because they practiced somewhat improved method and 53 % of the SHGs increased monthly income between Rs.1501 to 2000/-where as 18 % of the SHGs increased monthly income above Rs. 2000/- per month through the integrated activity of fishery with livestock farming. Survey indicated that around 45 % of the families assisted under SGSY could increase their income adequate to cross the poverty line. In the remaining 55 % of the cases, investment and return had not been adequate to make a dent on poverty.

Table 4 : Distribution of SHGs by increase in income per month through fishery project

Sl. No.	Monthly increase in income (Rs.)	No. of SHGs	% of total
1.	Up to Rs.1000/-	2	5
2.	Rs.1001 to 1500/-	9	24
3.	Rs.1501 to 2000/-	20	53
4.	More than Rs.2000/-	7	18
	Total	38	100

Table 4a: Fish production group wise loan repayment status

Fish production group	Income increment status (Mean ± SE)
1 (2)	$820.00^{a} \pm 70.00$
2 (9)	$1259.45^{b} \pm 29.74$
3 (20)	$1616.15^{b} \pm 51.07$
4 (7)	$2195.15^{b} \pm 41.74$

Figures in parenthesis indicate number of groups (SHGs)

Values with different superscript letters differ significantly ($P < 0.01$).

The highest producer group 4 (7) gained more income 2195.15± 41.74 and lowest producer group 1 (2) gained less income increment 820.00 ± 70.00 which was statistically significant ($P<0.01$).

Empowerment of Women through Entrepreneurship, Decision making, Economic and Social Participation in the community

Impact of empowerment of women on socio-economic variablesis presented in Table 5.Total women members in 38 SHGs were 644. It is shown in table that 70.65 % of the female SHGs take part in family decision making process because they have gained enough practical knowledge. Majority (73.29 %) of the SHGs possess leadership qualities whereas income increased in 52.80 %. Bank and market linkage pattern indicated that 34.16 % and 43.47 % of the SHGs were able to backward and forward linkage of their economic activities. 44.25 % of the SHGs protest against social injustice. Joining meeting, procession conference were done by 53.60 % of the SHGs. Poor women are now a part of the overall decision-making process, forming an organic part in the local development. Poor women are now treated as a potentially strong economic force. Statistical analysis of data collected during field survey indicated that the goal of empowerment by mobilizing women into Self Help Groups was partially achieved.

Table 5 : Impact of empowerment of women on socio-economic variables

Sl. No.	Variables	No. of Members	% of Total	Chi-Square(df[1])
1.	Participation in family decision.	455	70.65	12.72*
2.	Leadership qualities.	472	73.29	13.35*
3.	Increase in income.	340	52.80	20.58*
4.	Linkage with market.	280	43.47	23.33*
5.	Linkage with bank.	220	34.16	22.5*
6.	Protest against social injustice.	285	44.25	22.79*
7.	Joining meeting,procession,conference.	345	53.60	18.98*

*Significant at 1% level

The interventions to empower women members of the SHGs such as family decision making, leadership qualities, linkage with market and bank, protest against social injustice, joining meeting, procession, conference and increase in income through pisciculture and other income generating activities (IGA) have positively contributed towards socio-economic empowerment of women as indicated by high statistical significance Chi-Square values for variables.

CONCLUSION

From the above findings, it may be concluded thatSelf Help Group members utilized the credit and project loan along with Govt. subsidy for pisciculture which contributed substantially to their family development. The scheduled caste community was dominant over the other community for carrying out fishery as an economic activity. Majority of the SHGs after getting training followed somewhat scientific method for pisciculture and improved fish yield. Women SHGs envisaged a lot of constraints like financial difficulties, pilferage of fish stock, getting lease of ponds, pond excavation. Empowerment through pisciculture with the assistance availed under SGSY leads substantial gain to women SHG members in terms of employment, income, purchasing power, changed lifestyle, food protection and making fish culture more sustainable.

REFERENCES

1. Batiwala, S. (1994). "The meaning of women's empowerment: New concept from Action, in population policies re-considered: Health, Empowerment and Rights, Gita Sen, et al.,(Ed), Harvard Center for Population and Development Studies.
2. Jayalakshmi, R. (1998)."The empowerment of women in India-Grass rootwomen's network and the state, J.Rural Development, 17:5-9
3. Olagunju , F.I., Adesiyan,I. O. and Ezekiel, A.A. (2007). Economic viability of cat fish production in Oyo state, Nigeria. J. Human Ecology, 21(2):121-124.
4. Pinto, M.(1995). Development through empowerment of women in India, Kurukshetra, 10:11-13.
5. Sabri, A. M. (1998)."Ensure women's participatory power ", Social Welfare,19.
6. Samantary, B. R., Khuntia, C. P., and Das, B. K. (2009). Rural Pisciculture: A sustainable source of livelihood and food security. Kurukshetra, 57 (11): 30-32.
7. Satish, P. (2001). Some Issues in the Formation of Self Help Groups, Indian J. Agricultural Economics, 56 (3)410.
8. Snedecor,G.W., and Cochran,W.G.(1967).Statistical Method, 6thedn. Oxford and IBH Publ.Co. Calcutta, India.

CHAPTER - 27

Pattern of Internet use by Agricultural and Agricultural Engineering Technology Students

Mandavkar, P.M.[1] **and M.S.Talathi**[2]

[1]*Subject Matter Specialist (Agricultural Extension), Krishi Vigyan Kendra Roha, Dist.Raigad,Maharashtra*

[2]*Programme Coordinator, Krishi Vigyan Kendra, Roha, Dist. Raigad Maharashtra*

ABSTRACT

Information Technology includes computers and communication technology along with associated software. I.T. can be defined as the technology of creation, processing, storage, retrieval and transmission of data and information. Transmission of data and information between through local national, regional and global telecommunication channels are not new to us. Computer networks have been the real facilitators of data and information interchange in the individual and corporate levels. LANS and WANS are the good examples of it. Developing countries have a critical stake in how successfully they cope with the trade, environmental, energy, healthy, agricultural, industrial, population and other challenges facing them as they enter the information age. The World is experiencing an explosive growth of IT at the moment. There have been tremendous impacts of I.T. on the trade, commerce and industries of the globe.

Key words: Internet, use, IT, technology, information age

INTRODUCTION

The world is undergoing a revolution today. It is the Information revolution – the third major revolution after the agricultural revolution and industrial revolution, and the Internet is fueling it. Internet is the buzzword today and it will be for the foreseeable future. Internet is the fastest growing communication medium and college students are heavy users of Internet compared to general population. While to reach 50 million people, the Radio took 38 years Television 13 years, cable took 10 years and Internet has taken just 5 years. Never has an invention enabled so many people to do so many things. Use of Internet is a part of college student's daily routine because they have grown up with computers. Hence to know the degree of affinity and relation of students to Internet the study was designed with following objectives;

(i) To study the frequency of use of Internet among selected net users.

(ii) To study the type of net surfing by the selected net users.

(iii) To find out the after effect of Internet use on the selected net users.

METHODOLOGY

The present investigation was carried out in Konkan Agricultural University named as Dr.Balasaheb Sawant KonkanKrishi Vidyapeeth which is located in Dapoli tahasil of Ratnagiri District from Maharashtra State in April, 2012. The total number of Undergraduate students in College of Agriculture (COA) was 140, out of this 100 students were selected randomly and out of 58 students from College of Agricultural Engineering and Technology (CAET) , 50 students were selected. The selection of students from final year of U.G. has been done purposively for the study. The data were collected with the help of a schedule constructed for the purpose of the study. The data collected from the respondents were scored, tabulated and analyzed by using suitable statistical methods such as frequency distribution, percentage, mean and standard deviation.

RESULTS AND DISCUSSION

The findings and discussion has been presented in sections like, frequency of Internet use, type of Internet surfing and after effect of Internet use.

1. Frequency of Internet use among the selected internet users

This will explore the frequency of internet use among the selected students for investigating duration of Internet use, time spent on the internet, place of Internet access, money spent in a month for Internet and when started going online firstly. The results obtained in the survey have been presented in Table 1 to 4.

A) Duration of Internet use

Table 1 : Frequency distribution of students with respect to Duration of Internet use

Sr. No.	Duration of Internet use	Agriculture students (N=100)		Agril. Engg.& Technology (AET) students (N=50)	
		No.	Percentages (%)	No.	Percentages (%)
1.	Daily	20	20.00	30	60.00
2.	More than one in a week	31	31.00	12	24.00
3.	Weekly	27	27.00	06	12.00
4.	Once in a fortnight	12	12.00	02	04.00
5.	Monthly	5	5.00	-	-
6.	Occasionally	5	5.00	-	-
	Total	**100**	**100.00**	**50**	**100.00**

Table 1 presents the distribution of students on the basis of duration of Internet use by them. out of 100 Agricultural students,31% reported for using Internet more than once in a week, 27% for using weekly and 20% reported for daily use. Whereas,12% use Internet during once in a fortnight. Equal percentage (5%) of students reported monthly and occasionally use of Internet. Out of 50 AET students selected, 60 % reported that they use Internet daily and 24% reported for more than once in a week, 12% reported for using Internet weekly, 4% reported for once in a fortnight. These findings indicate that usually the Internet users use the Internet. Daily, once in a week and weekly use of Internet is also very common habit among the students. These findings are in line with the findings of Mendals (1998) and Gorden (2003).

B) Time spent per sitting on Internet

Table 2 : Frequency distribution of students with respect to Time spent per sitting on Internet

Sr. No.	Time spent per sitting on Internet	Agriculture students (N=100)		Agril. Engg.& Technology (AET) students (N=50)	
		No.	Percentages (%)	No.	Percentages (%)
1.	Less than one hr.	59	59.00	20	40.00
2.	1-2 hrs.	36	36.00	26	52.00
3.	2-4 hrs	05	05.00	03	06.00
4.	4-6 hrs.	02	02.00	01	02.00
	Total	100	100.00	50	100.00

Table 2 presents the distribution of students on the basis of time spent per sitting on Internet. Out of 50 AET students, 52% reported that they use Internet for 1-2 hrs., 40% use less than 1 hrs,6% use for 2-4 hrs, and 2% use 4-6 hrs Internet per sitting. Out of 100 Agricultural students, 59% reported that they use Internet less than one hrs, and 36% use for 1-2 hrs, only 5% reported for using Internet 2-4 hrs.

This infers that almost the selected Internet user spent from less than one hour to 1-2 hrs per sitting. These findings are in line with the findings of Gorden (2003) and Soren (2003).

C) Place of Internet use

Table 3 : Frequency distribution of students with respect to place of Internet use

Sr. No.	Place of Internet use	Agriculture students (N=100)		Agril. Engg. & Technology (AET) students (N=50)	
		No.	Percentages (%)	No.	Percentages (%)
1.	Personal	30	30.00	20	40.00
2.	Cyber cafe	56	56.00	17	34.00
3.	Other	14	14.00	13	26.00
Total	100	100.00	50	100.00	

Table 3 reveals the distribution of students on the basis of their place of Internet access. Out of 100 Agricultural students 56 % reported that they use Internet at Cyber Cafes, 30% reported for personal place. It is inferred that Cyber Cafes are the dominant place of Internet use. Out of 50 AET students, 40% reported for access to Internet by their personal computer. Whereas, 34% and 26 % reported for cyber café as a place of internet access and other place, respectively.

These findings are in line with findings of Joshi (2001) and Krishnatray and Kulshrestha (2002).

D) Started going online

Table 4 : Frequency distribution of students with respect to started going online

Sr. No.	Started going online	Agriculture students (N=100)		Agril. Engg. & Technology (AET) students (N=50)	
		No.	Percentages (%)	No.	Percentages (%)
1.	Last 6 months	11	11.00	-	-
2.	A year ago	24	24.00	05	10.00
3.	2 or 3 yrs ago	42	42.00	20	40.00
4.	More than 3 yrs. ago	23	23.00	25	50.00
	Total	100	100.00	50	100.00

Table 4 presents the distribution of students on the basis of when they firstly started going online. Out of 100 Agricultural students, 42 % reported for started going online 2 or 3 yrs ago, 24% reported they've started going online a year ago, 23 % started going online more than 3 yrs ago and 11% started from last 6 months.

Out of 50 AET students, 50% reported that they have started going online more than 3 yrs. ago, 40% reported that they have started going online 2 or 3 years ago and 10 % started a year ago.

It is clear from the findings that majority of Internet users started going online 2 or 3 yrs ago. A high number of Internet users also started going online more than 3 yrs ago. These findings is in line with findings of Jones(2002) and Soren(2003).

2. Type of internet surfing

Table 5 : Frequency distribution of students with respect to type of Internet surfing

Sl. No.	Type of Internet surfing	Agriculture students (N=100)		Agril. Engg.& Technology (AET) students (N=50)	
		No.	(%)	No.	(%)
1.a)	Sending mail regularly	32	32.00	43	86.00
b)	Sending mail Occasionally	68	68.00	07	14.00
2.	Use search engine	52	52.00	42	84.00
3.	Search for college and University	60	60.00	46	92.00
4.	Leisure & entertainment	52	52.00	36	72.00
5.	Visit Govt. website	45	45.00	34	68.00

Contd...

6.	Check sport score	55	55.00	20	40.00
7.	Information about jobs	60	60.00	35	70.00
8.	News/ information about politics	40	40.00	25	50.00
9.	Looking information about hobby	56	56.00	41	82.00
10.	Play video game	15	15.00	33	66.00
11.	Download files from Internet	52	52.00	32	64.00
12.	Chatting online discussion	34	34.00	31	62.00
13.	Participate online auction	07	07.00	09	18.00
14.	Just for fun/pass time	40	40.00	24	48.00
15.	Shopping online	05	05.00	16	32.00
16.	Make reservation for travel	45	45.00	40	80.00
17.	Visit cyber casino	-	-	05	10.00
18.	Visit pornographic site	31	31.00	20	40.00

Table 5 explores the type of surfing habit among the selected students on the basis of findings obtained in the survey for the study. The above findings infer that search for college and University, to get information about job, sending e-mail, looking information about hobby, make reservation for travel, chatting online discussion, using search engines, visiting Govt. websites, checking sports score, using net for just fun, pass time and visiting pornographic sites were the purpose in general for the students. These findings are in support with findings of Hunter(1997) and Krishnatray & Kulshreshtra (2002).

3. After effect of internet use

Table 6 : Frequency distribution of students with respect to after effect of Internet use

Sl. No.	After effect of Internet use	Agriculture students (N=100)		Agril. Engg.& Technology (AET) students (N=50)	
		No.	(%)	No.	(%)
1.	Enjoying the web irrespective to its capacity	41	41.00	20	40.00
2.	Web fanaticism	22	22.00	22	44.00
3.	Playing online game by interacting	18	18.00	10	20.00
4.	Like chatting online	24	24.00	24	48.00
5.	Participate in online discussion	22	22.00	30	60.00
6.	Like to display personal information on a web page/Facebook etc.	58	58.00	43	86.00

The above findings from Table 6 revealed that Internet user likes to display personal information on webpage/facebook, etc, like chatting, participate in online discussion and enjoying the Internet use even if they don't use it to its capacity. The entertainment resulting from the Internet use also helps them to learn new things and to teach themselves, so that they can have personality development in the fields where they feel laciness, in just little time. These findings are in line with the findings of UCLA(1999), Pew Internet and American Life Project(2001).

CONCLUSION AND IMPLICATION

The findings of this study reveals, an important feature of college life that students have come to expect in near ubiquitous internet access. Another feature of this study is that college student use internet as the knowledge and information source, they are a group primed for interactive entertainment. They may prove to be choosier consumer than previous generation, losing their purchase on pre-viewing media via file sharing. The college students are continuing adoption of Internet technology at work and at home. It is being habit in them to use computer and log on to the internet so, too, will this generation mix work and social activity online and further boundaries between work and home, work and leisure. Information technology has been proved as one of the most aspired field in all developed country and same can also be achieved in developing country like India.The online availability of information can serve as most useful system for agricultural education, research and extension development as there are many technologies available on it. The entire agricultural education pattern needs overhauling in the context of latest developments in Information Communication Technology available for educational system.

REFERENCES

Gorden, C. (2003). Computer use study contradicts conventional wisdom. San Francisco state University. Retrieved from http://www.sfsu.edu/news/2003/40.ntm

Hunter, D.C. (1997). "The uses and Gratification of Project Agora", University of Pennsylvania. Retrieved from http://www.asc.upenn.edu/usr/chunter/agorn_uses

Jones, S., (2002). "The Internet goes to college," Pew Internet and American life Project. Retrieved from http://www.pewinternet.org

Joshi,H., (2002) "Social class and Internet users" reported in Internet use among youth 2002, a study by Krishnatray and Kulshreshtha. Retrieved from http://www.mica_india.net/OUTPUT/pk.pdf

Krishnatray and Kulshreshtha(2002), " Internet use among youth-2002". Mudra Institute of Communications, Ahmedabad. Retrieved from http://www.mica_india.net/OUTPUT/pk.pdf

Mendals, P. (1998). Study shows students use Internet primarily for Research. Retrieved from http://www.nytimes.com/library/tech/98/04/cyber/articles/28education_html

Soren.M.K.(2003). A study on pattern of Internet use by Technology and Agriculture students. Unpublished M.Sc. thesis, Institute of Agricultural Sciences, BHU, Varanasi (Uttar Pradesh)

CHAPTER - 28

Role of ICTs in Family Farming: Experiences and Way Forward

Saravanan Raj[1] **and Bhattacharjee Suchiradipta**[2]
[1]*Associate Professor (Extension Education and Rural Sociology), Department of Social Sciences, College of Horticulture and Forestry, Central Agricultural University (CAU), Pasighat – 791 102, Arunachal Pradesh, India*
[2]*Ph.D. Scholar in Agricultural Extension, School of Social Sciences, College of Post Graduate Studies, Central Agricultural University (CAU), Umiam (Barapani) – 793 103, Meghalaya, India*

ABSTRACT

With the developing world still depending on agriculture to a great extent and rural people depending on it for their livelihood, it comes as no surprise that family farms have been the focus of development in recent times. Family farms have been legacy for most of the farmers and along with it being principal livelihood option for many, the increasing need of agricultural extension services for these farms become more important. But at the same time, the huge gap between the number of famers or farm households and the extension personnel is a reality. And to bridge this gap, Information and Communication Technologies (ICTs) can play a significant role. Another purpose it serves is making agriculture attractive to the young people engaged in farming, mostly by compulsion. ICTs like web portals, mobile phones and social media are the most popular and most widely used and can enhance the interaction among researchers, extension personnel and farmers. The youth today are hooked to the internet and their 1st major activity on the web is social networking. Also, mobile phones are the most popular form of ICTs as has been proven by the sheer number of subscribers worldwide. And these tools, used appropriately, can change the whole structure of agricultural communication. ICTs can be an important part of communication strategies, connecting the youth to peers and professionals.It is true that the 'apps won't plough the fields' but they will definitely help the youth do the job better. ICTs can prove to be a great catalyst in engaging agricultural community in family farms making agriculture attractive and sustainable again.

Key words: Mobile phones, Web portals, Social Media, Agricultural Extension, Family Farming

INTRODUCTION

Technology in agriculture has come a long way from animal drawn ploughs to mobile apps showing farmers how to plough. Development in agriculture has also been commendable, especially in the developed countries where agriculture is highly commercialized with resource rich farmers andlarge farms dominating the picture. But the scenario is completely different in developing countries where resource scarcity and subsistence farming is most prevelent with poor farmers following their family occupation. Agriculture has become non-remunerative in these areas and the new generations are losing their faith and interest in agriculture. But these family farmers are practically feeding the world. According to FAO (2014a), nine out of ten (of more than 570 million farms, above 500 are family farms) farms in the world are managed by family and are responsible for at least 56 per cent of agricultural production. Also, 85 per cent of the farmers in Asia are family farmers and are the main source production of rice, the staple food of the region.Family farms produce 80 per cent of the world's food and this makes them one of the most vital solutions to eradicating hunger (FAO, 2014b). As per FAO's State of Food and Agriculture 2014 (FAO, 2014c) report, around 75 per cent of agricultural resources of the world are under them making them important custodians of world's ecological sustainability. And at the same time, they are also responsible for feeding majority of the world, thus giving them the task of taking key role in eradicating food insecurity and hunger,judicious and sustainable use of resources available, and taking themselves out of poverty. This triple challenge is a lot to deal with especially for the smallholders who own 84 per cent of the family farms.

Extension has a huge role to play in helping the family farms face the challenges and achieve the target. The importance of communication in development has been long talked about. But the actual data about that is not so encouraging. From the agricultural census data available from ten developing countries shows that share of farms interacting with Government extension agents is maximum 25 per cent. In India, it is only a little higher than five per cent (FAO, 2014c). And when the information gap between farm families and extension agencies is so high, it poses a major threat to their sustainability and profitability. In this scenario, one aid to the extension agencies in aiding better communication with the farm households can be Information and Communication Technologies (ICTs). In 2014, 55 per cent of the active mobile broadband connections of the world come from developing countries;mobile-cellular market is approaching saturation with subscriptions reaching the world population and with 78 per cent from the developing countries; while 40 per cent of the world population are using fixed internet connections, in the developing countries its use has increased five times since 2009 (ITU, 2014) and all these statistics show the increasing importance of ICTs in the world around us. More importantly, ICTs are attractive to the new generations who find it modern and are more comfortable and used to communicating through them. When keeping the youth in the family farms is already a matter of concern for furthering their existence, ICTs can be of added benefit in this aspect too.

ICTs and Indian Agriculture

Coming to Indian agriculture, the utility of ICT tools can be best understood if we discuss the drawback of Indian agriculture. . As stated by the world Bank in their publication 'India Country Overview 2008', 'almost universal lack of good extension services are among the factors responsible' (Wikipedia[a], 2012). The statement is supported by the findings of Saravanan (2007) in Karnataka where the ratio for extension personnel to farmers for public extension system, NGOs and Agri Business Firms are 1:1819, 1.1061 and 1.13 respectively whereas in Japan the ratio is 1:230 (Colle, 2011). To bridge this gap, if the Govt. goes for employing extension personnel, it would need around 13,00,000-15,00,000 extension personnel (Saravanan, 2010) which is not very feasible in India's present economic status. A more practical approach would be to go for alternate communication or linkage process to tend to the increasing need for information of the farmers. Here come the ICTs in Agriculture.

Radio and community radio: India's experimentation with ICTs started with radio (Sharma, 2011). Radio broadcasting in India started in the year 1924 by Madras Presidency Club. All India Radio started as the Indian Broadcasting Company in 1927 at the then Bombay and Calcutta and was later renamed in 1932 as AIR after the broadcasting was undertaken by the Government of India. Starting with just 6 stations in the country in 1947, there are total 82 stations owned by AIR at present (Karthik, 2011). After more than 30 years of its introduction in the country, the first program with agriculture information dissemination in 1965 and was named 'KrishiPrasaran' (Chauhan, 2007). Some other initiatives taken in the country to use radio broadcasting for agricultural extension were inspired by the success of the same in USA and 'information centres' were launched to disseminate Intensive Agricultural District Program benefits to the rural parts of the country. The impact was so that the ADT 27 rice variety grown by the Tanjur farmers came to be known as 'Radio Rice'. Other experiments like 'Rural Radio Forum' by Ministry of Information and

Broadcasting and UNESCO and 'CharchaMandal' by AIR, Pune gained huge popularity in rural India (Kujur*et al*, 2009). Community Radio was launched in India in 1990s after the Supreme Court of India ruled that the 'airwaves are public property' (Wikipedia[c], 2014). UNESCO defined community radio in 2002 as 'Community radio is a type of radio service that caters to the interests of a certain area, broadcasting content that is popular to a local audience but which may often be overlooked by commercial or mass-media broadcasters' (Sharma, 2011). Some agriculture related community radios of the country are KumayonVani, Gurgaon kiAwaaz, TNAU Community e-Radio, *etc*. The first community radio of North East India 'Jnan Taranga' started in 2010 and operated from Handiqui State Open University, Guwahati. At present, there are about 148 community radio stations operating in India (Wikipedia[c], 2014). Recently, the PM of India gave great emphasis on community radio for reaching out to the farmers and actively engaging them (Parsai, 2014)and the Govt. has allocated Rs. 100 crores for a network of 600 community radio stations in the country (Choudhury, 2014).

Television: Television in India was launched in the year 1959 with Doordarshan, three decades after it was introduced in the world. Satellite television came in the 1990s and was soon followed by private television channels in 1995 (Wikipedia[d], 2012). It is the third largest industry in the world after China and US (Wikipedia[c], 2012). The first program for agricultural extension was broadcasted on 1967 and is one of the longest running programs in the world (Wikipedia[e], 2012). It is broadcasted by Doordarshan in collaboration with Ministry of Agriculture, Govt. of India, under the scheme Mass Media Support to Agricultural Extension and covers various aspects of Agriculture, Horticulture, Animal Husbandry, Dairy & Rural life of farmers (Doordarshan, 2011). Recently, on October 21, 2014, India launched its first agricultural TV channel Green TV and it has already penetrated 11 per cent of the TV owning homes of the country (Chakrapani, 2014). Also, the Government has allocated Rs. 100 crore towards an agricultural TV channel Kisan TV to be launched this year (Choudhary, 2014). Television in the country has gained popularity very fast and has been rising for the last decade (Census of India, 2011). Decreasing prices of the TV sets, increasing number of channels, TV serials and soaps are the main reasons behind the popularity.

While the older generation is not much into computer and internet,the younger generation is into more modern forms of ICT tools like the internet, mobile phones and social media. And so, the scope of this paper will be restricted to agricultural web portals, mobile based advisory services and social media platforms for agriculture.

Agricultural Web Portals and Rural Advisory

A web portal is a website that brings together information from diverse sources in a unified way. Each information source gets its dedicated area on the page for displaying information (a portlet) and often, the users can configure which one to display (Wikipedia[f], 2012). In India web portals are given their due importance for dissemination of agricultural information to the farming community and the extension mechanism. Utility of web portals in India lies in the fact that it can successfully complement the extension personnel in their work. Moreover, in a position where the need of extension personnel is fifteen folds of the existing number, web portals come in handy reducing the communication gap. Then again, the information in the web portals is regularly updated along with the changes in the dynamic field of agriculture. In such a scenario, web portals are going to be indispensible part of a healthy extension mechanism. Contents and special features of some major Indian portals are discussed hereunder:

Table 1 : Agriculture related web portals in India

Sl. No	Portal URL	Launched on	Launched by	Contents	Special feature	Target user
Portals for general information related to Agriculture						
1.	http://www.ikisan.com/	2000	Initiative by Nagarjuna Fertilizers and Chemicals Ltd.	Generic and crop specific info, weather forecast, e-commerce (Vyapar and Mandi – virtual market place)	-Services in Andhra Pradesh, Karnataka, Uttar Pradesh, Tamil Nadu, Punjab, Haryana & Maharashtra -Addresses information needs of farmers, trade channel partners, Agri output companies, etc., -Indias first agri business portal	Farming community
2.	http://www.ikissankerala.net	2003	-Department of agriculture, Govt. of Kerala and Indian Institute of Information Technology Management initiative	-Crop information, Market information, Expert answers on queries, GIS based Agro advisory, farming practices, Kerala Agri. Directory and Discussion forum	-Fertilizer Recommendation System (FRS) for registered farmers	Farmers of the Kerala state, extension personnel
3.	http://www.aaqua.org.	2003	-Developed by IIT-Bombay, KVK Baramati and Vigyan	-Success stories, Crop Doctor, Recommendations, aAqua news,	-Language interface : English, Hindi, Marathi, Kannada, Telugu	Stakeholders of Indian agriculture

Contd...

			Ashram -Owned by Agrocom Pvt. Ltd	Forums and Statistics -250 audio clips on 25 crops	(For questions, all Indian languages are supported) -Archived online discussion forum -Digital assistance -Services in 290 districts of India	
4.	http://www.iffco.nic.in	2003	-Initiated by Indian Farmers Fertilizer Cooperative Limited -Maintained by National Information Centre	Information on land preparation, seeding operations, agricultural inputs, irrigation facility, plant protection, harvesting and market information	-Covered 16 states -Accessed through the touch screen kiosks launched by IFFCO	Farming community
5.	http://www.hul.co.in(i-Shakti)	2003	-Initiative by Hindustan Unilever Ltd. -Content developed by CARE International and the AzimPremji Foundation;-Maintained by shakti Enterpreneurs	-Health Agriculture, Education, Employment, Legal Advice, Gender issues, Entertainment and Computer Game	-Special emphasis on gender issued -Accessed tjrpigj kiosks Focuses on entrepreneurship development	Farming community, especially women
6.	http://agropedia.iitk.ac.in/	2009	-Indian Agricultural Research Institute (ICAR)–National Innovation Project (NAIP) initiative -Maintained by IIT-Kanpur	-Crop info, blogs, chatroom, extension material, agroforum & digital assistance	-Multilingual interface -No. of pages: 14673 -Knowledge models for 17 crops -Agrotags version I and II	Farmes, extension personnel, researchers and all the stakeholders in agriculture
7.	http://www.rkmp.co.in	2009	-Developed by Directorate of Rice Research with &7 consortium partners (CRRI, IARI, ICRISAT, C-DAC, UAS-Raichur, IRRI, ICAR-NEH, BAU)	-Research, extension, farming info, general info, virtual platform for distant learning	-Web 2.0 -Registered users: 813 -Total no. of pages of the entire portal excluding 20 platforms: 13435 -News updates on rice from across the world -Cloud tags	Rice growers around the country, researchers, extension personnel and all other stakeholders
University Web Portals						
8.	http://agritech.tnau.ac.in	2009	-Tamil nadu Agrici;tira; University- Govt. of Tamil Nadu initiative	Information related to agriculture & allied sectors, DMI, weather forecasting, irrigation, news, e-govt. schemes and	-Links to Farmers' Associations -Available in English and Tamil	All stakeholders in agriculture with special emphasis on

Contd...

				other useful links		farmers
9.	http://www.pusavarsity.orgn.in	2008	-Funded by RashtriyaKrishiVikas Yojana Rajendra Agricultural University, Pusa, Bihar	Daily weather farecasting	-Individual forecasting for all the districts -Advice for crops depending on the weather -Summary of previous forecasts	Farming community

e-Commerce

10.	http://agmarknet.nic.in	2000	-Implemented by DMI, MoA, GoI: -Executed by NIC	-Daily commodity prices and arrival information from markets all over the country	-10 languages; 1000 markets -G2C portal -Date-wise, commodity-wise, market-wise price	Farmers, policy makers and other stakeholders
11.	http://www.echoupal.com	2000	Initiative of ITC Limited	Information in local language on weather, market price, scientific farm practice, risk management	-6500kiosks in 10 states (40,000 villages) -Operated by 'Sanchalaks'	Farming community
12.	http://www.megamb.nic.in	2003	Designed and developed by NIC, Meghalaya State Agricultural Marketing Board & Directorate of Market Information, Shillong	-prices of commodities from 62 regulated markets -District wise price information	-Commodity profile, market profile	Farming community and policy makers
13.	e-krishivipanan	2003	Initiated and executed by Madhya Pradesh Agril. Marketing Board, GOMP and MP Agency for Promotion of Information Technology	Market information of 400 locations around the state	-Aims for timely collection and dissemination of market information -Caters for 70% of the total Agricultural Produce Market of the State	Farming community and other stakeholders
14.	http://www.e-krishi.org.	2006	-Funded by UNTD -PPP mode; partnership of National Institute of Smart Government, Kerala State IT Mission	Agri-advisory services, market information and resource library	-Aims for 'grainlessmandis' -KISSAN online advisory -e-Trade	Farming community
15.	http://agriexchange.apeda.gov.in		Developed by Agricultural and Processed Food Products	Market news, country profile, service providers in agribusiness sector	-Data sources are UN Comtrade, FAO and DGCIS -550 commodities	In put dealers, traders and others

Contd...

			Development Authority (APEDA), Ministry of Commerce and Industry, GoI		related to trade
e-Governance					
16.	http://india.gov. in (Agriculture)	2005	Maintained by NIC	Information related to technology, irrigation, crop production, animal husbandry, horticulture and NRM	-Bilingual -English and Hindi -National portal of India	Farming community
17.	http://dacnet.nic.in	2002-2007 (under 10thFive Year Plan)	MoICT initiative, developed by NIC	Government schemes related to agriculture	-Objective: to bring e-Goverance in Directorates and field units -G2E, G2G, G2C (G-Government E-Employee C-Citizen)	All stakeholders of agriculture
18.	Agrisnet (Agricultural Systems Information Network)	2002-2007 (under 10th Five Year Plan) 2012 2013	-Implemented by Department of Agriculture and Cooperation and NIC -Mission Mode Project under National e-Governance Plan	-Internet/intranet services -National and international video conferencing -IT training -Decision Support System development -Disaster management	-Budget Rs. 49.47 crore -G2C service -Objective: improve dissemination of information on Govt. programme and services, agricultural resources and technology -Implemented in 26 states and Union Territories across the country	Agriculture departments and resource personnel
19.	http:// www.kiran.nic.in		-Implemented by ICAR-NEH RC -Maintained by NIC	-Practices in agricultural and allied sectors -Agromet advisory	-First of its kind in North East India -Gateway of information	All the stakeholders in NE India
20.	mKisan		-Implemented by Department of Agriculture & Cooperation, MI mo-stru pf Agriculture, Government of India	-Practices in agricultural and allied sectors	-the portal specifically if for farmers to register to mobile based advisory services through length and breadth of the country	Farming community across the country

Challenges faced: The bottlenecks in the implementation and utilization of the web portals in agricultural extension are many in the Indian context. The problem arises both in the developers' and users' front that needs to be effectively managed and removed to get a hassle-free service from them. The problems are discussed below:

a) *Illiteracy and lack of portals in local language:* Most of the farmers to whom the portals are directed are illiterate and if literate, have very little understanding English. But contents of most of the portals are mainly in English and in some cases, one or two major language of the region. The development of the portals in multiple languages sometimes becomes difficult which effect the users' perception and application of the content.

b) *Poor connectivity:* Poor connectivity in the rural areas is another problem preventing their use. According to the survey done by Akamai Technologies Inc. in 2011, India is the only country in the Asia-Pacific with average internet speed below 1mbps (Anwer, 2012). The rural areas are the worst hit and so the web portals in service are most of the time inaccessible.

c) *Timely updating of data:* Timely data update is another constraint in the content use of the web portals. Agriculture is a highly season dependent activity and the activity changes rapidly. Moreover, the temperature, humidity *etc.* determines the production to a large extent. The information update accordingly sometimes become difficult on the developers' part as it needs to be updated within short intervals.

d) *Developing location-specific information:* Agriculture is not only time-specific but also location specific as the choice of crop varies with soil type, topography, climatic conditions, *etc.* In a diverse country like India, creating location-specific content becomes difficult for obvious reasons.

e) *Integration with the web:* Organizations can pull various resources from the Web, using software tools and information specialists, and integrate them into the local resources (Chisenga and van Brakel, 2005). The portals in many cases are not properly integrated with the web with useful links and resources. This may sometimes stop the user to access the internet directly to get extra information on specific topics. It also limits the scope of the portal.

f) *Arrangement of links:* Too many information can easily confuse the users, especially farmers who have very less experience of using internet sources. An organized arrangement of the web links makes the portal simple to use and preferred by the inexperienced farmers.

g) *Ad hocweb development*: When using *ad hoc* approaches, most Web site and portal developers go directly into the implementation phase of the system, paying little or no attention to requirements acquisition and specification (Coda *et al.*, 1998). This leads to poor web site applications in the long run (Pressman, 2000). So use of formal approaches of web site development is very important.

h) *Integrated Knowledge system*: The resources of the web portals are often limited to the knowledge of the developers. This restricts the realm of the portal to the research system and does not extend to the local knowledge which often raises question about the credibility of the information provided.

Ways to improve: To make the web portals a real success and not a piece meal approach to use of ICTs in agricultural extension, more attention needs to be paid to the 'I' and 'C' of ICTs. Some of the steps that can be taken in this regard are:

a) *Wireless broadband in rural areas:* The information centres in the rural areas can be connected with wireless broadband to keep them free from the hassle of the wired connections and their frequent failure.

b) *Prioritizing an interface:* An interface between the users should be prioritized so that the web resources could be contributed by all the registered users to make a rich database. The extension personnel can share their experiences, the farmers their indigenous knowledge and the researchers their research finding immediately and thus reduce the hassle and time of passing the knowledge through the traditional route of institutional procedures. For the authenticity of the uploaded material or information, the developers can filter what is to be added to the database.

c) *Networking platform:* An interactive platform is always more lively and gives the user the opportunity to test his/her knowledge. The farmers can check with the experts instantly and take suggestions. It also makes the platform interesting and authentic.

d) *Video and audio interface*: As most of the target user of the portals, the farmers if the country is illiterate or semi-literate, the language poses a barrier in their understanding, the proper ways of spraying or weeding are sometimes really difficult to decipher from the texts. A supporting audio and video clip will make understanding easier and learning interesting.

e) *Partnership with local institutions*: To develop location specific data and for their timely update, partnership with local research stations, universities and other such local institution is a desirable step to take. The data generated by the institutes will not only help the farmers in getting up-to-date information, such link-ups will also make the work of data generation easier.

f) *Portal development*: The portal development should be based following the proper protocols so as to avoid any hassles at the later stage. A user-friendly web portal gives the user confidence in using it.

Mobile Based Advisory Services

Of all the ICT tools, mobile phones have created the maximum impact on farming community because of its easy availability, cheaper cost, easy handling of all the ICT tools. According to the International Telecommunications Union (ITU), the use of cellular telephones has grown faster than fixed-line telephones in developing countries (ECLAC, 2012). Mobile applications (m-apps) in general and mobile applications for agricultural and rural development (m-ARD apps) in particular hold significant potential for advancing development (Quing*et al*, 2011). And this opportunity has been taken up by many government and non-government agencies and service providers to help out the farmers and provide them timely and relevant information regarding the farming activities. An analysis of forty-three mobile based advisory services by Saravanan and Suchiradipta (2014) shows that majority of them are pilot projects and are operated in a small scale, and while some has been helpful in increasing the income of the farmers by a significant amount, some has also failed to have a measurable impact inspite of much hype surrounding them. Some such initiatives, providing farmer needs' based mServices through their technological innovative applications combined with creative business plans are discussed below:

Table 2 : Telecommunication based advisory services in India

Sl. No.	Mobile Service	Service provider	Network	Year	Service provided	Target user
1	Kisan Call Centre	DAC, GoI; TCIL and IFFCO	Mobile phones, land lines	2004	Accessible on identified telephone numbers (toll free) working on IVRS where any queries related to agriculture are answered at three levels	Accessible all over India
2	Fisher Friend	MSSRF and QUALCOMM, Tata Teleservices, Astute System Technology	CDMA	2007	Weather forecasting, market price, sea wave height information, fish shoal location via satellite scan data	Fishermen of Tamil Nadu
3	IFFCO Kisan Sanchar Limited	IFFCO and Airtel	GSM mobiles, landlines	2007	On a VAS platform, farmers can know about mandi prices, farming techniques, weather forecasts and fertilizer availability for free. Helpline for farmers from Green SIM Card	Service available in 18 states of the country

Contd...

4	Reuters Market Light	Reuters Group Plc.	GSM	2007	Localized pricing of onion, soybean, pulses, pomegranate, *etc.*, weather updates, agriculture news.	Farmers of Maharashtra and Punjab
5	SMS Broadcast Service	KVK, Babhaleshwar	GSM	2008	Weather forecast, disease forecast, market information,etc.; information on training to farmers' organizations,etc.	Farmers of Ahmednagar District, Gujarat. (Saravanan R, 2010)
6	Kissan Kerala	Deptt. Of Agriculture, Govt. of Kerala		2009	SMS based information service, voice based agro-advisory service, videos, SMS alerts regarding schemes, training programs, 'Krishideepam'.	Registered farmers in Kerala
7	Life Tools	Nokia and Idea Cellular	VAS from both GSM and CDMA	2009	Information on seeds, fertilizer and pesticide dosage, weather forecasting and commodity prices in Marathi and Hindi.	Can be accessed by anyone from anywhere in the country
8	mKrishi	Tata Consultancy Services	CDMA	2009	Information on microclimate, fertilizer dose, market info both in voice activated service and SMS	Registered farmers of Punjab, Uttar Pradesh, Maharashtra, Gujarat, Andhra Pradesh, Tamil Nadu and Rajasthan
9	Kisan Mobile Advisory Service (KMAS)	KVK, Ri-Bhoi (ICAR)and KVK-East Khasi Hills (Govt. of Meghalaya)	GSM	2010	Based on the cropping season and major crops of the area, SMS is sent to the farmers in local dialect on the package of practices, varietal selection, dosage of chemicals used, *etc.*	KVK-RiBhoi has around 500 farmers registered for the service and KVK-EKH has around 200
10	Market Price on Mobile	Rubber Board of India	GSM	2010	Subscription based service (annual subscription Rs. 300) through SMS. Domestic and international prices at 4 p.m and 11 a.m respectively.	Accessible all over India
11	Mandi on Mobile	BSNL tied up with state Govts. and IIT Kanpur	GSM	2011	Market rate of vegetable, grain, pulse or fruit on software that work on voice command. To access, user needs to call up a specified BSNL number.	Farmers of Uttar Pradesh, Himachal Pradesh

Contd...

12	SMS service	Govt. of Haryana	GSM	2011	Queries on all aspects of agriculture can be sent on a specific number	Accessible to farmers of Haryana
13	SMS service	Tamil Nadu Agricultural University	GSM	2011	Day to day market price of vegetables in the state; expert advice based on the pictures of their crops sent by the farmers	Registered farmers of Tamil Nadu
14	Kisan Help Line	Bihar Agricultural University	Mobiles, landlines, *etc.*	2012	Queries related to farmer's field and crops are answered	Farmers of Bihar
15	m4agri NEI	Media Lab Asia (MoC&IT) & CAU, Barapani, Meghalaya	GSM	2012		5000 farmers from Ri-Bhoi and East Khasi Hills district of Meghalaya
16	SMS Weather Alert	IMD	GSM	2012	Weather related information and forecasting	Farmers of Maharashtra can access the service
17	aAqua mini (redesigned for mobile browser)	Agrocom	GSM and CDMA		Localized remote crop diagnostic solution, audi prompted guide application in English, Marathi and Hindi, remote crop and land based diagnostics, microclimate, SMS enabled register and query mechanism.	Registered users of the country
18	Hello Uttam	Chambel Fertilizers and Chemicals Limited	Mobiles, Landlines, *etc.*		Any queries related to farming can be asked using the helpline numbers.	Farmers of Rajasthan, Punjab, Uttar Pradesh and Haryana can access the service
19	Via SMS	Chambel Fertilizers and Chemicals Limited	GSM		Any crop and market related queries	Farmers of Rajasthan, Punjab, Uttar Pradesh and Haryana can access the service
20	mKisan	Government of India	GSM and CDMA	2013	Location based information on all agricultural activity; mobile app for all mobile phones	Registered users of the country

(*Source*: Saravanan and Suchiradipta, 2014, Saravanan et al., 2011)

Challenges faced: Mobile phones are speeding up ways in which farmers get, exchange, and or manipulate information (Nyamba and Mlozi, 2012). Literature, for example, show that, a developing country with an average of 10 or more mobile phones per 100 people between 1996 and 2003 would have enjoyed per capita Gross Domestic Product growth of 0.59% higher than an otherwise identical country with a mobile density of less than 10 phones per 100 people (Verheye, 2000). But in India, in spite its popularity in parts of the country, use of mobile phones in agricultural extension is facing a number of bottlenecks as discussed below:

Language barrier and illiteracy: Language is a big hurdle to cross in all sphere of ICT implementation in India. Be it SMS service or the IVRS, language remains a problem in understanding of the information provided. High rate of illiteracy in parts of the country only adds more to the problem.

Poor signal: Though the mobile service has reached nearly every nook and corner of the country but a lot of work needs to be done on the signal strength provided. Optimal utilization of the tool is prevented by the poor signals in the villages limiting its scope.

Electricity: It is another problem in using mobile phones. Many parts of the country are still to get proper electricity connections and even where it is there, power cut is more frequent than power availability. It makes difficult to use mobile phones in those areas.

Lack of smart phones: Most of the users in the rural areas have no idea what a smart phone is and even if they buy it, they have no clue how to use most of the features. Internet access through mobile phones is also negligible in rural areas and so the best features of mobile phones in agriculture is left unexplored.

Lack of awareness and promotion: Most of the time, the farmer are not aware of the services they can avail through mobile phones. Sometimes, they do not know whom to call. Poor promotion is a reason for this. The private players take enough measures to promote their services but it is still a weakness of the public sector.

One mobile per household: Most of the households have only one mobile phone which is carried by different persons at different point of time and as a result the messages always do not reach the intended farmers.

Expensive service: Many rural farmers say that it is expensive for them to afford a mobile phone or if they have a phone, they can't afford the service provided (Suchiradipta, 2012).

Confidence on the system: In many cases, lack of confidence in the system creates a barrier in the use of mobiles. The farmers are not fully convinced by the information provided and so the medium loses its credibility.

Areas to improve: To improve the service quality and make mobiles an important medium of information delivery in extension service, some steps that can be taken are:

Transparency and feedback: Transparency and feedback mechanism needs to be improved. Most of the service provided are more involved with providing information but not interested in the usage pattern of the service. This should be done to evaluate the success of the service.

Add on feature: Expert advice through mobile should be added to the existing services so that the process becomes more interactive and authentic.

Network independency: Network independent operations can give better service in areas with poor signal strength or high cost charged.

Consolidated info: Together with the daily basis information provided, consolidated weekly or fortnightly information can increase the credibility and usefulness of the information provided. Moreover it will be useful for the users to learn the trend and act accordingly.

Solar chargers: Since electricity is a problem in parts of India, solar chargers can be installed to provide an alternate source of power to the villagers. This will help them to stay better connected along with other related facilities.

Social Media

Social media refers to the internet-based tools for sharing and discussing information among people. It refers to the user generated information, opinion, video, audio, and multimedia that is shared and discussed over digital networks (Andres and Woodard, 2013). Merriam-Webster (2013) defines Social Media as forms of electronic communication through which users can create online communities to share information, ideas, personal messages and other content. The definition of Ahlqvist*et al.* (2008) is focussed on three basic components – content, communities and Web 2.0 and operationalises social media as the interaction of people and also to creating, sharing, exchanging and commenting contents in virtual communities and networks. According to Michelle Chmielewski, Social Media is not about what each one of us does or says, but about what we do or say together, worldwide, to communicate in all directions at any time by any possible digital means (Cohen, 2011). Social Media tools include social networking sites (Facebook, LinkedIn, MySpace); Video and photo sharing websites (Flickr, YouTube, Tumbler, Pinterest); Blogs and microblogs (Blogger, Twitter, Instagram); Forums, discussion boards and groups (Google hangouts); Wikis (Wikipedia); Video and podcasts; Video conferences and web conferences (webinars); Socially integrated mobile text messaging (Line, WhatsApp, Viber, BlackBerry Messaging, WeChat); Websites with social plug-ins and layers; Social bookmarking (Delicious, Blinklist); Social news (Reddit, Propeller, Digg); Social gaming and many more (Wikipedia[g], 2014).

Social media has become the number one activity on the internet and the major communication media for the youth. There are 1.5 billion social media platforms in the world (Wikipedia[g], 2014), Facebook user base is second to only the population of China, *i.e.*, if it was a nation, it would be the second most populated in the world; YouTube is the second most popular search engine after Google; LinkedIn was found to be 277 per cent more effective in lead generation than Facebook and Twitter. Young people are the main consumers of social media and they can be easily attracted to embrace agriculture by digital technology more effectively. And so, social media, when effectively and strategically used, can be highly effective in family farming and the youth concerned with it. Some examples of social media in agriculture are given below:

Table 3 : Social media in Agricultural Extension in India

Social media platform	Roles
Facebook	
Turmeric Farmers' Association of India	-Information sharing among turmeric farmers of India-Both B2B and B2C communication
Agricultural Extension India	-Information sharing among agricultural stakeholders in India
Indian Council of Agricultural Research	-Communicating recent developments in Agriculture by the institution
CRISP India	
Agriculture in India – AG's Group	-An open forum to discuss about Indian agriculture for and sharing ideas,thoughts and latest methodologies
Agricultural Extension in South Asia	-Information sharing among the agricultural extension professionals in South Asian countries
LinkedIn	
Agricultural Extension in South Asia	-Information sharing among the agricultural extension professionals in South Asian countries
YouTube	
Indian Council of Agricultural Research	-Sharing videos of various agricultural developments/ technologies
Blogspot	
Agricultural extension in South Asia	-Information sharing among the agricultural extension professionals in South Asian countries
Agropedia	Wikipedia of agriculture with the same privileges
Digital Green	
Farmbook	-Facebook for the farmers where the comments and likes and dislikes of the farmers on the videos they watch are uploaded. The farmers can also be followed on Facebook when linked up with Facebook
Wondervillage	-Social gaming with the real farmers on Farmbook

(Source: Saravanan and Suchiradipta, 2013)

Challenges faced: Social media is non-arguably the next best thing in communication but inspite of its vast popularity, the question of its applicability in agriculture still remains. No matter how incredible social media is for personal communication, before using it for professional communication and making it important in family farms seeking agricultural information, some major hurdles still needs to be cleared.

Age: Social media platforms like Facebook may be becoming more and more used by 45+ populations for personal communication, but so is not the case in agriculture, especially for farmers of the developing countries. Mostly residing in the rural areas, they are not tech savvies yet and so the intricacies of smartphones or social media are something not much familiar to them.

Awareness: While the use of social media, especially Twitter is more and more used by farmers of developing countries (a good example is @Agchat in Twitter) to discuss their problems, the developing countries are yet to see that. Awareness about use of social media for agriculture and its practical use is very less in countries like India.

Infrastructure: Even though the price of smartphones are becoming cheaper by day, there are still places in the country where cell phone towers haven't reached and if they have, internet data packs are costly for a farming family to use, especially if the amount of information they are supposed to gain from it is negligible.

Information explosion: Social media has made information gathering and discussions with peer groups real easy but exposure to so much information on a single topic sometimes may be confusing and act as a deterrent for its further use for getting advice on agriculture.

Areas to improve: This minimal number of users of social media in extension in a country like India can mean only one thing – the huge scope that it has in front of it. Some steps that can be taken for increasing the use of social media in agricultural extension are:

Breaking the mould: In India, inspite of numerous ICT projects functioning and completed, the number of sceptics in use of ICTs is still not less and social media is still not seen as something that can be used for agriculture. But a little awareness and inspiration from those who are already using it,especially among the young farmers, can break this mould and make social media a great medium of communication among professionals and farmers.

Regulating the information: Regulating the information shared on a discussion by one or more professional can make the discussion really enriching rather than confusing for the users. This also will interest the farmers to use it when the response is so quick. Also, the amount of money and time that is saved through these platforms can be enthusiastic and act as an incentive in making agriculture 'cool'.

Infrastructure development: As any other ICT tool, using social media also requires some basic infrastructural facilities like electricity and cell phone towers which are not just required for using social media but are also the basic building blocks for development. Also, some regulation of data tariff and special offers at cheaper rates for rural areas can also be helpful for the potential users.

CONCLUSION

Family farming is not just a way of living but is a tradition or a legacy passed on for many. Family farmers are not just responsible for feeding majority of the world; they are also responsible for the protection of majority of the world's biodiversity and ecology. And most importantly, they are the way out of world poverty and hunger. So, making agricultural information easily accessible to them as and when required is something that cannot wait for the future. And so, it is much easier and faster to make a phone call or a Facebook post or a tweet rather than building a road to reach everyone personally. It is true ICTshave their own drawbacks but what doesn't? It is our job to learn from failures and find a way around the hurdles if not jump over them and using the boon that ICTs has to offer can always be a help and a great complement to the personal efforts of the extension mechanism.

REFERENCES

Andres, D. and Woodard, J. (2013) Social media handbook for agricultural development practitioners. USAID and FHI 360.

Anwer, J. (2012). India among countries with worst internet connections. The Times of India. Retrieved from http://articles.timesofindia.indiatimes.com/2012-05-02/internet/31537373_1_256kbps-speed-internet-speed-akamai. (Published on 02.05.2012; Accessed on 17.10.2014).

Census of India. 2011. Mode of Communication: 2001-2011. Office of the Registrar General and Census Commissionar, India. Ministry of Home Affairs, Govt. of India, New Delhi, India. Retrieved from http://www.censusindia.gov.in (Accessed on 21st November, 2012).

Chakrapani, V. (2014). Green TV to ho the AFP way, soon. Retrieved fromhttp://www.indiantelevision.com/television/tv-channels/specialised-and-niche/green-tv-to-go-the-afp-way-soon-141105. (Posted on 05.11.2014; Accessed on 10.11.2014).

Chauhan, Jitendra. (2007). Agriculture Extension Education: Communication in Agriculture. Retrieved from *nsdl.niscair.res.in/.../517/1/PDF+Communication+in+Agriculture.pdf* (Accessed on 07.11.2014).

Choudhury, V. (2014). Budget 2014: Govt. allocates Rs. 100 crore for community radio. Retrieved from http://www.livemint.com/Politics/HATaZIsYut0SCALrSUYfQO/Budget-2014-Govt-allocates-Rs-100-crore-for-community-radio.html. (Posted on 10.07.2014; Accessed on 10.11.2014).

Chisenga, J. and van Brake,l P.A. (2005). Guidelines for developing agricultural information portals in the Southern African Development Community Region (SADC). *South African Journal of Information Management.* Retrieved fomhttps://ujdigispace.uj.ac.za/bitstream/handle/10210/5296/Van%20Brakel,%20Chisenga.pdf?sequence=1. (Accessed on 17.10.2014).

Coda, F., Ghezzi, C., Vigna, G. and Garzotto, F. (1998). Towards a software engineering approach to Web site development. Retrieved from http://www.cs.ucsb.edu/~vigna/publications/1998_coda_ghezzi_vigna_garzotto_iwssd98.pdf (Accessed on 17.10.2014).

Cohen, H. (2011) 30 social media definitions. Retrieved from http://heidicohen.com/social-media-definition/ (Accessed on 29th September, 2013).

Colle, Royal D. (2011). R. Saravanan (Ed), ICTs for Agricultural Extension. Book Review *Journal of Development Communication, 22:1* Retrieved from http://www.saravananraj.net/index.php?option=com_content&view=article&id=53&Itemid=60 (Accessed on 22.11.2012).

Doordarshan. (2011). PrasarBharti: Agriculture (KrishiDarshan). Retrieved from *http://www.ddindia.gov.in/Kendra/Jaipur/Program+Column+2/Movie.htm*. (Accessed on 07.11.2012).

ECLAC (Economic Commission for Latin America and the Caribbean). (2012). Facts and Figures. Retrieved from http://www.eclac.cl/socinfo/noticias/paginas/3/44983/newsletter18ENG.pdf. ECLAC Newsletter: 18. pp-12.

FAO. (2014a). Family farmers: Feeding the world, caring for the earth. Retrieved from http://www.fao.org/resources/infographics/infographics-details/en/c/230925/ (Updated on 05.05.2014; Accessed on 30.10.2014).

FAO. (2014b). Putting family farmers first to eradicate hunger. Retrieved from http://www.fao.org/news/story/en/item/260535/icode/ (Updated on 16.10.2014; Accessed on 30.10.2014).

FAO. (2014c). State of food and agriculture 2014. Retrieved from http://www.fao.org/3/a-i4040e.pdf (Accessed on 30.10.2014).

GoI. (2014). Press in India Highlights for the Year 2013-14. Registrar of Newspapers for India, Government of India. Retrieved from http://rni.nic.in/ (Updated on 7.11.2014; Accessed on 7.11.2014).

HeinonenAhlqvist, T., Back, A., Halonen, M. and Heinonen, A. (2008) Social Media Roadmaps – Exploring the futures triggered by social media. Retrieved from http://www.vtt.fi/publications/index.jsp (Accessed on 14th December, 2013).

ITU. (2014). The world in 2014: ICT facts and figures. Retrieved from (Accessed on 30.10.2014).

Karthik, P. (2011). Essay on the Brief History of Radio Broadcasting in India. Retrieved from http://www.preservearticles.com/2011112417838/essay-on-the-brief-history-of-radio-broadcasting-in-india.html. (Updated on 24.11.2011; Accessed on 17.11. 2012).

Kothari, Brij. (Undated). Let a Billion Readers Bloom: The Power of Bollywood, TV and Subtitles. Retrieved from http://www.indiafutureofchange.com/indiastory_Education_Same Language.htm (Accessed on 22.11. 2012).

Kujur, G., Jha, M. N., Chaudhary, B.N., Kabdal, D. C., Deepkumar, V. S., Singh, R. C. and Deshmukh, P. A. (2009). Media Support to Agricultural Extension: Success Stories of All India Radio. PrasarBharti, Directorate General, All India Radio, New Delhi – 2009. Retrieved from http://allindiaradio.gov.in/NR/rdonlyres/...EB8B.../Kisinvani_Books.pdf(Accessed on 28.11. 2012).

Merriam-Webster. (2013) Social media. Retrieved from http://www.merriam-webster.com/social-media/ (Accessed on 28th September, 2013).

Nyamba, S.Y. and MloziMalongo, R.S. (2012).Factors Influencing the Use of Mobile Phones in Communicating Agricultural Information: A Case of Kilolo District, Iringa, Tanzania. *International Journal of Information and Communication Technology Research*. 2:7.

Parsai, G. (2014). Farmers deserve a standing ovation: NarendraModi. Retrieved from http://www.thehindu.com/news/national/farmers-deserve-a-standing-ovation-narendra-modi/article6260560.ece. (Posted on 29.07.2014; Accessed on 10.11.2014).

Pressman, R.S. (2000). What a tangled Web we weave. *IEEE Software* (January/February). pp 18-21.

Quiang, C.Z., Kuek, S.C., Dymond, A. and Esselaar, S. (2011). Mobile Applications for Agriculture and Rural Development. ICT Sector Unit. World Bank.

Saravanan R. (2010). 'India' In: *ICTs for Agricultural Extension: Global Experiments, Innovations and Experiences.* Saravanan R (Eds.).Retrieved from http://www.saravananraj.net/index.php?option=com_content&view=article&id=51&Itemid=57(Accessed on 21.11.2012).

Saravanan, R., Kathiresan, C., and Indra Devi, T., (2011) (Eds.) *Information and Communication Technology for Agriculture and Rural Development,* New India Publishing Agency (NIPA), New Delhi

Saravanan, R and Suchiradipta, B. (2014). Mobile phones and social media for Agricultural Extension: Getting closer to hope or hype? Lead paper presented in Third International Conference, Extension Educational Strategies for Sustainable Agricultural development – A Global Perspective, December 5-8, 2013, University of Agricultural Sciences, Bangalore, India.

Saravanan, R. (2007). Effectiveness Indicators of Public, Private and NGOs' Agricultural Extension Programs in Karnataka State, India. Retrieved from http://www.saravananraj.net/index.php?option=com_content&view=article&id=51&Itemid=57 (Accessed on 17.11.2012).

Saravanan, R. and Suchiradipta, B. (2014). Mobile phone applications for agricultural extension in India. In: Saravavan, R. (ed.) Mobile phone for agricultural extension: Worldwide mAgri innovations and promise for future. New India Publishing Agency, New Delhi, India.

Sharma, A. (2011). Community Radio as an Effective Tool for Agricultural development (Research). Retrieved from http://www.youthkiawaaz.com/2011/07/community-radio-as-an-effective-tool-for-agricultural-development-research/ (Updated on 07.12.2011; Accessed on 18.11. 2012)

Suchiradipta, B and Saravanan, R. (2013). Youth and ICTs for agricultural development. In: Gowda, K.N., Nataraju, M.S. and Veerabhadraiah, V. (eds.) Youth in Agriculture and Rural Development. New India Publishing Agency, New Delhi, India.

Suchiradipta, B. (2014). Mobiles for mobilizing Agricultural Extension in India. Credit Seminar presented in School of Social Sciences, College of Post Graduate Studies, Central Agricultural University, Umiam, Meghalaya – 793 103.

Thurstone, L. L., (1946) Comment. *American Journal of Sociology*, 52: 39-50.

Verheye, W. (2000). Food Production of Food Aid: An African Challenge. "Customer perceived value in banking services". *International Journal of Bank Marketing*,24(5): 266-283.

Wikipedia[a]. (2014). Agriculture in India. Retrieved from http://en.wikipedia.org/wiki/Agriculture_in_India. (Updated on 06.10.2012; Accessed on 16.10. 2014)

Wikipedia[b]. (2014). Communication in India. Retrieved from http://en.wikipedia.org/Communications_in_India_files/ (Updated on 12.10.2014; Accessed on 02.11.2014).

Wikipedia[c]. (2014). Community Radio. Retrieved from http://en.wikipedia.org/wiki/Community_radio#India Updated on 24th November, 2012 (Accessed on 28.10. 2014)

Wikipedia[d]. (2014). Television in India. Retrieved from http://en.wikipedia.org/wiki/Television_in_India Updated on December 7th, 2012 (Accessed on 08.11.2014)

Wikipedia[e]. (2014). KrishiDarshan. Retrieved from http://en.wikipedia.org/wiki/Krishi_Darshan. Updated on 5th December, 2012 (Accessed on 08.11.2014).

Wikipedia[f]. 2014. 'Web Portal' Retrieved from http://en.wikipedia.org/wiki/Web_portal, (Accessed on 20.10.2014).

Wikipedia[g]. (2014). Social Media. Retrieved from en.wikipedia.org/wiki/Social_media (Accessed on 8.11.2014).

CHAPTER - 29

Agricultural Extension Services System Supported by ICTs in India: Progress and Challenges

Raghuveer Singh Meena, Mohammad Hashim, Pramod Kumar Prajapati and Richa Sachan

Research Scholars, Department of Extension Education, Institute of Agricultural Sciences, Banaras Hindu University, Varanasi- 221005, Uttar Pradesh

ABSTRACT

The conventional agricultural extension service in India is mainly provided by extension workers visiting farmers to provide agricultural advisory services. This system of extension service provision faces a number of challenges including the few number of extension workers and limited resources. The effectiveness of an impact-driven, radio-based extension service delivery system that has been introduced in some rural areas of India. The system aims to enable extension workers to reach many farmers with minimum efforts. However little is known about the effectiveness of this new extension service delivery system. In rural areas, farmers have started sharing agricultural related information and best practices. Some farmers have also started to change their farming practices. This brings a new finding that farmers can adopt and practice easily what is aired by their fellow farmers in the community radio stations.

Keywords: Extension Services, ICT, Farmer Radio Voice, Community Radio Station

INTRODUCTION

ICT is having capability to make the farmers up to date with the recent and relevant information and/or services that may be information on new agricultural technologies, best agronomic practices, demands, price trends and so on. Extension has the responsibility to bring information and new technologies to farming communities, allowing them to improve their production, income and standard of living. Some extension performance are also there like the cost of reaching large, geographically dispersed and remote smallholder farmers, high transport costs, weak accountability to clients and financial unsustainability. Only very few farmers can be dealt directly due to very high farmer to extension agent and agents tend to focus on the larger, better resourced and more innovative farmers. Here comes the importance and role of ICTs which has great potential to transform the way public extension is organized and delivered -including interactions with farmers.

Trends and Issues

There are a plenty of research findings that ICTs can revitalize research-extension in ways that fulfils the demands and needs of the farmers. The traditional top-down, technology-driven extension system is getting a tough competition from time and input effective and efficient ICTs tools that are making the issues more transparent and reliable. ICTs have four broad functions in the context of rural advisory services that support innovation. First, they need to deliver or provide access to information. Second, ICTs help in documentation and store information for future use. Local and indigenous know-ledge is often transmitted orally, records are often unavailable, and information is dispersed only to nearby family and friends.Third, ICTs connect people and networks. Fourth, ICTs need to empower rural communities. A key element is to use ICTs to give rural people the skills and tools to tell their own stories, in their own words and languages, in ways that reach and influence others.

ICTs That Preserve Farmers' Knowledge

Some of the ICTs tools are in use of the rural community since long time ago like radio while digital videos are new and innovative to rural community. These are now helping in conserving and disseminating the abundance of indigenous technological knowledge lying with the farmer.

Using ICT to Share and Elicit Local Knowledge

Seeing ICTs just as tools to bring information and modernity to rural areas is under estimating the capacity of the ICTs. It may also be applied to empower rural people to store, preserve and disseminate their own regional and local knowledge by making the documental proof of the same.

Documenting and Mobilizing Indigenous Knowledge

A related information-sharing effort documents indigenous knowledge. As experienced farmers migrate to urban areas, as the local farming population ages, or as climate change and social upheavals uproot agricultural communities, much knowledge can be lost.

Using Radio and Video to Reach Rural Farmers

Although there are a number of communication media available but till now we don't have a substitute of the radio and even now it is most pervasive, cheap, popular and socio-culturally compatible. Radio is still the only medium for disseminating information rapidly to large and remote audiences, including critical information about markets, weather, crops, livestock production, and natural resource protection.

Challenges: The challenges of access to ICT can be divided into two:

(i) access to ICT infrastructure and

(ii) access to ICT services.

Access to ICT infrastructure is the basic challenge to have the physical presence of the tools to be used for agriculture purpose while just the presence of the ICT infrastructure can not serve the purpose till the ICT services are not accessible.

Opportunities

The ICT sector in India has reflected a tremendous growth in the past two decades.India is now setting an example in the IT market and exemplifying itself as a global leader in the ICT sector.The Indian IT sector has changed the image of the nation from a nation of slow moving bureaucratic economic economy to a land of successful entrepreneurs. The market size of information technology in India is expected to touch US$ 44.8 billion in 2014 as compared to US$ 35.1 billion in 2012, as per International Data Corporation (IDC).The Indian IT infrastructure market (comprising servers, storage and networking equipments) is projected to grow to US$ 2.1 billion in 2013, an increase of 9.7 per cent as compared to 2012, according to Gartner. The mobile services market in India is expected to grow to Rs 1.2 lakh crore (US$ 19.97billion) in 2013, an increase of 8 per cent from Rs 1.1 lakh crore (US$ 18.28 billion) in 2012, as per Gartner.

CONCLUSION

The rural community can be benefited in real sense only when the right and the relevant information or services are delivered to the farmers and other stakeholders at right time and in right format. But these should not be obtained on the price of the indigenous and tactical knowledge repository lying with the farmers. Policy and investment priorities that government and its partners should consider in order to promote cost-effective knowledge management in agriculture have been highlighted. As the mobiles have become a part of the normal human life and the farmers are not an exception of this, the mobile phones can be effectively used for delivering the relevant information and services to the farmers at right time. So there is a need of making the mobile phones compatible with the local language and there should be a team of the extension experts who can take the responsibility of training the local people about how to use the mobile phones for coding there problems and getting the relevant remedial information. The government can encourage such types of efforts by providing subsidies on such ICT tools These initiatives, we believe, will assist the government to rationalize its expenditures in the sector, streamline the agricultural extension system, speed up agricultural transformation and attain the objective of doubling agricultural production and productivity by the end of the year 2015.

REFERENCES

Adam, L. 2010. Ethiopia ICT Sector Performance Review 2009/2010: Towards Evidence-based ICT Policy and Regulation. Volume Two , Policy Paper 9.

Ballantyne, P. G, A. Maru, and E. M. Porcari. 2010. "Information and Communication Technologies Opportunities to Mobilize Agricultural Science for Development." Crop Science 50(2):S1–S7. ILRI, http://mahider.ilri.org/handle/10568/653, Accessed September 2014.

Ballantyne, P. G, A. Maru, and E. M. Porcari. 2010. "Information and Communication Technologies Opportunities to Mobilize Agricultural Science for Development." Crop Science 50(2):S1–S7. ILRI

http://www.fao.org

http://www.moia.gov.in

http://www.oifc.in

Nyirenda-Jere, T. 2010. "Unlocking the Promise of ICTs for Transforming Agriculture in Africa." Wageningen:Technical Centre for Agricultural and Rural Cooperation.

Swanson, B., and R. Rajalahti. 2010. "Strengthening Agricultural Extension and Advisory Systems: Procedures for Assessing, Transforming, and Evaluating Extension Systems." Agriculture and Rural Development Discussion Paper No. 44. Washington, DC: World Bank. UNCTAD (United Nations Conference on Trade and Development). 2010.

CHAPTER - 30

Role of ICT in Betterment of Youth Farming

Richa Sachan, Sneha Singh, Swati Pahuja and Pramod Kumar Prajapati

Research Scholars, Department of Extension Education, Institute of Agricultural Sciences, Banaras Hindu University, Varanasi - 221005

ABSTRACT

Small holdings in developing countries have long been associated with poor farming practices, intensive labour and low income, which make the sector unattractive to a new generation of farmers. Lack of youth interest and involvement in agriculture affects the sustainability and survival of family farms. This interest in farming is further strengthened by ICT providing access to markets and information to boost production. Researchers found that youngsters easily grasp the ICT applications, acquire information, and jump onto new market opportunities. Considering the above facts the present paper will address the issue of the role that ICT can play in invoking the interest of youth in farming.

Keywords:ICT, youth, farming, sustainability

INTRODUCTION

Engaging youth in agriculture has been a prominent topic recently and has risen up the development agenda, as there is growing concern worldwide that young people have become disenchanted with agriculture. With most young people – around 85% – living in developing countries, where agriculture is likely to provide the main source of income it is vital that young people are connected with farming.

ICTs play an increasingly important role in agricultural value chains. Though important, cellphones aren't the only ICT being used to improve agriculture.

ICTs encompass radios, digital cameras, geographic information systems (GIS), cloud computing, tracking mechanisms, etc.

Five ways in which ICT can help tackle key challenges in agricultural value chain development are:

1. Pricing and weather information systems
2. Applications (apps) to help buyers manage transactions with the thousands of small-scale farmers who supply to them
3. Mobile banking and apps that facilitate quick payments
4. Initiatives to expand the reach of farm extension services through phone, radio, video and sometimes all three
5. SMS or text messaging campaigns for enabling environment advocacy

The increasingly important role of ICTs in agriculture can help change the face of the sector. In fact, it should form part of the larger thrust to attract more young people to the sector. ICTs offer employment opportunities in the sector that are both attractive to young people and are in demand.

Agriculture has an image problem. There is a decreasing interest among youth in entering agricultural related fields due to the persistent perception of agriculture as an outdated field with minimal financial returns. But it remains true that youth face large challenges accessing land and finance without collateral.Information and Communication Technology (ICT) not only improves the status of young person's using it, but also of the farming sector in general. Youngsters who used to see farming as a type of last resort source of income without much perspective now see it as a potentially strong source of rewarding business.

ICTs can play a role in countering youth migration to urban areas by enhancing access to market information, production techniques, new technologies and financing opportunities. The use of ICTs enables choice, the option to stay on farms and take full advantage of new technologies and farming techniques, while incorporating valuable traditional practices and knowledge. Efforts should be aimed at further fostering youth involvement in agricultural activities and decision-making processes. These efforts can seize on the youth's affinity for using ICTs, their capacity to innovate and their propensity for taking higher entrepreneurial risks.

"There is a lot of enthusiasm among the youth to embrace the latest farming techniques such as greenhouse technology and use of certified improved planting material and I believed this is being triggered by the recent wave of ICT platforms emerging that are making free information which is of a high quality available to them even in their mobile phones."

Eric Sirengo, Syngenta

Attracting Youth

Working with youth and media to provide more interesting portrayals of agricultural careers is important, as is working with ICTs and social media to reach a broader respondent.The use of ICT increases the social status of young persons. This acts as a motivating factor for ICT use. ICT creates an image of entrepreneurial young farmers who are keen to adopt new innovations. This results in the perception of young farmers changing in the eyes of external actors' such as government extension officers. Because of increased yields and new farming practices applied, young people also became technical resource persons to other farmers. ICT is also used to keep proper records for improved farm management, compare different crops and seasons, calculate profit margins and assess risks.

Education

The skills and competencies of agricultural graduates do not meet the needs of today's agricultural sector. Curriculum reform in tertiary level agricultural education is required to meet the demands of the labour sector, with particular attention to agribusiness and entrepreneurship, where many youth demonstrate an interest. Youth require a range of skills and competencies beyond their technical discipline, particularly those 'soft' skills such as communication, leadership and business skills. Reform of the agricultural curriculum must be a fully inclusive process which involves a wide range of stakeholders including youth.

With high unemployment rates and the lack of interest in traditional agricultural pursuits, a greater focus on entrepreneurship in agriculture is emerging. Entrepreneurship has the potential to contribute to the rejuvenation of the industry, creating more employment opportunities, increasing the potential for profit and moving away from the perception of agriculture as a low prestige career.

Innovation

There is a lot of concern about engaging youth in agriculture, in many ways, young people are not very much interested in continuing in agriculture because they don't see much prospect in the future of agriculture, they don't see it is as an active profession in the long-run, so many of the smallholder farmers are quite aged. ICTs could provide new opportunity for making agriculture more interesting for young people.

The increased use of mobile phones in farming can also help deter young people away from stereotypes of traditional farming and help change their perceptions on agriculture, helping them to view it as an exciting and innovative industry.

A Chance to Make a Difference

Farming offers the young generation a chance to make a difference by growing enough food to feed the world.Those who become farmers now have the opportunity to be the generation that end hunger and alleviate malnutrition as well as helping the sector adapt to climate change.

There are many challenges ahead for the sector but if young people are offered education in agriculture, a voice at policy level, and in the media, and are engaged with innovations then the agriculture industry can attract youth again. A responsible agriculture, meeting food security needs without depleting its resources, can only become a reality if young professionals are actively engaged in shaping the sector's future.

CONCLUSION

ICT has been find to catalyse and genisational development especially in relation to involvement and motivation and commitment of youth to improved management and development as a farmer.

Now what we need is to Focus on young smallholder families who already madea conscious choice for farming and are likely to apply ICTsto their farming activities and Pay attention to removing obstacles that prohibit young men and women from fully participating in, and taking advantage of,ICT-related farming activities There is also a need to furnish ICT centres as learning spaces to support informal exchanges and networking by young farmers.

CHAPTER - 31

Information and Communication Technology in Youth and Rural India: An Overview

Lokesh Kumar Meena, Shoji Lal Bairwa, Kerobim Lakra and Chandra Sen
Department of Agricultural Economics, Institute of Agricultural Sciences Banaras Hindu University, Varanasi (U.P.) - 221005

ABSTRACT

Information and Communication Technology (ICT) has become one of the basic building blocks of modern society. Young people are the world's largest resource in development. Forty percent of the world's population is under 20 years of age while 66 per cent of the total population in the country is below the age of 35 and this youth play very important role in adoption of ICTs. Present study is concerning with Information and Communication Technology, importance of ICTs on youth and rural peoples of our country, the importance of ICTs is not the technology as such, but its enabling function in facilitating enhanced access to information and communication across large distances. It involved development and need of Information and Communication Technology. It also focused on strategies for sustained ICTs growth in India and development ranking of some selected countries. India has been ranked 121st among 157 countries in terms of progress in the realm of information and communication technology.

Key words: ICTs, Youth, Network, LCCs and ITU

INTRODUCTION

Definition of Information and communication technologies (ICTs) consist of hardware, software, networks, and media for the collection, storage, processing, transmission, and presentation of information (voice, data, text, images), as well as related services. Communication technologies consist of a range of communication media and devices, including print, telephone, fax, radio, television, video, audio, computer, and the internet (Neto *et. al.,* 2005).

India has the largest population of youth in the entire world — 66 per cent of the total population in the country is below the age of 35.Nearly 40 per cent of the Indian population is aged 13 to 35 years. The median age in India in 2010 was just 25.2 years, compared to 34.5 years in China (Ghia, U.2014).Young people are the world's largest resource in development. Forty percent (40%) of the world's population is under 20 years of age and over 1 billion people are between the ages of 15 and 25. In some developing countries, youth constitute more than two thirds of the population

Young people who are given early access to information and communications technologies (ICT) tend to become early adopters and adapters of the technologies, skills valued for spurring innovation and economic growth. Youth are at the forefront of societal transformation through ICT because technology is what they have grown up with, what they know more about than their parents, and what gives them an edge. With the Internet, young people have acquired a powerful new tool to connect, communicate, innovate and take action on things that matter to them on a scale that transcends their locality, making them global actors (Prelis,2013).

Young people's high level of social consciousness are often underestimated, but they care about many things including climate change and the environment, the spread of HIV and other diseases, economic inequality and the lack of employment opportunities, social justice and human rights. Because of their ideals, those among them who are proficient with ICT and are motivated to create change have a propensity for linking the use of the technologies to development goals. However, without entrepreneurial skills, they lack the necessary ability to sustain their efforts and overcome challenges that stand in their way.

As enablers of ICT-facilitated development and as social entrepreneurs who want to make a difference in the world, empowered youth can be a potent force for change. Yet there are still few strategies and initiatives that truly empower young people to become leaders in creating sustainable social, environmental and economic impact worldwide. There are many initiatives that engage youth through conferences and connect them for virtual networking, but few go on to provide a comprehensive support structure that enables youth to create development impact for themselves and for others (Prelis, 2013).

The World Summit on the Information Society in 2003 acknowledged that young people are the future workforce, leading creators and early adopters of ICT, and that they must be empowered as learners, developers, contributors, entrepreneurs and decision-makers. The reality is that some young people are already part of the global workforce due to poverty and circumstances. A large number of them are unable to find employment. The International Labour Organization reported 76 million youth were unemployed in 2008, comprising 40% of total world unemployment. Little progress has been made in improving the position of youth in the labour market and young people still suffer disproportionately from a deficit of decent work opportunities. Moreover, youth who are employed often work long hours under informal, insecure and intermittent work arrangements characterized by meager earnings, low productivity and reduced labour protection. The issue of youth employment is and will continue to be an important variable when considering effective youth empowerment strategies and it makes a focus on entrepreneurship critical (Prelis 2013).

Information is the key to democracy. Information empowerment is fundamental to a successful democracy. The advent of information technology (IT) has nurtured the swift emergence of a global "information society "that is changing the way people live, learn, work and relate. Even after 60 years of India's independence, the most pressing problem for the country is still how to deal with its rural poor. Out of 1027 million people, 742 million (72.2%) live in rural areas and 285 million (27.8%) in urban areas as presented in Table 1 (Census of India, 2001).

Nowadays, when people think of India, the IT industry andBangalore comes to mind. Conversation is about how IT is modernizing India's economy and transforming the lives of millions of Indians. But this discussion refers only to one India—urban India. There is still the other India-composed of those living in rural India who has largely been left out of the IT revolution. India's agricultural sector employs over 65% of its workforce but contributes less than 25% to GDP. Indian farmers remain amongst the poorest in the world. The average annual income for a farm household in India was just under Rs.20 000.00p.a. ($420.00 p.a.) (NSSO, 2003 & Gollakota, 2008).

Investments in information and communication technologies (ICT) around the world have grown at a staggering rate over the past two decades. Information technology (IT) industry consultant, Gartner reported that worldwide spending in this area was on the order of US$3.4 trillion in 2010, rising by about 5.3% to US$3.5 trillion in 2011, reflecting a new emphasis on consumer mobile devices, virtualization solutions and security software (Burt ,2010).

Importance of ICT

Importance of ICTs for Rural Development given the importance of Information and Communication Technologies (ICTs) in national development, countries across the globe have put in place mechanisms such as Universal

Service Funds and other forms of Government intervention to achieve Universal Access to ICTs.These focus inter alia on bridging the digital divide between urban areas/populations and rural areas/populations.

The importance of ICTs is not the technology as such, but its enabling function in facilitating enhanced access to information and communication across large distances; improved access to governmental and quasi-governmental resources and services; opportunities to trade or bank online through kiosks; new opportunities to design, manufacture and market products through internet or intranet systems; increased and improved education through computers or about computers or both; superior medical advice and diagnostic information; information about local resources, opportunities to earn a better living by learning a new skill in the knowledge-based economy, improving agricultural productivity, etc., that are essential elements in today's economic and social interaction, conducting businesses, compete in markets and shape developmental priorities (Siriginidi, ,2002).

ICTs play a major role in a nation's politics, economy, social and cultural development. Over the years, role of ICTs both established (radio, television, video, compact disc) and emerging (wireless, internet, broadband)as a powerful if not indispensable tool in the massive scaling up and inter-linkage of development interventions and outcomes inherent in this objective has become recognized. ICTs offer enormous opportunities to narrow social and economic inequalities; support sustainable local wealth creation by overcoming obstacles of geographic isolation, lack of access to information and challenges in communication. Thus, ICTs fuel the global economy and relate to human rights, helping at best, to support freedom of expression and right to information according to Article 19 of the Universal Declaration of Human Rights. About 1.2 billion people are experiencing extreme poverty that is considered by many to be the worst human rights violation in the world (UNDP, 2005; United Nations, 2005).

Importance of ICTs by Hazra (2012)

1. Strengthening Rural Governance
2. Enhancing people's participation in nation building process
3. Encouraging social transformation
4. Ensuring a better quality of life
5. Strengthening the Information-base of rural communities
6. Intensifying Effort towards implementation of the rural development initiatives

Strategies for Sustained ICT Growth in India (Chowdary, J.A.)

1. Improving the quality of Education
2. Increasing the number of Universities and colleges
3. Industry Academic linkages
4. Providing High bandwidth Data-com Network
5. Tax incentives for R&D
6. Setting-up dedicated VC funds for ICT dev
7. Promotion of IP development

ICT Development Ranking of Selected Countries

India has been ranked 121st among 157 countries in terms of progress in the realm of information and communication technology (ICT) in a newly-released report of the International Telecommunication Union (ITU), which makes an annual assessment based on a wide range of parameters and data.

The ranking has been made on the basis of an ICT Development Index (IDI), a benchmark made up of 11 indicators. It comes in the wake of the Broadband Commission for Digital Development, in a recent report, ranking India 145th among nearly 200 countries in terms of the percentage of individuals using the Internet and 106th in the case of mobile broadband penetration. Though India's IDI score had gone up from 2.13 in 2011 to 2.21 in 2012 in the report 'Measuring the Information Society 2013 released on October 7, its ranking declined from 120 to 121 during this period (Ramachandran. T., 2013)

India's rankings in three sub-indexes which relate to access, skills and use, on which the IDI is based, are also in a similar range. In the skills sub-index, which evaluates ICT capability or skills in terms of indicators that include adult literacy as well as gross secondary and tertiary enrolment, India was ranked 117 in 2012, the same position it had occupied in 2011.

In terms of the use sub-index, which broadly measures the uptake of ICTs and the intensity of usage based on parameters such as Internet users per 100 inhabitants, fixed (wired)-broadband subscriptions per 100 inhabitants, and wireless-broadband subscriptions per 100 inhabitants, India has been ranked 121st, up from 124 in 2011.

And the access sub-index rankings place India in the 122nd position, down from 114 a year ago.

This sub-index has to do with ICT readiness in terms of fixed-telephone subscriptions, mobile cellular telephone subscriptions, international Internet bandwidth per Internet user, percentage of households with a computer, and percentage of households with Internet access. Among countries in the region, the report takes note of the gains Bangladesh has made, up by four places to 135th in the latest IDI, with maximum gains in the access sub-index.

The report includes India among a group of 39 least connected countries (LCCs) with low IDI 2012 values. "In these least connected countries, most ICT access and use is limited to basic voice and low-speed data services. While a number of least connected countries have reached relatively high levels of mobile-cellular penetration, more advanced ICT services, including broadband Internet access, remain very limited," the report said. Most of the countries in this group are the least developed ones in Africa.

In terms of overall global rankings, the Republic of Korea, with a value of 8.57, leads the world, a track record it has set in recent years. The Nordic countries Sweden, Iceland, Denmark, Finland and Norway follow with the Netherlands, United Kingdom, Luxembourg and Hong Kong (China) also finding a place among the top ten (Ramachandran. T., 2013).

CONCLUSION

This study should be seen only as a snapshot in the analysis of using ICT in youth and rural India. With any evolving technology, any study done is not predictive of future use, but is a review of progress at a certain point in time. Information and Communication Technology (ICT) has become one of the basic building blocks of modern society. It play very important role in rural India and the youth is synonymous with the need for change in this country. India has been ranked 121st among 157 countries in terms of progress in the realm of information and communication technology. It has importance in national development; countries across the globe have put in place mechanisms such as Universal Service Funds and other forms of Government intervention to achieve Universal Access to ICTs. Improving the quality of Education and Industry Academic linkages are main strategies for sustained ICT growth in India.

REFERENCES

Burt, J. (2010) Global IT spending to grow 5.3 percent in 2010: Gartner. eWeek.com, April12,2010.<http://www.eweek.com/c/a/IT-Infrastructure/Global-IT-Spendingto-Grow-53-Percent-in-2010-Gartner-839373.

Gulati, A. G. (2012) Role of ICTs in Rural Development. The Monthly Journalkurukshetraministry of rural development.Vol. 60, No. 3, Pages 3-7, January 2012.

Hazra, A. (2012) ICT: A Catalytic Intervention for Empowering Rural India. The Monthly Journal.Kurukshetra.Ministry of Rural Development.Vol. 60, No. 3, Pages 9-11, Jan 2012.

J A Chowdary (2102).ICT development in India.President and Managing Director, Portal Player Pvt. Limited.President of HYSEA.

Kamala Gollakota (2008) ICT use by businesses in rural India: The case of EID Parry's Indiagriline. International Journal of Information Management 28 (2008) 336– 341.

Neto, Isabel, Charles Kenny, SubramaniamJanakiram and Charles Watt. (2005) "Look before you leap: The bumpy road to E-development." In Robert Schware, (eds.), E-development: from excitement to effectiveness. Washington, D.C.: World Bank.

NSSO, Government of India (2003) Income, expenditure and productive assets of farmer households /http://www.mospi.nic.in/mospi_nsso_rept_pubn.htms.

Ramchandran.T. (2013) India 121st In Global Information and Communication Technology Rankings. The Hindu 8 October 2013.

Saji Prelis (2009) Empowering Youth to Change the World with ICT and Entrepreneurship.The United Nations Educational, Scientific and Cultural Organization, this 62-page radio toolkit.www.network foryouthintransition.org.

UnnatiGhia (2014) Statistics show youth to play major role in 2014 elections31 March 2014.http://www.dnaindia.com/india/report-statistics-show-youth-to-play-major-role-in-2014-elections-1971300.

Siriginidi, S.R. (2002) Knowledge management in India's rural community projects. In: Online Information 2002 Proceedings, Learned Information, Oxford, pp. 29–38.

UNDP (2005) Millennium development goals - MDGs. Retrieved April 4, 2007 from http://www.undp.org/mdg

United Nations (2005) What are the millennium development goals? Retrieved April 4, 2007 fromhttp://www.un.org/millennium goals

CHAPTER - 32

Leveraging the Potential of ICTs for Training and Development of Rural Youth

Rupesh Ranjan[1], Birendra Kumar[2] and Aditya[3]
[1]*P.G. Student (Agricultural Extension), GBPUA&T, Pantnagar, Uttarakhand*
[2]*Dean & Professor, Department of Agricultural Communication, GBPUA&T Pantnagar, Uttarakhand*
[3]*Asst.Professor-cum-Junior Scientist, Extension Education, BAC, Sabour, Bihar*

INTRODUCTION

According to the World Bank, youth present an opportunity for a sustainedeffort to participate in nation's development process because they possess greaterenergy and potential and have the capacity to drive positive change. Stimulating growthof employment in the agricultural sector remains paramount priority in countries with alarge agricultural sector, as improvements in agricultural productivity can generatemore and better jobs in most developing countries (World Bank, 2008).Youth who possess non-farm skills are 1.4 times more likely to move out ofagriculture. Their age has significant role in favor of shifting out of agriculture, becauseoccupational mobility was higher among younger farmers (Sharma and Bhaduri, 2012).India has the largest youth population in the world. About 70 per cent of India'spopulation is below the age of 35 years, making India the youngest nation in the World.According to some estimates, the proportion of population under 25 years in India is 51per cent. Some experts refer to this large proportion of youth as "demographicdividend" because greater proportion of the population is young and in the workingage-group which can lead to economic growth. However, out of the youth population of460 million, only 333 million youth in India are literate and unemployment rate ishighest (10.6 per cent) among youth (CIA World Factbook, 2012).

By 2020, the average Indian will be only 29 year old, whereas in China and theUnited States of America the average age is estimated to be 37 years. This high percentage of youth can be utilized for taking Indian agriculture to new heights by channelizing the creative energies of the youth through development of appropriate skills, knowledge and attitude. This predominance of youth in the population is expected to last until 2050. Hence, the National Youth Policy (2012) emphasizes youth empowerment in different spheres of national life. There are several advantages of involvement of youth in agriculture as they have the latent energy, capacity and ability to produce, propensity to learn and grasp new ideas or technologies faster and they areexcellent source of ideas and innovations (Government of India, Ministry of Agriculture, 2012).Youth are dynamics section of society full of vigour, hope and zeal to try out new ideas. In agriculture rural youth have been inspired for entrepreneurship in an agriculture, horticulture, livestock and allied sectors. In Bihar a lot has been done to propagate integrated farming, fishing, mushroom, bio-fertilizers', floriculture etc. It has been observed that in twenty KVKs of Bihar Agricultural University many youths have come forward as innovative farmers. However, scenario in general is not very sanguine. Rural youths are become averse to farming and wish to go to urban areas in search of greener pastures in many states rural youths due of lack of education and proper opportunities of training are engaged in passive leisure, addicted to drugs, alcohol and gambling. What need to be done to attract them to agriculture and enhance production to a level that it is remunerative.Recommendation of the Planning commission of India's working group on agricultural extension for XI five year plan (2007-2011) states that the agricultural growth is stagnating and sluggish. Hence, there is an immediate need of vibrant, dynamic and innovative approach to be adopted for agricultural extension in order to achieve targeted growth rate and serve the farmers better. To ensure the future viability of the agricultural sector, tackle rural poverty and generate employment opportunities, it is crucial to equip the farmers of tomorrow. ICT in agriculture is an emerging fields focusing on the enhancement of agriculture and rural development. It involves applications of innovative ways to use ICT in rural domain. The advancement in ICT can be utilized for providing accurate, timely, relevant information and services to the farmers, thereby facilitating an environment for more remunerative agriculture. The paper aims to discover how Information and Communication Technology can spread the message of innovative remunerative entrepreneurial ventures among rural youth to build their competency and inspire them.

Significance of ICT in Agriculture

There is lot of criticisms about the way extension personnel of the department of agriculture disseminate the technologies and messages to the farmers through various extension methods. Most national governments and agricultural stakeholders are trying to revitalize their extension systems. However, estimates in India indicate that to disseminate advanced agricultural technological information to its 120 million farm holdings requires at least 1.3 to 1.5 million extension personnel against which present availability is only 0.1 million (Planning Commission, GOI, 2007). In India, during 2004, public extension (including the KrishiVigyan Kendra centers) reached only 6.4% of the farmers, who instead mostly got their information from other progressive farmers and input dealers (17 and 13%, respectively) and mass media (Binswanger-Mkhize and Zhou, 2012). Further, even after continuous efforts of extension organizations, 60% of the farmers do not access any source of information for advanced agricultural technological information, resulting in a huge adoption gap (NASSO, 2005). The diversity of agro-ecological situations of the country adds to the challenge. Farmers' needs are much more diversified and the knowledge required to address them is beyond the capacity of the grass root level extension functionaries. Hence, extension reforms are underway and integration of new Information and Communication Technologies (ICTs) are rapidly transforming agricultural extension.

There is need of such technologies in rural areas for increase the production of agriculture especially in Indian countries it has provided average access to agricultural information where farmers are getting many problems in connectivity of communication technologies (Meera, Jhamtani & Rao, 2004). ICTs have brought significant changes in agriculture development and transfer information and knowledge through various technologies among farmers (Birkhaeuser, Evenson & Feder, 1991). ICTs have potential to disseminate the agricultural systematic information among smallholder farmers similarly the mobile phones, television, Internet and radio have the facility to transfer related and timely information that helps to make decisions to use resources in the most productive and profitable way (Ekbia & Evans, 2009; Ommani & Chizari, 2008).It is only through the introduction of ICT that information can also be updated and extended at the lowest cost. Similarly those who have used the e-services, e-commerce applications also increased their income (Chapman and Slaymaker, 2002 B Sideridis, 2010). There are several ICT models in Indian agriculture, which have made significant difference to agricultural operations (Meera, Jhamtani & Rao, 2004).

ICT for Agricultural Extension Work in India

The reports indicated that 45 per cent of the world's ICT projects implemented in India(Manzar, 2004). And also Asia's highest number of information kiosks has been implemented across rural India. However, the most rural ICT projects are implemented in the socio-economically developed states of South and North India (Paul *et al.*, 2004). Numbers of ICT initiatives are documented in the literature (Saravanan, 2010a; Saravanan *et al.*, 2011; Sulaiman, 2012).

ICT initiative include creation of Web portal for specialized agriculture knowledge like aAQUA , KISSAN Kerala, TNAU AGRITECH Portal , AGRISNET, e-Krishi, India Development Gateway (InDG) portal, ASHA, Rice Knowledge Management Portal (RKMP) and Agropedia.Web Portals for Market Information and Agri-Business Firms' Portal to Farmers includesAGMARKNET, ITC-e-Choupal, EID Parry- Indiagriline, Indiancommodities.com/, Mahindra KisanMitra, IFFCO Agri-Portal, Agrowatch Portal AND iKissan.Some notable projects have been in village communication use ICT for provided relevant information to farmers on demand likeVillage Knowledge Centres (VKCs)-M.S. Swaminathan ResearchFoundation (MSSRF), Village Resource Centres (VRCs) –Indian Space Research Organization (ISRO), Community Information Centres (CICs) and Common Service Centres (CSCs).Used Telephony/ Mobile Telephony in providing information to youth includes Farmers Call Centre (Kissan Call Centre), Lifelines India, IFFCO Kisan Sanchar Limited (IKSL), Fisher Friend, Reuters Market Light (RML) and Mobile Advisory Services by KrishiVigyanKendras KVKs) of Indian Council of Agricultural Research (ICAR). However there have been some Hybrid Projects (Mix of ICTs, Info-mediaries& Conventional Extension Methods) such as e-Arik, e-Sagu, Digital Green and Knowledge Share Centers. So, in past …one decade, intensive efforts have been made in India to use ICT for serving knowledge information needs of the farmers. In India, by different type of organization such as ICAR has established IP-Telephony and Video Conferencing at 23 ICAR Institutes with Head Quarters. VSAT Connectivity is provided to 200 KVKs through ERNET, KVK and ICAR Network. Up gradation of ICAR-ERNET network involving all ICAR Institutes and SAUs is on-going under NAIP Project Component.

ICAR has to focus on development of situation specific (district wise) virtual repository of information thus taking farmers technological contributions on a world/ national platform. Sharing farmers' experiences from region to region through internet/intranet, online availability of advisory services and integration of PC and mobile services also has to be initiated in all the KVKs. This can be achieved through development of situation specific content/ information, ensuring satisfactory connectivity, adequate training to all SMS, developing static multimedia content, developing ethics, codes and standards for content hosting, registering exclusive common domain and web space for KVKs for web hosting.

ICTs can play catalyst role in providing new training and employment opportunities for youth. There are a number of successful initiatives from both developed and developing countries that endeavor to provide support for young people in developing skills and employment opportunities. Some of these initiatives are clustered and presented in four groups. These groups include initiatives on (a) providing ICT and skill training (b) ICT employment generation through entrepreneurship (c) using ICT-based employment opportunities for youth (d) Mobile phones foster new ways for business linkage.

One of the many successful initiatives that provide Training and Agriculture Technology information Through ICTs in the Bihar Agricultural University (BAU), Sabour, Bhagalpur, Bihar. Under AKDS, electronic media center have been established at BAU, Sabour with facilities with audio and video multimedia production system. Agriculture Technology and Knowledge generated by the university have been stored in audio and video in multimedia format. A website www. Kishangyan.com. has been specially hosted for the farmer to provide appropriate information suitable for agro ecological situation of bihar.All the important Agriculture Technology information have been uploaded on the site and there updated from time to time. Beside all the 20 KVKs under the jurisdiction with KVKs have been linked with electronic media center at the head quarters to provided standardized training. It possible to the provide training and media support from the center to the different KVKs. beside SMSs facility have been provided to farmer through mobile services.

Gandhi et al., (2008) indicated that the Digital Green project (participatory digital video for agricultural extension). The project increased the adoption of certain agriculture practices sevenfold over classic extension approaches. The Digital Green project was shown to be 10 times more effective per dollar spent and created positive social side effects for agricultural extension. Digital Green also facilitate knowledge exchange between community engagement for partners looking to learn, contribute and connect on social innovation practices toward improving lives in rural. It video technology with community organization to train them on sustainable agriculture. In four years, Digital Green has been able to achieve the following:

- Reached 2,000 villages across India, Ethiopia and Ghana
- Produced more than 2,600 videos in 20 different languages
- Improved lives of more than 160,000 community members (70% women) in South Asia and Sub-Saharan Africa
- Increased the cost-effectiveness of development efforts by 10 times.

The aim to reach 10,000 villages in India and Sub Saharan Africa, enabling the communities we reach to live with dignity.

Village Knowledge Centre (VKC) was set up in Yagrung village in East Siang district of Arunachal Pradesh State. After three years of implementation, around 42% of e-Arik(e-Agriculture) beneficiaries reported increased production of rice and 29% reported increase in Khasi mandarin crops. It was estimated that on an average, the income increased by Rs.1689 for Rice crop and Rs. 5251 for KhasiManadarin trees, among 500 e-Arikbeneficiaries. Besides improvements in farm, the e-Arik advisory services helped farmers save a lot of time and cost in accessing information. In comparison to the conventional extension services, farmers in this system saved around Rs.2400 per year on their travel cost. (Saravanan, R.2013).

The ICT based agricultural education has the potential of empowering farming communities by improving access to information and knowledge sharing.The e-Arik project staff regularly undertook field visits to observe crop conditions and to diagnosis pests, diseases, nutrient deficiencies and physiological problems. They could then digitally document these issues using ICTs in the field and, via email and webcam, communicate them to staff at the e-Arik Research Laboratory at the Central Agricultural University.Problems were analyzed by the experts (who themselves sometimes also undertook field/advisory visits) and recommendations were passed on to the e-Arik Village Knowledge Centre by email and then to the concerned farmers by phone or personal facetoface communication by the farmerfacilitators.Dissemination of information and good practice was also addressed by innovative approaches such as farmertofarmer communication and local selfhelp groups.

SEVAa successful initiative from India has shown that actually trade unions can play important roles in providing ICT training to their young members. This would lead to job security and income protection too many young entrepreneurs.SEWA (Self employed women's association) has been one of the first organizations in India to realize the potential for harnessing ICT to help women in the informal sector. It has sought to develop the organization's capacity to use computers by conducting awareness programs and imparting basic computer skills to its team leaders, SEWA's main goals are to organize women workers for full-employment whereby workers obtain work security, income security, and social security (at least health care, child care and shelter). SEWA organizes young women to ensure that every family obtains full-employment and self-reliance both economically and in terms of their decision-making ability.

In the view of the discusses made in the proceeding section about state of rural youth and use of ICT in agriculture, it is now imperative to motivated rural youth to take up constructive and productivity activity since of large majority in rural area entrepreneur activity in agriculture can be very good solution .there is growing literature about use of ICT for capacity building of youth adoption of used of mobile phone increased in rural area and there by internet accessed also growing. Just it is possible advance of technology to provide need based, skill oriented, entrepreneurial activity in agriculture and related field like horticulture, livestock, home science, fisheres, etc. It is provided fact youth can easily learned the tidbit of ICT and apply them day to day life to accessed information advertised projected, advertised product, tract marketing information and expand their enterprises.

There is a great scope to implement ICT in order to communicate and integrate the complete agri-food supply chain, as the e choupals are doing in Madhya Pradesh to procure soyabean. The other beneficiaries of ICT can be food-processing companies, and suppliers within the agri-food sector. On the other hand, the need to market the agricultural produce at reasonable prices will also change the farmers' attitude, and they will be more dependent on ICT. ICTs will, thus, definitely help the KVKs in sustaining the Indian agriculture.In that perspective the following generalized recommendations might be useful in forming policies and in addressing some of the youth education, skill and employment problems of the developing countries.

Firstly, promoting youth employment and employability requires important integrated effort that includes actions in the areas of education, skills development, job supply and support for young low-income entrepreneurs, particularly in the knowledge intensive sectors. For being successful the initiatives must provide participation by various stakeholders, from state to private entrepreneurs, including but not limited to international and national NGOs, local authorities, youth leaders, the media, and the public.

Secondly, it is clear that there is an extensive potential for ICTs to generate employment for young people. However, this potential will not be realized unless a country has a range of supporting strategies in place, including an enabling environment. The enabling environment includes opportunity to participate in skill training in technology, access to data and information, economic support to young people who need to develop or improve skills, fair distribution of jobs and wages, free education for the ones that cannot afford, and quality work and training environments.

Thirdly, ICTs offer developing countries the opportunity to close the gap with developed countries and narrow the global digital divide. Applying ICTs in education is key to provide young people with ICT skills. Therefore, governments need to ensure that they provide quality education to all, and ICTs are integral components of education planning and school curricula.

Fourthly, the participation of young people in the development and implementation of initiatives involving the use of ICTs to generate employment is likely to be a key factor in the success of such initiatives.

Fifthly, mentor support for starting ICT-related enterprises is a key service that governments,

NGOs or international organizations could organize to provide advice and guidance to young entrepreneurs.

Sixthly, partnership with international organizations such as United Nations and its agencies may help in implementing new best practices. These organizations assist developing countries and regional institutions of developing countries in building global, regional, national and local networks of partnership.

Lastly, investing only on technology is not the solution of the young population's problems, investing on social and human capital makes a better change. As Mansell and When (1998) indicated investing in mere technology and not human capital will only lead to increase in energy consumption and not in economic growth and employment opportunities for young people.

REFERENCES

A, R, Chhachhar. et al. (2014) Impact of Information and Communication Technologies in Agriculture Development. J. Basic. Appl. Sci. Res., 4(1), 281-288.

Andersen, B., Howells, J., Hull, R., Miles, I. and Roberts, J. (2000). (Eds.), Knowledge and Innovation in the New Service Economy. Edward Elgar: Cheltenham.

Castells, M. (2000) End of the Millennium: The Information Age Economy, Society and Culture.Volume 3.Blackwell: Oxford.

Curtain, R. (2000) Identifying the basis for a youth employment strategy aimed at transitional and developing countries. Report commissioned by the United Nations Social Development Division.

Curtain, R. (2003) Creating more opportunities for young people using information and communications technology. The World Summit on Information Societies, Geneva, 12-14 December 2003

E. Gelb et.al. (2008) Adoption of ICT Enabled Information Systems for Agricultural Development and Rural Viability..ICT Adoption Workshop at the IAALD-AFITA-WCCA Conference 2008.Atsugi Japan.

Ezeh Ann Nnenna (2013) Access and application of information and communication technology (ICT) among farming households of south east Nigeria.Agric. Biol, J. N. Am., 2013, 4(6): 605-616.

G. Adamides. And A. Stylianou (2013) ICT and Mobile Phone Use for Agricultural Knowledge Sharing by Cypriot Farmers.Agris on-line Papers in Economics and Informatics, 2(5), 3-10.

Guttman, C. (2003). Education in and for the Information Society. United Nations Educational,Scientific and Cultural Organization, Paris.

Lever, W. (2002).Correlating the Knowledge–base of cities with Economic Growth. Urban Studies. 39(5/6): 859–870.

Mansell, R. and When, U. (1998). (Eds.), Knowledge Societies: Information Technology and Sustainable Development. Oxford University Press: Oxford.

Mobbs, P. (2002) Glossary and Cross-Reference Index.GreenNet Civil Society Internet Rights Project. Accessed on 6 June 2006 from www.internetrights.org.uk.

Morris, P. (2000) World Wide Work: Knowledge-intensive Globally Distributed Work.Department of Industry Science and Resources, Canberra

Naamwintome, B.A and Bagson, E (2013) Youth in agriculture: Prospects and Challenges in the Sissala area of Ghana.Net Journal of Agricultural Science.1(2), 60-68

Najafabadi,M.O., Poorsadegh, M and Mirdamadi, A.M.(2013) Challenges of Application ICTs in Technical and Vocational Training from Students' and Instructors' Perception in Maragheh.International Journal of Advanced Science and Technology, 54:105-111.

Raj, S. (2013) e-Agriculture Prototype for Knowledge Facilitation among Tribal Farmers of North-East India: Innovations, Impact and Lessons. Journal of Agricultural Education and Extension, 19(2), 113-131.

Saravanan, R., e-Arik: Using ICTs to Facilitate "Climate-Smart Agriculture" among Tribal Farmers of North-East India, ICTs and Agricultural Adaptation to Climate Change Case Study, 2011, Centre for Development Informatics, University of Manchester, UK. www.niccd.org/NICCD_AgricAdapt_Case_Study_eArik.pdf.

Shaik N. Meera, Anita Jhamtani, and D.U.M. Rao, (2004). "Information and Communication Technology in Agricultural Development: A Comparative Analysis of Three Projects from India" Agricultural Research & Extension Network, Network Paper No.135, 1-20.

V. Venkatasubramanian. et.al. (2009) Concepts, Approaches and Methodologies for Technology Application and Transfer - A Resource Book for KVKs. Zonal Project Directorate, Zone – iii, ICAR, Umiam, Meghalaya

World Conference On Agricultural Information and IT World.

Yigitcanlar, T. and Baum, S. Providing Youth with Skills, Training and Employment Opportunities through ICT Initiatives

CHAPTER - 33

Gender Issues and Access to ICTs for Agricultural Farming and Rural Development in Gazipur District of Bangladesh

Foyez Ahmed Prodhan[1] and Md. Safiul Islam Afrad[2]
Lecturer[1] and Professor[2], Department of Agricultural Extension and Rural Development, Bangabandhu Sheikh Mujibur Rahman Agricultural University Gazipur-1706, Bangladesh

ABSTRACT

This study critically examined the access of women farmers to Information and Communications Technology (ICTs) as against their male counterparts. The study was conducted Kapasiaupazila of Gazipur district. A total of 120 respondentswere selected using proportionate random sampling technique and interviewed through structured questionnaires. Data collected were subjected to descriptivestatistics, Chi-square and rank test. The results indicated that the women folk, despite their dominancein agricultural activities, were dominated in terms of access to and full participation in the use of ICTs.In terms of access to agricultural information through ICTs, it was evident that, people in the study area capture the advantages of increased number of ICTs to access information related to their farming business. Most of respondents valued ICTs as easy, fast and convenient way of communicating agricultural information and also allow more contacts among the farmers for family farming along with increase income of the family people in the community. Chi-square analysis on the relationship between socio-economic characteristics of the respondentswith their access to ICTs showed that age, education and innovativeness have significant relationship with therespondents' access to ICTs. Kruskal Wallis rank test showed that there was a significant difference inthe gender access to and the uses of ICTs

(P d" 0.005) with the male dominating the female gender. Chi-squareanalysis revealed that age and educational background and innovativeness of the respondents had significantrelationships with their access to ICTs (P d" 0.005).On the other hand religious/cultural barriers, language barriers and cost of possessing and accessing ICTs were the main reasons for gender disparity in gaining access to ICTs.

Keywords: Gender, ICT, agricultural farming, rural development

INTRODUCTION

Gender is an international issue operating in all fields of life. Gender refers to the roles and responsibilities of men and women and the relationship between them. Gender does not simply refer to women or men, but to the way their qualities, behaviours and identities are determined through the process of socialization (Saghir *et al.*, 2009).It is often argued that the concerns for gender and ICTs in developing countries are not a compelling one, and should be secondary to meeting basic needs. This contention, however, doesn't quite hold, as it is not and either' 'or' situation, and both need to be addressed simultaneously. The major problems lie with developing countries is poverty, illiteracy, lack of access to ICTs infrastructure. However, ICTs can offer significant opportunities for girls and women, including those in rural areas; their ability to take advantage of these opportunities is contingent upon many things. Extension of infrastructure, particularly wireless and satellite communications, to rural areas and semi-urban areas is vital to increasing women's access to information technology (Singh, 2003). Communication interventions play the key role for gender development in various fields of life. Like any other field, ICTs has various recent advances, such as: mobile phone, cable, internet, computer, digital devices, and cyber space (Elijah and Ogunlade, 2007). These interventions are fruitful in working of several fields of life like medicine, research, education, industry and manufacturing. But the usage of latest interventions in agriculture technology transfer is very poor. Agricultural research in one part of the world can give benefit to the farmers of other country through effective utilization of these technologies (ObayeluandOgunlade, 2006). In agriculture, like in many other sectors, information is becoming a major input, whilst, knowledge and information plays a central role for farmers to respond to opportunities that could improve their agricultural productivity. Information communication technologies (ICTs) are therefore continued to be the best hope in developing countries to accelerate their development process (Nyamba and Mlozi,2012).The history of ICT (Information and Communication Technology) use in the Agriculture of Bangladesh is not so rich. In 2003, Support to ICT, task force program launched by the Ministry of Agriculture (MOA BD). In Bangladesh, private sector operators are the main providers of ICTs (mobile phones, computers and internet, television channels, radio, and fixed-line telephony on a limited scale), whereas the state controls the fixed-line telephony and two national TV channels and 10 radio centers. The government also formulates and implements ICT policy. Recognizing the role that information can play in improving the livelihoods of the poor, NGOs began to look at tele-centers as a means of information sharing (Saiful, 2013). The economy of Bangladesh is mainly agricultural. Agriculture is the largest sector in Bangladesh, comprising almost 80% people's involvement by contributing 21-23% of the GDP. About 80 percent of the total population lives in rural areas, and 62 percent of them are directly, and others are indirectly engaged in a wide range of agricultural activities.The female contribution to the overall economy, particularly in agriculture is 59 per cent.In Bangladesh, being a traditional Muslim society, women in Bangladesh hardly participate in

agricultural activities outside home (Jaim and Hossain, 2011). Women's agricultural activities were confined to homestead production and post-harvest operations; however, in recent years they are mostly involved in livestock and poultry rearing activities besides crop production activities.For Information and Communication Technologies (ICTs) to benefit women in agricultural production and to challenge existing gender imbalances in rural livelihoods, it is necessary to understand women's status and the gender roles and responsibilities in the society.Women are more likely to work in agriculture than in other sectors in most regions.Yet, accordingto the Food and Agriculture Organization of the United Nations (FAO), even though women are major producers of food crops, in most of the world, women lag wellbehind men in ownership of agricultural land and access to income from land. This lack of access can also be applied to the area of information access and the abilityto share and learn from this knowledge.These factors should be considered when understanding the complexities of the multiple gender dimensions that impact women's accessto and use of ICTs in the areas of agriculture andrural development (Anonymous, 2010). Hence, this research is designed to investigate the gender access to ICTs and problems women encounter in gaining access to ICTs.

Objectives of the Study

The objectives of this study are to:

1) determine the socio-economic characteristics of the respondents in relation to their access to ICTs;
2) identify gender access to ICTs in the study area and
3) inspect the reasons for disparity in gender access to ICTs

METHODOLOGY

This research was carried out in Kapasiaupazila under Gazipur district of Bangladesh. The choice of this area for this research is drawn from the fact agricultural activities are done based on gender division of labor. The people of this area are predominantly small-scale farmers. They engaged themselves in both crop and animal production for domestic and commercial purposes. A structured questionnaire was administered while oral interview complemented the questionnaire. A total of 120 respondents were selected from a population of 480 following proportionate random sampling technique.Descriptive statistics was used to analyze some of the data, Kruskal Wallis rank test was used to investigate whether there was a significant difference in gender access to ICTs and Chi-squareanalysis was used to test the relationship between the socioeconomiccharacteristics (age, education, farming experience,innovativeness and cosmopoliteness) of the farmers and their access to ICTs. The formula for computing Krukalwalis rank test is:

$$H = \frac{12}{N\ (N+1)} \frac{\Sigma R2}{ni} - 3(n+1)$$

Where, N = Total number of observations over all samples, R^2 = square of the sum of rank for samples i, n_i = Sample size of sample i, Ó = summation sign

RESULTS AND DISCUSSIONS

Personal Characteristics of the Respondents

Characteristics profile of the farmers were determined and presented in Table 1.

Table 1: Distribution of the respondent according to their socio-economic characteristics

Character	Measuring unit	Categories	No. of respondents	Percent	Mean	SD
Age	Actual year	1-20	40	33.33	34.10	5.4
		21-40	50	41.67		
		41-60	20	16.67		
		>61	10	8.33		
Sex	Actual	Male	64	53.33		
	number	Female	56	46.67		
Education	Year of	Non-formal	35	29.17		
	schooling	Primary (5)	40	33.33		

Contd...

		Secondary (6-10)	30	25.00		
		Higher (>10)	15	12.50		
Farming experience	Year of farming	1-8	38	31.67	10.75	6.01
		9-16	45	37.50		
		17-24	22	18.33		
		>24	15	12.50		
Innovativeness	Scores	Up to18 (Low)	30	25.00	24.49	6.97
		19-30 (Medium)	55	45.83		
		31 and above (High)	35	29.17		
Cosmopoliteness	Scores	Up to 15(Low)	43	35.83	19.57	4.37
		16 to 25(Medium)	47	39.17		
		26 and above (High)	30	25.00		

Age of respondents

Table 1 revealed that 33.33% of the respondents were within the ages 1 to 20 years, 41.67% said they were within ages of 21 to 40 years, 16.67% were found to have fallen within the ages 41 to 60 and those that fell within the range of ages over 60 years were only 8.33% of the respondents. This implies that majority of the respondents were in their youthful and active age which will enable them participate in ICTs efficiently.

Sex of the respondents

Table 1 showed that the respondents were dominantly male (53.33%) while 46.67% were female. This result implies that all things being equal, both genders should have access to ICTs.

Educational background of the respondents

Table 1 indicated that 29.17% of the respondents never obtained any form of formal education, 33.33% of them said they attended primary school, 25.00% of the respondents said they attained secondary school education while only 10.00% of the respondents attained secondary education and 12.50% attained higher education. Most of the respondents attained one form of educationor the other at least a primary school education. If their educational background can be improved, it will go a long way in enhancing their efficiency in the use of ICTs like Computer/ Internet require some form of education or skills to operate them.

Farming experience

Table 1 reflected the farming/marketing experience of the respondents of which 31.67% had the experience of 1 to8 years, 37.50% had 9 to 16 years of experience, 18.33%had 17 to 24 years while only and 12.50% had 25 years and above. Several studies have shown that the more experienced a farmer are in farming business, themore exposed he is to sources/channels of information (Ibrahim *et al.*, 2009). This will avail the farmer of the opportunities in accessing agricultural information.

Innovativeness of the respondents

Table 1 revealed the innovativeness of the respondents of which 25.00% of the respondents were within the ranges of up to 18 (low) innovativeness, 45.83% within ranges of 19 to 30 (medium) and 29.17% were found to have fallen within the ranges of 31 and above (high) innovativeness. The highest proportion (75.00%) of the respondents in this study had medium to high innovativeness which will enable them participate in ICTs efficiently. Because for a respondent, innovativeness is a characteristic that helps him consider and except new technologies earlier than other members of the society.Respondents who are low in innovativeness are likely to face difficulties in adopting modern technologies.

Cosmopoliteness of the respondents

Table 1 reflected the cosmopoliteness of the respondents of which 35.83% of the respondents were within the ranges of up to 15 (low) cosmopoliteness, 39.17% within ranges of 16 to 25(medium) and 25.00% were found to have fallen within the ranges of 26 and above (high) cosmopoliteness. This implies that highest proportion (75.00%) of the respondents in this study had low to medium cosmopoliteness. Through cosmo politeness quality an individual becomes aware of the recent information which will help in adopting and accessing new ICTs.

Access to ICTs of the respondents

In Table 2, the male gender as against their female counterpart, dominantly have access to mobile phone (75.00%), television (33.33%), radio (29.17%),CD/ DVD (19.17%), camera (10.87%), and computer/internet (5.03%). This result generally indicated a low access for both male and female alike tothat of contemporary ICTs except for that of mobile phone werethere was a relatively appreciable access to ICTs. This implies that if the enabling environment is provided for contemporary ICTs like computer/internet, it will also blossom and spread even to the rural areas as in thecase of mobile phone. Information as reflected in Table 3, Kruskal Wallis Rank Test H =71.92, X^2 = 13.08, P d" 0.05 showed that there was asignificant difference in the access of male and female gender to ICTs with the male gender dominating. Chi-squareanalysis on the relationship between the farmers'socio-economic variable and access to ICTs as indicatedin Table 4 revealed that age, education and innovativeness are have significant relationship (P d" 0.05) with their access toICTs. However, farming experience and innovativeness does not have significant relationship (P d" 0.05) with their access to ICTs as seen in several studies. This is due to the factthat farmers are seen to have lower access to particularly contemporary ICTs (computer/internet, camera and except mobile phone)than they do to the conventional ICTs (radio and television).

Table 2: Distribution of respondents according to access to ICTs

ICTs	Young male	Adult male	Young male	Adult female	Total male (%)	Total female (%)	Dominant gender
Mobile phone	35 (29.17)	55 (45.83)	10 (8.33)	15 (12.5)	75	20.83	Male
Television	25 (20.83)	15 (12.5)	18 (15)	13 (10.83)	33.33	25.83	Male
Radio	20 (16.67)	15 (12.5)	5 (4.17)	-	29.17	4.17	Male
CD/DVD	14 (11.67)	9 (7.5)	10 (8.33)	5 (4.17)	19.17	12.5	Male
Camera	11 (9.17)	2 (1.7)	3 (2.5)	7 (5.83)	10.87	8.33	Male
Computer/ Internet	4 (3.33)	1 (0.83)	2 (1.7)	-	5.03	1.7	Male

Table 3: Kruskal Wallis rank test results on gender access to ICTs

ICTs	Young male	R_1	Adult male	R_2	Young male	R_3	Adult female	R_4	Over all mean rank
Mobile phone	35	20	55	13	10	22	15	21	19.75
Television	25	15	15	17	18	11	13	13	13.67
Radio	20	14	15	11	5	9.5	-	-	13.12
CD/DVD	14	8	9	9.5	10	13	5	7	9.68
Camera	11	12	2	7	3	4.5	7	9	7.25
Computer/ Internet	4	7.5	1	6.5	2	6	-	-	4.00
Total	109	76.5	97	64	48	66	40	50	

H=71.92 X^2=13.08 Pd"0.05

The significant relationship in age with accessto ICTs as indicated in this study agrees with Omotayo (2011) who said that younger people are more disposed to ICTs especially the modern ICTs (computer/internet and camera) than the older ones hence, age of the farmers determine to some extent their accessibility to ICTs. Ibrahim *et al.* (2009) confirmed that education does notonly affect women accessibility to credits and other factors of production but also, affect their accessibility tosources/channels of agricultural information (radio,television, extension agents, newspaper).

Table 4: Chi-square analysis showing the relationship between socio-economic characteristics of respondents and access to ICTs

Socio-economic variables	Access to ICT	
	X^2-calculated	X^2-tabulated
Age	24.31**	
Education	33.17**	
Farming experience	12.23	27.58
Innovativeness	37.19**	
Cosmopoliteness	8.56	

*= significant (Pd"P)

Reasons for disparity in gender access to ICTs

Result presented in Table 5 showed that female were the dominant gender in the reasons for low participation in ICTs which include: Religious/Cultural barrier (Female), cost of possessing and accessing ICTs (Female), Gender prejudice (Female), Illiteracy (Female), and only in language factor that the male became the dominant gender. This implies that women have more constraints in gaining access to ICTs than their male counterpart and these are the reasons for gender disparity in access toICTs.

Table 5: Reasons for gender disparity in gaining access to ICTs

Reasons	Male	Female	Dominant gender
Religious/cultural	25 (20.84)	79 (65.84)	Female
Language barrier	58 (48.34)	36 (30)	Male
Cost of possessing and accessing ICT	41 (34.17)	79 (65.84)	Female
Literacy	31 (25.84)	63 (52.50)	Female

CONCLUSION AND RECOMMENDATIONS

Findings of the study reveals thatthe male gender as against their female counterpart dominantly have access to mobile phone,television, radio,CD/DVD, camera, and computer/internet. This result generally indicated a low access for both male and female alike to that of contemporary ICTs except for that of mobile phone werethere was a relatively appreciable access to ICTs. It was also found that there was a significant difference in the access of male and female gender to ICTs with the male gender dominating. Chi-square analysis on the relationship between the farmers'socio-economic variable and access to ICTs as indicated in Table 4 revealed that age, education and innovativeness are have significant relationship (P d" 0.05) with their access to ICTs.Female and male farmers in the study area face similar problems, but they affect the female farmer more adversely. Information is essential for facilitating agricultural and rural development and bringing aboutsocial and economic change. Unfortunately, the female gendersin Kapasiaupazila of Gazipur Districtwho participate more in agricultural activities haveless access to Information and Communication Technologies(ICTs) compare to their male counterpart as reflected results of this research. This inequality in gender access to ICTs in the study area can be addressed by the following recommendations:

(i) Involvement of women in decision making and policy formulation in the development of agricultural sector especially at the grassroots (Local Government Areas) where, particularly the issue of their access to ICTs and other factors of production.

(ii) Empowering women in the rural areas so as toimprove their agricultural production; and this will in turnbetter their socio-economic status to afford some of these ICT facilities.

(iii) Education of the girl-child and adult women in thestudy area is important, because it is a function of women empowerment and development toward gaining accessto ICTs. Especially in the area of computer literacy programs.

(iv) Government should intensify its effort on the policies of liberalization and deregulation of the economy to incorporate ICT infrastructure, so that the prices of ICTfacilities like the television, telephone sets, computer andother accessories will be affordable by both male and female farmers alike at the grassroots.

REFERENCES

Anonymous (2010) Gender, Information and Communication Technologies (ICTs) and Rural Livelihoods. Available at: http://www.fao.org/docrep/017/ap856e/ap856e.pdf

Elijah, A. O.and I. Ogunlade (2006) Analysis of the uses of information and communication technology for gender empowerment and sustainable poverty alleviation in Nigeria, International Journal of Education and Development using ICTs 2 (3).

Ibrahim, M. K., Adejoh S.O. and Edoka M. H.(2009) Sources and Uses of Agricultural Information among Rural Farmers in ijumu Local Government Area of Kogi State, Nigeria. Proceedings of the 43rd Annual Conference of the Agricultural Society of Nigeria (ASN) held on the 20th–23rd October, 2009, at Abuja, Nigeria.pp. 558 -561.

Jaim, W. M. H. and H. Mahabub (2011) Women's Participation in Agriculture in Bangladesh 1988-2008: Changes and Determinants. Paper presented in the pre-conference event on "Dynamics of Rural Livelihoods and Poverty in South Asia", 7th Asian Society of Agricultural Economists (ASAE) International Conference Hanoi, Vietnam, October 12, 2011. Available: http://research.brac.net/publications/final%20hanoi%20paper.pdf

Nyamba, S. Y. and Mlozi Malongo R. S. (2012) Factors Influencing the Use of Mobile Phones in Communicating Agricultural Information: A Case of Kilolo District, Iringa, Tanzania.International Journal of Information and Communication Technology Research, 2(7):559-563. Available at: http://esjournals.org/journaloftechnology/archive/vol2no7/vol2no7_3.pdf

Obayelu, A. and Ogunlade, I. (2006). Analysis of the Uses of Information Communication Technology (ICT) for Gender Empowerment and Sustainable Poverty Alleviation in Nigeria. International Journal of Education and Development using ICT, 2(3). Available at: http://ijedict.dec.uwi.edu/viewarticle.php?id=172%20amp;layout=html&layout=html

Omotayo, O. M. (2011) ICT and Agricultural Extension: Emerging Issues in Transfering Agricultural Technology in Developing Countries. In: Madukwe, M.C. (ed). Agricultural Extension in Nigeria.Published by Agricultural Extension Society of Nigeria (AESON). pp. 200–215.

Saghir, A., M. Ashfaq and A. Noreen. (2009) Gender and Information and Communication Technologies (ICTs). The Journal of Animal & Plant Sciences,19(2): 94-97. Available at: https://www.apc.org/en/system/files/Genardis_EN.pdf

Saiful. (2013) Use of ICT in agriculture of Bangladesh. Available at: http://agriculture and farming. wordpress.com/2013/10/05/use-of-icts-in-the-agriculture-of-bangladesh/

Singh, R. (2003) Gender Issues are Central to the empowerment dialogue, Comments to Digital Opportunity Channel: Providing opportunity to All, Ministry of Agriculture, Government of India.

CHAPTER - 34

ICT Tools in Agriculture: Opportunities and Challenges

Mohammad Hashim, Raghuveer Singh Meena and Pramod Kumar Prajapati

Research Scholars, Department of Extension Education, Institute of Agricultural Sciences, Banaras Hindu University, Varanasi- 221005

ABSTRACT

ICT innovations are bringing new information services to rural areas on which, farmers, as users, will have much greater control than ever over current information channels. A greater opportunity to access the services of ICT innovations is crucial requirement for the sustainable development of the farming system, which is the demand of the time. ICT when applied to the rural and agricultural conditions can improve the linkages between research and farmer sub-systems. The variety of new ICT tools for agriculture is impressive, but the tools need to be placed in an overall context of agricultural information and service needs. By looking at the critical information needs of agriculture and farming communities, the focus should be on applicable and relevant ICT tools for the specific purpose to serve the rural community without compromising with their individual applications towards understanding of their overall role in promoting productive, equitable, and sustainable agriculture. Extension workers well equipped with the ICT can prove more efficient in gathering, storage, retrieving, adaptation and dissemination of broad range of information needed by farmers, thus transforming them from extension workers into knowledge workers. Emergence of such knowledge workers will result in the realisation of the much talked about bottom-up, demand driven technology generation, assessment, refinement and transfer.

Key words: ICT Tools, Farming System, Information Services, Sustainable Development

INTRODUCTION

ICT is an emerging area which is having immense potentials to change the world agriculture by providing relevant information and services at the time when it is needed the most. Many farmers and rural entrepreneurs generally lack access to accurate and timely information on markets, production processes, crops and livestock disease; as well as access to services such as land registrations, permits, government regulation forms, and internet.But now the Indian IT infrastructure market is promising contribution in the field of agriculture as it is projected to grow to US$ 2.1 billion in 2013, an increase of 9.7 per cent as compared to 2012. The market is driven by hardware refresh, optimisation and consolidation efforts along with the data that show that the telecom sector in India has registered a phenomenal growth during the past few years and has become the second largest telephone network in the world, only after China.During the past decade, the power of Information and Communication Technology (ICT) to improve productivity and reduce costs has increasingly been applied to the enhancement of agricultural and rural development and to sectors which have been slow to adopt and benefit from technology. Since about 70% of the world's population targeted by the Millennium Development Goals (MDGs) lives in rural areas and predominantly depends on agriculture for their sustenance, improving agriculture and rural development is key to meeting the MDGs. Also the Consultative Expert Meeting of CTA's observatory on ICTs formulated a large number of recommendations as awareness raising, capacity development, publishing and packaging, developing methodologies and approaches, research, gender and ICTs, donor coordination.

OPPORTUNITIES

In recent year the ICT initiatives in the country has multiplied to several times with the hope that ICT can help developing countries reach the Millennium Development Goals. Yet, rigorous field-tested knowledge about "what works and why" have been relatively scarce, as well as a deeper understanding of the enabling conditions and success factors in ICT for development initiatives.As all the stakeholders of the agriculture are going to be affected with the quality and timing of the information provided to them that in turn affect their decision taking ability and best output of their inputs.The opportunities of ICTs in the field of agriculture can be discussed as follows:

Drivers of ICT in Agriculture

Five main trends have been the key drivers of the use of ICT in agriculture, particularly for poor producers: (1) low-costand pervasive connectivity, (2) adaptable and more affordable tools, (3) advances in data storage and exchange

(4) innovative business models and partnerships, and (5) the democratization of information, including the open access movement and social media. These drivers are expected to continue shaping the prospects for using ICT effectively in developing country agriculture.

Adaptable and More Affordable Tools

The proliferation of adaptable and more affordable technologies and devices has also increased ICT's relevance to small holder agriculture. Mobile-based applications are also becoming more suitable for poor and isolated communities, especially though feature phones. Drawing on simple, available technologies such as SMS, service providers can offer mobile banking, other transactional services (selling inputs, for example), and information services (market price alerts).

Cloud services

Adoption of these will see a new life and greater momentum in the context of application/feature upgrade and application modernization. The influence of the business managers can be expected both in the selection of the solution and to ensure future alignment to business.

New Business Models and Public-Private Partnerships

The development and use of many ICTs originated in the public sector but were quickly dominated by the private sector when their profit potential became clear.

Trading on the Internet

If ever there was an example of expectation in IT being ahead of reality, then e-commerce must be it. In agriculture there was great enthusiasm for on-line trading of inputs and outputs and several multi-national projects were established to exploit this perceived opportunity. Now days the online marketing is dominating the market due to its ability to provide quickservice with plenty of options without any physical mobility. This potentiality can also be utilized for the field of agriculture.

Improving Decision Making

Using computer systems to assimilate information and provide advice is perhaps the most exciting opportunity for the future. The timely availability of the information improves the decision taking ability of the farmers.

Enhancing Agricultural Production

Increasing the efficiency, productivity and sustainability of small scale farms is an area where ICT can make a significant contribution. Farming involves risks and uncertainties, with farmers facing many threats from poor soils, drought, erosion and pests. All these can be tackled by applying relevant remedial procedures at the critical time which is possible by utilizing the potentialities of the ICTs.

Improving Market Access

Awareness of up-to-date market information on prices for commodities, inputs and consumer trends can improve farmers'livelihoods substantially and have a dramatic impact on their negotiating position.

Capacity-Building and Empowerment

Communities and farmer organisations can be helped throughthe use of ICTs to strengthen their own capacities and better represent their constituencies when negotiating input and output prices, land claims, resource rights and infrastructure projects. ICT enables rural communities to interact with other stakeholders, thus reducing social isolation.

Automated Capturing

Automated data capture through technologies like GPS, sensors,satellites, and remote sensing has been around longer than other models. However, only recently has it become affordable, accessible, and tailored to development specialists and developing-country governments.

Challenges

The information provided by the system must be in user-friendly form, easy to access, cost-effective and well protected from unauthorized accesses. Recording of text, drawings, photographs, audio, video, process descriptions, and other information should be in digital formats. Achieve greater interactivity in communicating, evaluating, producing and sharing useful information and knowledge. In the changed perspective, farmers need access to updated and exact information for improving the quality and quantity of the agricultural products marketing. ICT also have a moderate results showed that the situation of agricultural marketing is fairly desirable.

Some challenges to be discussed here include:

- Providing electricity to the rural areas.

- Ensure low prices for broadband internet in rural areas
- Understanding current trends in ICTs as they pertain to agriculture and the contributions that ICT can make to enhance agricultural strategies and their implementation.
- Making the available tools compatible with the local language and formats suited the most.
- There should be a formal training both for the farmers and the extension workers to how to use the available ICT tools effectively for dealing with the problems related to the agriculture.
- The availability and access to relevant information should be made by imposing cyber laws and cyber.
- The transparency in the public dealing along with the proper provisions to safeguard the information should be there.
- Foster public-private partnerships to make market andbusiness information accessible.Enhance rural access.
- Foster combined public-private efforts and cost sharingarrangements to ensure sustainability of rural information centres
- Foster awareness-raising and capacity-building of ruralcommunities to use and maintain ICT
- Support technical innovations for rural connectivity, such as wireless broadband connections or solar powered systems.

REFERENCES

De and Jirli (2013). A Handbook of Extension Education.

http://www.e-agriculture.org/newsletters/new-online-forum-challenges-and-opportunities-capturing-impact-ict-initiatives

http://en.wikipedia.org/wiki/ICT_in_agriculture

http://www.oifc.in/information-and-communication-technology-ict-india-individual

http://www.businessworld.in/news/business/it/for-indian-ict-industry-year-2014-is-the-transition-year/1289915/page-1.html#sthash.7xzT6syz.dpuf

http://www.worldbank.org

www.ictinagriculture.org

http://www.infodev.org/infodev-files/resource/InfodevDocuments_1162.pdf

www.ictinagriculture.org_sites_ictinagriculture.org_files_m%26E in ICT in ag.pdf

Chapter - 35

Efficacy of Mass Media in Contributing to Knowledge Empowerment of Farmers for Better Family Farming

Gopika M.H.
Assistant Technology Manager (ATM), ATMA Project, Dept. of Agriculture, HoskoteTaluk, Bangalore Rural District- 562114

ABSTRACT

The main function of extension is to bring a desirable change in human behavior. The extension workers, thus are essentially educators and their most important job is to communicate useful and practical research findings to the farmers in such a way that they understand, accept and adopt. In order to achieve this objective, the extension agencies are disseminating new technologies through different means including mass media like radio, T. V, and print media. Considering the wide areas of coverage, exposure and frequent use by people, the mass media such as television, newspaper and radio play an important role in the transfer of technology. Mass media are increasingly becoming a veritable instrument and tool for transforming agriculture. Hence, mass media play a vital role in agricultural extension. There is a gap between the agricultural researchers and the farmers. A lot of findings from the research institutions and laboratories are not used by the farmers, either because of the information did not reach them, or the received information was not implemented by them. To overcome this problem, mass media play a vital role in diffusing scientific knowledge among the mass. This is used to link researchers and extension workers and collecting the feedback from farming communities. Further, agricultural technology information acquired by farmers through mass media. Generally, the agricultural technology information includes techniques of applying fertilizer, insecticides and fungicides to crops, improved methods of cultivation and conservation techniques of planting, maintenance, harvesting and storage of crops. There are also new technologies of animal husbandry as well as processing and marketing various agricultural products. Hence, mass media is the most effective means of reaching people with new ideas and persuading them to utilize these innovations. The appropriate use can create knowledge and catalyze behavioral change among farming communities for better family farming.

Key words: Mass media,transfer of technology, agricultural technology information, scientific knowledge, agricultural extension

INTRODUCTION

Agriculture is the backbone of India's economy. Above 65 percent of population is engaged in agriculture business. During the last 25 years, the total production of food grains, vegetables, fruits and milk has been increased.Now a days we are paying attention to production. India ranks second in human population. The annual growth rate in India is also high. In order to meet the requirements of food grains and non food grains of India's population, it is essential that the cultivators use modern techniques in production of crop and maintenance of livestock. In the transfer of technical knowledge of agriculture among the cultivators, the Agriculture Universities, center and State Departments are playing an important role.

The main function of extension is the production of desirable changes in human behavior, the extension workers, thus are essentially educators and their most important job is to communicate useful and practical research findings to farmers in such a way that they understand, accept and adopt. Effective communication of new research findings and technologies in agriculture to rural farmers for better family farming remains a promising strategy for increasing agricultural productivity. Generally, the agricultural technology information includes techniques of applying fertilizer, insecticides and fungicides to crops, improved methods of cultivation and conservation techniques of planting, maintenance, harvesting and storage of crops. There are also new technologies of animal husbandry as well as processing and marketing various agricultural products. For the farmers to adopt the new technologies and put them to use, the new ideas must reach the farms and homes through effective extension and mass media channels.

Therefore, there is a direct need to apply science and technology in the field of agriculture. In this context, farmers need adequate information exposure to the latest technologies.Past researches have shown that by and large farmers' information exposure is most likely to be an important factor influencing their adoption behaviour. In order to achieve this objective, the extension agencies are disseminating new technologies through different means including mass media (radio, TV and print media). The various tools and media are used by different agencies in transfer of agricultural technology among the cultivators in India. Mass media are increasingly becoming a veritable instrument and tool for transforming agriculture. People will derive pleasure from learning how the food they eat daily is produced and they may be encouraged to develop an interest in growing something themselves. In India, over the last few years Mass media has seen a phenomenal growth both in terms of reach and advancements in technology. Considering the wide area coverage, exposure and frequent use by people, the mass media such as television, newspaper and radio play an important role in the transfer of technology.

Role of Mass Media in Agriculture Contributing to Knowledge Empowerment of Farmers

Mass media of communication play a very important role in progress and development of agriculture in the country. Thus mass media forms the corner stone of communication channels utilized by farmers during the adoption process. The farmers use different mass media and their combinations to obtain the latest information of their interest.

Mass media such as radio, TV and printed materials are used to reach large number of people quickly. The cost of extension advice through mass media comes to be considerably low as compared to individual and group contact methods. Mass media can serve as an important and valuable function in stimulating farmers' interest in new ideas. By these, farmer gets right information in right time through right form by right person which contributes for better family farming.

1. Newspapers

Newspapers in India are playing the expected leading role in agricultural information dissemination. India ranks between the World top ten countries with highest newspapers circulation. News papers in addition to their roles of informing and educating also serve as purveyors of change, watchdogs and custodians of the people's interest. Newspaper a product of the print media is considered as:Collective organizer; A gate keeper of information; An advocate; Purveyors of information; Vehicle for change. Since extension agents are change agents and Newspapers are vehicle of change thus, newspapers are meant to be extension counterpart. Potentials of Newspaper are:Create awareness; Serve as resource material; Prepare reader for social change. Some of the qualities of Newspaper such as : Can be stored and repeatedly referred to when necessary; Can be used for announcing different events and meetings that are important to the farmers; Can be used to raise level of awareness of farmers and their interest; Can serve as resources material to extension personnel and farmers;

Excellent expository resource material from where farmer can gain insight into agricultural research findings; Can present timely information on crop and livestock operations; Featuring success stories; Surveying the effect of new improved farm practices.

2. Electronic media

In India, farm and home broadcasting systems were introduced in 1966, to enlighten farmers on the use of various technologies to boost agricultural production. Radio and T.V have proved the most effective media in promoting agriculture.Farm Radio programme has a wide popularity and a high credibility in the minds of farmers. They reach large number of people quickly and inexpensively; Overcomes illiteracy; Do not depend on physical facilities; Reaches people who are unable to attend extension meetings; Especially suited to give Emergency and timely information; Radio/ T.V helps to Inform, Alert Suggest, Direct, Creates interest, Motivate and Stimulate the people; Radio/T.V programmes can easily be made to include latest information and findings to operate the knowledge of the local people as well as extension workers.

Radio and T.V can be used to disseminate agricultural information in the following ways.

1. Research findings can be distributed through Radio and T.V to Non Governmental Organizations (NGOs) dealing in agriculture, Extension workers, Farmers themselves, Academic Institutions.
2. Radio and T.V can make the link between researchers and extension workers by offering information on: Where research can be obtained and used;How to pass it on to users; Communities' feedback regarding research.
3. Radio and T.V can be used to collect feedback from Communities Through programmes, farmers give their responses to research; Farmers share experiences on use of research;rural radio airs out farmers' views/recommendations on how to improve the research; Communities provide alternatives depending on their experiences.
4. Radio can also be used to announce processes of research and extension work: Give venues for meetings with extension workers; Advise on where to get inputs/services; Advise on where to get technical support.

Effective utilization of mass media in agricultural extension

1. *Shift from individual to group approach*

Important issues may be segregated and focused through proper media strategy. Shifting from individual to group approach is inevitable considering the cost, time and existing extension scarce resources. To ease the growing burden of extension worker, an exclusive TV channel, more time on radio for agricultural programmes, more space in print media, innovations in information technology utilization in agricultural extension, capacity building of extension personnel, infrastructure and more importantly proper media strategy and liaison with mass media are need of the hour.

2. *Triple role of mass media*

Mass media has triple role of informing, entertaining and educating the masses. Politics, current affairs and entertainment have lion share in mass media coverage leaving inadequate space and time for agriculture. Unlike entertainment, agriculture does not enjoy the back up of agribusiness companies industry. Hence, it is necessary for the agri-business companies to utilize mass media effectively in promoting their products and sponsoring agriculture programmes. Production of innovative and quality programmes by professionals will help in sustaining the interest of farmers.

3. *Professional approach*

Agricultural journalism is the field where tools and techniques of journalisms are applied in the area of agriculture. Journalism is the profession of conducting or writing for the journal/ newspapers/magazine/radio/TV and IT in the present context. Very few organizations are training the personnel in agriculture journalism, which is highly inadequate. It is also most important for the mass media to utilize only qualified professional.

4. *Mass media co-ordination committees*

This concept of mass media co-ordination committee aims at bringing concerted action between various stake holders in agricultural extension and mass media to benefit the farmers. It could be group of all stake holders in agricultural extension like farmer, scientist, traders, extension workers and media people. They could meet frequently to decide the areas to be covered for the next one month or quarter and also should decide who has to contribute article/ programme and screen before going to masses. This process will ensure right news, in right form, at right time, by right people through right channel.

5. Network of correspondents/ reports

Network of resource persons to contribute to mass media on various issues of agriculture is the most important issue to be tackled first. Either the existing reporters may be trained, or trained professionals may be recruited or a network of progressive farmers, extension workers and scientists has to be maintained.

6. Innovation, key for success

Some newspapers have question-answers column which covers important field problems raised by farmers. Hence, more the innovation more will be popularity of mass media which is essential to sustain economically. Creative input is crucial in extension though mass media.

7. Inadequate coverage

Majority of the mass media focus on urban based issues leaving little space and time for agricultural information. Chandra Shekara*et al* (1998) reveled that newspapers covered very little information (2.51%) on agriculture. Hence, there is need to dedicate more space and time for agriculture issues handled by professional agriculture journalists.

8. Motivate extension functionaries to use mass media

It is necessary to make utilization of mass media in extension, a mandatory act to be performed by extension personnel. Necessary training in this regard may be imparted.

9. Cover demerits

It is tendency of the writers to focus only on merits of technology, not the limitations. Thus after realizing the field level difficulties, farmers lose faith in source of information. Hence, it is necessary to provide all information on technology were farmer's makes final decision.

Agricultural Technology Information Acquired through Mass Media By Farmers

The radios, televisions and farm magazines are the commonly used tools by the extension personnel in diffusing scientific knowledge among the masses. In a country like India, where the literacy level is low, the choice of communication plays very important role in transferring modern agriculture technology to literate farmers.

1. Mass media used

The coverage of different subject matter by radio, television and print media are not only used in agricultural extension but also in Agriculture, Horticulture, Animal Husbandry and Agriculture Marketing. Thus radio, television and newspapers play an effective role in the transfer of technology to the largest mass in short time. Each of them has their effect in the field of agriculture through communication appropriate technology for economically, educationally and spatially disadvantaged rural, resource poor, farming communities. As a result farmers are able to obtain timely technical advice from participating agricultural workers and researchers. Likewise, the researchers are able to draw from the traditional wisdom of farmers. State Agricultural Departments also encourage the publication of farm magazines particularly through farmers association.

2. Information Acquired through mass media

Effective communication of farm information to the farming community is a key to raise the socio-economic condition of the people of the nation. It is established fact that through mass media it is possible to reach large number of people to raise their interest in farming. Progress and change in farming can only emerge by timely and rapid communication of the farm information. Mass media can cater the need to a greater extent for the farmers of less accessible and isolated villages both on plains and hills.Various mass media are in use; out of those important are television, radio, farm literature, kisanmela and exhibitions. These media of communication possess the potentiality to disseminate the information on new agricultural and allied technology quickly, promptly and effectively.

According to Nirmal (2003) Technology information acquired through mass media includes:

- Techniques of applying fertilizers
- Insecticides and fungicides to crops
- Improved methods of cultivation
- Soil conservation
- Techniques of planting
- Maintenance
- Harvesting and storage of crops
- New technologies of animal husbandry
- Processing and marketing

Technical/Scientific Information

This type of information arises from research and development work carried out in various agricultural research institutions including universities.The result of such research is aimed at increasing agricultural production by providing high yielding seedlings and control of major pests, and diseases.This type of information is need by researchers, extension officers, agro based industries staff including illiterate farmers.

Commercial Information

Farmers need commercial information in order to maximize their profits, research findings based on the type of seedling, fruits and varieties of crops to be cultivated.Plus information that covers credit and cooperatives, ways of getting loans from government to boost their productivity, national prices for export commodities.Extension officers and agro-based industries and services staff also benefit from this type of information, as do policy makers and planners.

Social/Cultural Information

This encompasses traditional information on agricultural practices, local cultures, background information on farming communities on the types of crops to be grown, modern systems of farming, new innovations in storage facilities for harvested crops etc.This type of information will be useful to policy makers and planners, extension staff farmers themselves.

Legal Information

All legislations that affect agriculture such as land tenure, the production,distribution and sales of agricultural produce will come under this category. There should be legislation which allows illiterate farmers the opportunities to sell their harvested crops in any of the urban centers of the country without interruption.

CONCLUSION

Mass media is the most effective means of reaching people with new ideas and persuading them to utilize these innovations. The appropriate use can create knowledge and catalyze behavioral change among farming communities for better family farming. Hence, in the present context, mass media can play very important role in agricultural extension. They can reach more people in less time and less cost. Here, farmer gets right information in right time through right form through right channels by right person. Work load on extension worker reduced by using mass media in extension. Scientist gets the feedback from fields, traders and farmers can make good profit, mass media will increase its popularity and circulation. Hence, mass media in agriculture is a game where all the players are winners. But, most important element is co-ordination among all stakeholders in deciding what time, what information, by whom, how and to whom. Making use of mass media it is possible to bring desirable action for better family farming.

REFERENCES

Chidanandappa G., (1985). A study on mass media utilization by extension personnel. Unpublished M.Sc. Thesis in Agricultural extension, University of Agricultural Sciences, Bangalore.

Chandrasheekara, P. et al., 1998. Role of newspapers in coffee education. India coffee, april. Coffee Board, GOI. Pp 38-44

Kumar, N. (2003). Dissemination of farm information through mass media.Maharashtra Journal of Extension Education.XXII(2), 2003

Upap. P. J., (2010) Role of mass media in Agricultural Extension, Journal of Agriculture and Rural Development., Vol 32. No 10.

CHAPTER - 36

Mobile Phone Technology- An Eminent ICT Tool for Better Family Farming

Minu Singh[1], Adhiti Bhanotra[2], Niketha L[1], Sajad Ahmed Wani[1] and Mukesh Kumar[1]

[1]PhD Scholar, Division of Dairy Extension Education, NDRI, Karnal

[2]PhD Scholar, Division of Veterinary & A.H. Extension Education, SKUAST-Jammu

ABSTRACT

In 21st century, the world has turned into a global village. This herculean task has been made possible with the advent of ICT tools. When we think of ICT, the most omnipresent tool that comes in our mind is mobile phones. With the increasing rate of mobile phone penetration in the urban as well as in rural areas, it is now possible to connect the farmers leaving in the far-flung areas with the institutes, who are working for the betterment of farming community. Today, mobile phones play a significant role in everybody's life, besides being an efficient way to stay connected with the family; it also paves a way to bridge the gap between the availability and delivery of agricultural inputs and agricultural infrastructure. Mobile phones provide basic connectivity, mobility and security to owners, it works using the radio spectrum, so no need to rely on physical infrastructure, requires basic literacy only, also its services are available in local language; short message service (SMS), multi-media service (MMS) and voice communication are the added benefits. It allows transfer of data which can be used in the context of applications for the purpose of health, education, commerce and governance. Mobile phones are increasingly affordable due to factors like increased private sector competition and innovative payment methods. Due to high access, its reach, good adoption and real-time interaction mobile phones offer efficient solutions to governments' communication challenges. Mobile phones helps the farmers in agriculture decision making, as they provide information related to crops, weather forecasting, seeds, fertilizers, bio-pesticides, soil fertility, pest and disease diagnosis, demand and supply of agricultural products, different schemes and technologies. Mobile based initiatives like SMS services related to weather forecast, disease forecast, and market information are issued to farmers by different organizations like IFFCO Kissan Sanchar Limited (IKSL) and Reuters Market Light (RML). Farmers can raise queries related to agriculture and allied sectors using their mobile phone to a farmer call centre which has been operating in every state of India. There are many such initiatives, which are working to provide valuable services, and also it provides opportunities to further explore the potentialities of mobile phones as a tool for transforming the farming scenario of India.

Key words: ICT Tools, Mobile Phones, Farmers, Communication Agriculture Decision Making

INTRODUCTION

Today, the world has turned into a global village with the emergence of information and communication technologies (ICT). And in India the mobile subscription penetration as on July, 2014 has been reported to be around 70% (Srivastava, 2014), which suggests that people are rapidly advancing in mobile phone usage. So it offers the Extension personnels ample opportunity for leveraging the potentials of mobile phones. Additionally, mobile phones are equipped with some of the useful features like touch screen, key pad sensor, camera, microphone, geopositioning system (GPS), a light sensor, gyroscope (to sense device orientation), radio, send, receive, and connect store sense audio/ video. It also has a SIM card which will identify a phone number. All these facilities provided in mobile phones make it a competent device for dealing with many agriculture related issues like communication of market prices, dissemination of new ideas of agriculture, information related to crops, weather forecasting, seeds, fertilizers, bio-pesticides, soil fertility, pest and disease diagnosis, demand and supply of agricultural products, different schemes and technologies.

In a research done by the World Bank's independent evaluation group found that- (1) Farmers who regularly connect with extension staff are more likely to adopt technology sooner. (2) It is a fact that the progressive farmers get connected to extension agents faster than other farmers.(3) The technology that provides returns quickly is accepted quickly and its use is more wide spread. If these three observations are taken as facts, modern ICT can be used as an efficient tool to get non-progressive farmers connected to extension information and to make them adopt technology faster. ICT play an important role in adoption of technologies that are in an early stage of adoption (Fischer *et al*., 2009).

Relevance of Mobile Phones in Technology Transfer

The relevance of mobiles for the agricultural development in general and for Agricultural Extension in particular is extremely high for our country, where majority of the population is rural and depends on agriculture. Despite large, well-educated, well trained and well-organized Agricultural Extension manpower, the ratio of Extension personnel to farmers is 1000:1. At this point, the mobile phone with its massive penetration among the masses can play crucial role in reaching the unreached, and can help in disseminating desired information in right time.

Benefits of Mobile Phones

Mobile phone provides communication link even in isolated circumstances, and facilitates in reaching the farmers leaving in far flung areas. It requires basic literacy; also its services are available in local language. However, with the advent of smart phones it has become possible even for the illiterate farmers to use mobile phones with ease. The immediacy provided by touch screen technology in conjunction with audio-visual feedback can enable illiterate people to engage with digital information (Banks, 2012). It provide basic connectivity, mobility and security to owners, it works using the radio spectrum, so no need to rely on physical infrastructure; short message service (SMS), multi-media service (MMS) and voice communication are the added benefits. It allows transfer of data which can be used in the context of applications for the purpose of health, education, commerce and governance. Mobile phones are increasingly affordable due to factors like increased private sector competition and innovative payment methods.

In a study on mobile phones contribution to growth it has been shown that farmers use mobile phones to get connected with people like traders and other farmers who have agricultural activities related information. They also reported that there is some increase in convenience and cost savings from using their mobile phones as basic communication devices to seek information, such as input availability or on market prices. Farmers also benefited from improved access to information including seed variety selection, best cultivation practices, protection from weather-related damage and handling plant disease. (Mittal and Mehar, 2012).

Mobile phones helps the farmers in agriculture decision making, as they provide information related to crops, weather forecasting, seeds, fertilizers, bio-pesticides, soil fertility, pest and disease diagnosis, demand and supply of agricultural products, different schemes and technologies. Jensen (2007) found that the introduction of mobile phones decreased price dispersion and wastage by facilitating the spread of information for fishermen in Kerala. This made markets more efficient and enhanced both consumer and producer welfare.

Programmes Based on Mobile Phones

Mobile based initiatives like SMS services related to weather forecast, disease forecast, and market information are issued to farmers by different organizations like IFFCO Kissan Sanchar Limited (IKSL) and Reuters Market Light (RML). Some of the farmer's need based mobile services are aAQUA mini, Fisher friend, mKrishi, eSagu, Nokia Life tools and CERES. Farmers can clear their queries related to agriculture and allied sectors by simply making a call with their mobile phones through Kisan Call Centre helpline-1551 which has been operating in every state of India. mKrishi is a initiative through which farmers can translate information into action so that they can mitigate agricultural risk. Tata consulting services has started mKrishi service which allows two way interactions. A farmer uses the platform to access best practices and query agricultural experts through low-cost mobile phones, mostly using SMS (Banerjee, 2010). Similarly, in Kerala, the fishermen with the reach of mobile networks were now able to contact the markets in various ports for best offer for their daily catch of fish, on one hand it helped to sell the fishers in the place where it was more highly valued and on other hand it reduced wastage and equalized the prices throughout the regional ports (Jensen, 2007).

CONCLUSION

The mobile phones have emerged as an important tool for the development of the farming community in general and agricultural production in particular. It has a tremendous scope of contributing in the development of various sectors because of the special features which are being possessed by it such as- wireless connectivity, affordability, mobility, voice as well as text transfer services, reliability and above all its ability to reach the unreachable have shown us the path of success in the development of rural people who actually face the problems due to lack of infrastructure. Many studies have highlighted the significant role played by the mobile phones in establishing connectivity among the various stakeholders in agriculture and also by increasing the production by saving the time and travel cost in search of desired information, and helped in decision making by providing on time information to the farmers. It has the potential to bridge the gap between the Extension functionaries and the farmers, there by strengthening the Extension system.

REFERENCES

Banerjee, D. (2010). Mobile phone farming. The wall street journal, August 24, 2010. http://bit.ly/cB2Ofn, accessed September 2010.

Banks, Ken. (2012). Mobile Learning: How Smartphones Help Illiterate Farmers in Rural India – News Watch. Retrieved from: http://newswatch.nationalgeographic.com/2012/06/05/mobile-learning-how-smartphones-help-illiterate-farmers-in-rural-india/

Fischer, R.A., Byerlee, D.and Edmeades, G.O. (2009). Can Technology Deliver On The Yield Challenge To 2050? Expert Meeting on "How to feed the World in 2050", Food and Agriculture Organization of the United Nations and Economic and Social Development Department, Rome.

Jensen, R. (2007): The Digital Provide: Information (Technology), Market Performance and Welfare in the South Indian Fisheries Sector. In: Quarterly Journal of Economics 122 (3): 879-924.

Mittal, S. and Mehar, M. (2012). How Mobile Phones Contribute to Growth of Small Farmers? Evidence from India. Quarterly Journal of International Agriculture 51 (2012), No. 3: 227-244.

Srivastava, Bhavaya. (2014). Mobile And Internet In India 2014: 349 Million Unique Mobile Phone Users [INFOGRAPHIC]. Retrieved from: http://www.dazeinfo.com/2014/07/11/mobile-internet-india-2014-349-million-unique-mobile-phone-users-70-traffic-mobile-india-shining-infographic/

CHAPTER - 37

Community Radio Station: An Effective Tool for Communication

Binita Rani[1], Rakesh Kumar[2] and R. K. Sohane[3]
[1]KVK, Patna;[2]KVK, Arwal and [3]Directorate of Extension Education
Bihar Agricultural University, Sabour, Bhagalpur

ABSTRACT

The role of community radio has been conceptualized as an important and powerful tool for informing, educating, entertaining and influencing the masses. Community Radio Station at KVK, Barh, Patna has brought about useful contributions to the farmers of surrounding villages and broadcasting different programmes related to agriculture, health, hygiene, nutrition and entertainment in the local language. Present study was carried out to see the impact of organic farming related talks.We found it a very effective tool to communicate with small and marginal farmers of surrounding villages for delivering useful information related with agriculture and organic farming.

Key words: Community radio, farmer, villages, organic farming information

INTRODUCTION

We are living in a world that is saturated by mass media with our environment brimming with data and information (Hutchings and Matthews, 2008). In India, where literacy remains a substantial barrier to development, radio especially Community Radio, can reach a large number of poor people because it is affordable. Community Radio as a mass media is a radio service offering a third model of radio broadcasting in addition to commercial and public broadcasting. The rational for Community Radio in India is strong on legal/ constitutionl, social, cultural and developmental ground but is an inexpensive medium with comparatively simple technology and more suitable for illiterate and peasant communities and societies characterized by oral and folk traditions (Pavaral, V., 2003).

Community Radio Station can be critical enablers of information, voice and capacities for dialogue, in social accountability and public voice through community radio programming. While community media can take many forms – for example, using cassette, tape, video, public address systems and print – it is community radio which has particularly come to the fore as a medium which is accessible, low cost and, often, both popular and participatory. The growth of community radio is a story of people and communities striving to speak out and to be heard community radio has provided a means of empowerment and of self reliance.

According to the ministry of information and Broadcasting Government of India, the status on 25 April 2013 of community radio in India was:

- No of applications received : 1200
- Grant of Permission of Agreement (GOPA) signed : 191
- Operational Community Radio Station : 148

Modern Community Radio Station serves their listeners by offering a variety of content that is not necessarily provided by the larger commercial radio stations. As community radio is confined to a small geographical area and depends on low power transmission covering not more than 20-30 km radius, it serves a community which uses common resources for livelihood has common development issues and concerns. Which are relatively localized, nevertheless connected to national and regional development goals. Agriculture has always been a highly knowledge- intensive sector requiring continuous information flow. Farmer's quest for authentic, credible and usable information both from established systems and traditional practices is ever increasing in this fluctuating global environment, to operate efficiently and compete economically. Effective extension education and communication services are probably some of the key strategies for sustaining agricultural growth, strengthening food security and combating hunger and malnutrition. However, diverse socio- cultural backgrounds, linguistic barriers, geographical remoteness and differential incentives make the task of information dissemination challenging. Community Radio is the vehicle or system for delivering useful information to farmers and assisting those farmers to develop requisite knowledge, skills and attitudes to make use of this information or technology effectively. As the awareness about organic farming is increasing day by day and farmers do not know the potential of organic input to be used in organic farming, keeping in view, this study aimed to see the impact of talks related to organic farming.

METHODOLOGY

The study was carried out in the need by villages where farmers listen different programmes broadcasted by Community Radio Station at KVK,Barh,Patna. Famers listener group were formed and the content were developed keeping in view the need of the local farmer and the available natural resources existing. Farmers after taking training on vermicompost started its production at household level. A well structured interview schedule was prepared with relevant questions and for data collection, trainees as well as radio listener were interviewed personally.

RESULTS AND DISCUSSION

Community Radio Station established in 2011 at KVK, Patna covers 8 blocks of the district (Table 1). Formation of listeners group was also started to see the impact of radio talks which is presently 16 in number having 160 farmers.

Table 1: Area and listener groups covered by CRS , Patna

Sr.No.	Year	Area (No of block)	Listeners Group	No of farmers
1	2012	8	6	60
2	2013	8	16	160

Community Radio Station at KVK, Barh, Patna has brought about two useful contributionsto the farmers of surrounding villages, firstly, those living in remote corners of the district have been pulled out of their pockets of isolation and merged into the nation's mainstream and second it has provided a 'window' on agricultural scene in other parts of the country and beyond.Community Radio Station, Barh, Patna started broadcasting different programmes related to agriculture, health, hygiene, nutrition and entertainment in the local language. Duration of broadcasting increased from 2 hrs daily in 2011 to 3 hrs daily in 2014 (Table 2). Radio content are developed and broadcasted for 3.0 hr daily in two sessions i.e. morning and evening including talks of 30 minutes exclusively on agriculture related topics (Table 3).

Table 2: Duration of transmission

Sr. No.	Year	Timing	Duration
1	2012	07:00 AM to 08:00 AM (Morning) 06:00 PM to 07:00 PM (Evening)	20 Min
2	2014	07:00 AM to 08:30 AM (Morning) 05:30 PM to 07:00 PM (Evening)	80 Min

Table 3: Composition of different programmes broadcasted, under CRS, KVK, Patna

Sr.No.	Type of programme	Duration (min.)	
		Morning	Evening
1	Krishakmanch	30	30
2	Mahilajagat	15	-
3	Balmanch	-	15
4	Swasthyacharcha(Radio Natak, Gramwarta, YogBhagayeRog)	30	30
5	Lok Rang	15	15

Farmers responding to the Community Radio talks belonged to age group between 25-45 years out of which 35 % of respondents had education upto high school level. Thus, it is inferred that even illiterate farmers were interested to get knowledge about vermicompost after listening talk by radio, simplybecause they were interested to gain knowledge by attending training to use it in actual field situation to generate better crop production and improve soil health. Out of 160 registered farmers, 54 farmers responded to the talks by phone and 28 visited to KVK during 2013 to know details about vermicompost production. After attending one week of training they collected verms from the practicing farmers and started production of vermicompost at their farm. Out of 133 farmers of surrounding villages, 54 farmers are producing vermicompost at small scale for their personal use and more than 60 farmers have applied for subsidy provided by the Government to extend the production at larger level (Table 4).

Table 4: Impact of CRS on farmers' response and vermicompost production

Sr. No.	Farmers (No.)	Response (Media)	Farmers started vermicompost production	Farmers applied for subsidy
1	54	Telephone	54	60
2	28	Visit		

This impact of Community Radio talks is very encouraging as it has created awareness among the farmers about modern farming system in terms of organic farming, off season vegetables, selection of seeds and marketing of the agriculture products. This implies that Community Radio is playing significant role in the surrounding villages in diffusing technical information on various aspects of agriculture and allied sector.We found it as a very effective tool to communicate useful information with small and marginal farmers of surrounding villages.

CONCLUSION

From the above study it can be concluded that community radio can be used to communicate information related to modern agricultural issues and organic farming. There is need to create a common room in the village where important issues regarding agriculture, nutrition, health and hygiene can be discussed by villagers for the development of content for Community Radio broadcasting. Broadcasting expert, extension worker, panchayat member and villagers can meet and share their ideas and opinions for the purpose to have better impact on the community.

REFERENCES

Hutchings, K and Matthews, C. (2008) Teching and Learning Guide for: Ecocriticism in British Romantic Studies. Blackwell Publishing Ltd.

Pavarala, V. (2003) "Building Solidarities-A case of Community Radio in Zharkhand" Economic and Political Weekly. 22: 2188-2197.

CHAPTER - 38

Agribusiness: The Way of Attracting and Retaining Rural Youth in Farming

Shoji Lal Bairwa[1] and Saket Kushwaha[2]
ICAR-Zonal Project Directorate, Zone II
Kolkata 700097, West Bengal

ABSTRACT

Rural youth are the future of the agricultural sector in the country. A rising number of rural youth are migrating to urban areas from villages in search of better employment as well as better life. They are migrating towards cities due to limited resources, weak infrastructure, and unemployment in rural areas. Thus, the number of young people in rural areas is falling and the exodus of rural youth means fewer small-scale farmers for today and tomorrow. Agribusiness may prove important tool to stop the exodus of rural youth from rural to urban areas. Agribusiness provides the various employment opportunities in the areas of agricultural inputs, crop production, agro chemicals, fruit and vegetable cultivation, agricultural marketing and processing of agricultural produce. Agribusiness helps in remunerative farming which increases the level of income, employment and living standard of people in rural India. The present paper will discuss the role and importance of agribusiness in attracting and retaining rural youth in farming.

Key words: Agribusiness centre, rural youth, urban migration, employment and ARYA

INTRODUCTION

The growing population of India is dependent on agriculture for their livelihood but Indian agriculture is low in productivity with large number of disguised unemployment (Bairwa *et al.*, 2014d). Engaging youth in agriculture has been a prominent topic and growing concern worldwide that young people have become disenchanted with agriculture. There is a lot of concern about engaging youth in agriculture, in many ways, young people are not very much interested in continuing in agriculture because they don't see much prospect in the future of agriculture, they also don't see it is as an active profession in the long-run (Gandhi *et al.*, 2001). The scenario of migration is very dangerous and thinkable because 19 percent of the Indian population was in urban areas in 1951 and expected to jump to around 42.5 per cent of the total population by 2025 (Kurukshetra, 2012). In the last 50 years the rural population has decreased from 82.0 to 68.9 per cent in India.

Recently, Indian Council of Agricultural Research (ICAR) has recommended a proposal on "attracting and retaining rural youth in agriculture (ARYA)" to the Planning Commission of India which has agreed to include it in the 12th five-year-plan in order to tap the youth potential in farming and also prevent their migration to urban areas. The proposal regarding the pilot project had been recommended to the ICAR by its committee headed by K. Narayana Gowda formed to suggest the measures to be taken to attract youth in farming. The committee had recommended that a Rs. 200-crore pilot project should be taken up in 30 different places across the country to motivate the rural youth to remain in agriculture. "The prospects of agriculture are getting affected because of the youth shunning farming and migrating to cities in search of menial jobs. The only way of attracting them to agriculture is to turn farming into a profitable venture. The proposed project aims at not only making farming a profitable venture, but also increasing the farmers' incomes by three folds in five years. According to committee report, the efforts will be made to increase the farmers' incomes under the proposed project through various technological interventions depending upon the local requirements and bringing about convergence among the local research institutes, administrative bodies and farmers. This would also involve organising the youth on the basis of crops being grown by them and promotion of integrated farming. To begin with, about 500 farm youth would be chosen from each centre, which would comprise a cluster of three to four village panchayats. They would be given training in various aspects of farming including hi-tech farming and value addition to crops. The project would also link them with banks to get loans for setting up processing units and custom-hiring services related to farm machinery. The main intention of the project was to instil confidence among rural youth that they can lead a decent life through farming rather than chasing menial jobs in cities (Kumar, 2014).

Role and Importance of Agribusiness in Retaining Youth in Farming

India is a reservoir of agro biodiversity and other natural resources which fulfils the requirements of all the types of raw material for various small scale industries such as handmade items, sugar mills, food processing and dairy industries. Agribusiness opportunities can be explored through research and development at various states such as production, harvesting, processing, marketing, storage and transportation of agricultural produce (Bairwa *et al.*, 2014c). Agribusiness sector have the potential to contribute to a range of social and economic development such as employment generation, income generation, poverty reduction and improvements in nutrition, health and overall food security in the country (Bairwa *et al.*, 2014b). Agribusiness growth propels even the industrial growth by supplying huge quantities of raw material to agro-based industries like – sugar, textiles food processing, cigarettes, paper etc. The food production in India is likely to grow two-fold in the next ten years. Thus, there is ample of opportunities for investments in food and food-processing technologies, equipments, especially in areas of canning, dairy and food-processing, specialty processing, packaging, frozen food and thermo processing, cold chains and in the area of food retailing (Gandhi *et al.*, 2001 and Bairwa *et al.*, 2014a). Agro industry opportunities in India currently tend to favour growth and development of medium-scale industries that match the current production levels and the distortions in marketing of produce (Bairwa and Kushwaha, 2012).

The potential for agro based industrialization, especially labour intensive industrialization, is truly enormous. In India, to promote agro-industries and offer people living in rural areas new avenues of employment close to the place they work and live. However, unlike in the West, where much of this was highly mechanized due to labour shortages, we need labour using technologies. In many developed countries, the strategy of food processing and agro industry was focused essentially on increasing farmers' incomes without a focus on generating rural employment (Singh, 2007). In a labour surplus economy like ours, we need solutions that increase producers' incomes but also generate new employment opportunities. The food processing sector must have these objectives in mind (Prime Minister of India addresses Global Agro Industries Forum, 2008).

Major Potential Areas of Farming for Agribusiness Development

- Agriclinics and Agribusiness Centres – Agriclinics are envisaged to provide expert advice and services to farmers on technology, cropping practices, protection from pests and diseases, market trends, prices of various crops in the markets and also clinical services for animal health which would enhance productivity of crops/animals and increased income to farmers (http://www.agriclinic.net). Agribusiness Centres are envisaged to provide farm equipments on hire, sale of inputs and other services for crop production (Bairwa *et al.*, 2013).

These centres will provide a package of input facilities; consultancy and other services with the aim of strengthen transfer of technology and extension services and also provide self employment opportunities to technically trained persons (Chandrashekara, 2003).

- Dairy Farming – dairy farming is considered as village industry in India. Increasing population and growing consumption of milk and milk products like *Dahi, Cream, Paneer* indicate that there is a great potentiality for dairying in India. India's milk demand is likely to increase to 170 million tonnes by 2020 from the present 105 million tonnes (Bairwa *et al.,* 2014c). Dairying is the most suitable area among many livestock enterprises, to small and novice agripreneurs. Individuals with less risk taking nature but high obsession for agripreneuership may opt for setting up a small dairy farm that can generate him/her income from the day one (Srivastava, 1989 and Bairwa *et al.,* 2014d).
- Poultry and Piggery Farms– The Poultry sector has a large growth and expanding market in changing business environment. The egg production has just 1.6 per cent share in the livestock value of the state but poultry are gaining importance in recent times because income elasticity of demand is high for these products (Singh, 2007). Although Piggery is not a significant subsidiary occupation of the farmers in the country. Generally, landless labourers and some marginalized poor households in the urban areas are engaged in piggery (Bairwa *et al.,* 2014d).
- Post Harvest Management Business – It includes supply chain and value addition management. Supply chain management is most crucial for every industry especially for agro processing industry (Gandhi *et al*., 2001). As evident, large number of middlemen are present in the supply chain of the agro processing industry. Still, the number of agripreneurs at every nodal point in distribution network is scanty. Supply chain is hardly observed associated with value chain. Value addition is the fact which needs attention to strengthen supply chain. Value addition and creation at every stage of supply chain can create multiple opportunities for agripreneurs and thereby ensure supply of more designed and ready to use agro food products and other agricultural products to consumers (Srivastava, 1989). This does not end here rather it create circular effect as habits of eating processing food increases and make stride in the overall development and expansion of food processing and agro processing sector.

- Fishery and Fish Processing – Fish and fish processing sector is an emerging tool to agriculture. Fish and shrimp have good export potential but there is an immense lack of cold storage and modern processing facilities. For instance fish production is around six million tonnes a year and the frozen storage capacity spread over 500 units is only one lakh tones. India, with its 7,500 km long coastline and an exclusive economic zone of 2.02 million square km; 191,024 km of rivers & canals and 4.4 million hectares of reservoirs and fresh water lakes has an enormous potential for fisheries. In 1999, the country had an estimated 1, 81,284 traditional fishing crafts; 44,578 motorized traditional crafts, 53,684 mechanized fishing boats and about 200 deep-sea vessels in operation (Singh, 2007 and Bairwa *et al.,* 2014c).

Strategies for Successful Agribusiness Development

Considering the changing and complex present business situations, it is necessary to create a congenial atmosphere in the agriculture field. There should be a unanimous option among government officials and farmers about the need and benefits of promoting self-employed youth or private entrepreneurs to facilitate the farmers to enhance agricultural production and profitability (Bairwa *et al.,* 2014a). The Government should discontinue the practice of providing free services in those sectors where the work has been assigned to private entrepreneurs. The technical skills and ability of the entrepreneurs should be evaluated to ensure high standards. There should be a monitoring agency to check the quality of the services and the charges collected from the farmers to avoid exploitation. To popularize the services of the entrepreneurs, the agricultural extension agencies and farmers organisations should give wider publicity about the services available to the farmers (Bairwa *et al.,* 2014c). Such publicity can enhance the credibility of the services provided by the entrepreneurs. The Government should encourage the entrepreneurs by introducing various concessions and incentives. Networks of entrepreneurs may be established to share their experiences. These networks can also establish a close link with research institutions and universities to become acquainted with the latest research findings and seek solutions for their field problems.

CONCLUSION

Agripreneurship is solution to many economic problems like urbanization, poverty, unemployment and low economic growth. This situation can be changed by generating employment opportunities in rural areas through agribusiness. Agribusiness has much potential for the betterment of rural areas from the economics point of view. It will be play a vast role not only in the eradication of major challenges such as poverty, hunger, malnutrition and unemployment from the country but also in the development of living standards of people, income source which helps in security of basic livelihood. Government should give special attention on the seven pillars of agribusiness development(Improve agricultural productivity, Upgrading value chains, Exploiting local and national demand, Technological development, Innovating financing, stimulating PPP model and Energy and Infrastructure development) during agribusiness development policy making in order to take the advantages from the agribusiness opportunities in the country.

REFERENCES

Bairwa, S. L. and Kushwaha, S. (2012) Agro Industry Scenario in India In Edited by Prof. S. P. Singh "Agricultural Research and Sustainable Development in India", Bharti Publications, New Delhi, 110093, pp 159-182.

Bairwa, S. L., Kushwaha, S., Meena, L. K., Lakra, K. and Kumar P. (2014a) Agribusiness Potential of North Eastern States: A SWOT Analysis. Edited by Singh et al., 2014 "Agribusiness Potentials in India: experience from hill states". EBH Publishers (India) Guwahati – New Delhi. PP 544 – 556.

Bairwa, S.L., Kalia, A., Meena, L.K., Lakra, K. and Kushwaha, S. (2014b) Agribusiness Management Education: A Review on Employment Opportunities, International Journal of Scientific and Research Publications, 4(2): 536 - 538.

Bairwa, S.L., Lakra, K. and Kushwaha, S. (2014c) Agripreneurship for Rural Development. Edited by Rai and Goyal "Agriculture and Rural Development in India". M.R.F. Publication, Varanasi. PP 422 – 425. ISBN 978-81-926935-6-9

Bairwa, S.L., Meena, L.K., Lakra, K. and Kushwaha, S. (2014d) Agriclinic and Agribusiness Centres Scheme as Poverty Reduction Tool. Edited by Munnilal et al., 2014 "Poverty in India". pp. 17-24, ISBN 978-81-927285-1-3

Chadha, G. K. and Ashok Gulati (2007) Performance of agro-industry in India: Emerging issues and prospects, in Agricultural Diversification and Smallholders in South Asia, Academic Foundation, New Delhi, pp. 349-84.

Chandrashekhra, P. (2003) Third Wave in Indian Agriculture: Introduction to Agriclinics and Agribusiness Centre Scheme. MANAGE Ext. Res. Rev. January-Jun, 2003, pp. 10-20.

Gandhi, V., Kumar, G. and Marsh, R. (2001) Agro industry for Rural and Small Farmer Development: Issues and Lessons from India. International Food and Agribusiness Management Review, 2(3/4): 331-344.

Kumar, B. S. S. (2014) Pilot project to attract rural youth to farming in the pipeline, the Hindu, 20th July 2014. Retrived from www.thehindu.com/.../pilot-project-to-attract-rural-youth-to-farming-in

Singh, Sukhpal (2007) Agribusiness in South Asia: A Fact Sheet, Make Trade Fair Campaign http://www.maketradefair.com

Srivastava, U. K. (1989) Agro-processing Industries: Potential, Constraints and Tasks Ahead. Indian Journal of Agricultural Economics, 44(3): 242–256.

Chapter - 39

Innovation and Creativity in Rural Retail Marketing for Indian Scenario

Bibhu Santosh Behera[1], Anama Charan Behera[2] and Rudra Ashish Behera[3]

[1]Ph.D Research Scholar, OUAT Bhubaneswar [2]Professor cum Principal in Economics,Department of Economics,Dadhi Baman College,Turumunga,Keonjhar [3]Student,MBA,The Techno School,Bhubaneswar

ABSTRACT

Marketers make consistent attempts to innovate tools and strategies to overcome the challenges they face in the business arena. Business innovations are broadly classified under two heads, namely Product / Service innovation. Marketers need to design creative solutions to overcome challenges typical of the rural environment such as physical distribution, channel management, promotion and communication.India's rural markets offer a sea of opportunity. The urban rural split in consumer spending stands at 9: 11, with rural India accounting for 55 percent of private retail consumption. Indeed the market can be tapped with focused attention and strategy. Currently the Indian retail market is estimated at Rs. 13,30,000 crore and almost half of this growing retail market at present lies in rural India, which is a tremendous growth sector that needs to be tapped with care. This paper examines how the corporate sectors with their innovation and creativity tapping the Indian rural market with their retail marketing.

Key words: Innovation, retailing, rural retailing, rural India, creativity

INTRODUCTION

According to the great management guru Peter Ducker "the organizations have only two functions, one is marketing and other is innovation." Rural markets are characterized with huge potential for marketers, but at the same time pose several challenges to serve them with similar set of marketing mix used in urban set- tings. The main challenges in rural marketing are:

- Physical distribution
- Channel management
- Promotion and communication
- Poor infrastructure
- Uneconomical market size
- Diverse socio-economic Consumer profile

Against the backdrop of such a market environment, marketers need to design creative solutions to achieve success in rural markets. The problems of physical distribution and channel management adversely affect the quality of service (delivery) and cost. With poor or even no means of communication to exchange in- formation with rural consumers directly, the success of a brand depends largely on the village retailers. Therefore, rural marketers have felt a great need to overcome the existing limitations across business stages in general and distribution in particular of which retailing is the final stage.

Need of Innovative Rural Market

As the competition increases in the rural market there might be the need for competitively priced products that are developed as per the needs of the rural consumers. Non-consumers of yester- years are entering into the rural market as first time buyers for a large number ofproducts in a large numbers. This calls for shift in management thinking from gross margin to higher profit from high value unit sales to game of high volumes, capital efficiency and from one solution fits all thinking to market innovation.

Principle of Innovations for Rural Market

The principles and practices of innovation to be adopted in rural market have to take into consideration: needs, lifestyles and consumer behavior of the rural population. It is extremely important that the product, pricing, promotion and distribution strategy are not just innovative alone but they must make product value proposition attractive and relevant for rural consumers.

Process Innovation

Process innovations are critical in rural markets. Innovation must focus on building a logistics infrastructure, including manufacturing that is in accordance with the prevailing conditions and can deliver solutions in a cost effective manner.

Product Innovation

Product development must start from a deep understanding of functionality, not just form. Marginal changes to products developed for customers in the towns might not be that effective in rural market. The infrastructure and environment, in which the rural consumers live and work in, demand a rethinking of the functionality a new. Poor consumer's problems cannot be solved with old technologies. New technologies need to be developed to make the product relevant to the rural consumers whose product use environment is very different from urban consumers.

Price Innovation

Conserving resources is the mantra; the product meant for rural market must eliminate or reduce the various recurring costs and thus should reduce resource intensity. The option of reuse, refill and recycle are critical principles in product development for rural market as they reduce the overall cost of the product. India is a value for money society- while the value may range from a few rupees tomillions;the basic instinct remains the same. Market strategies must ensure the prices of the product are reduced either through re-engineering, cost saving in operations, reduction in pack sizes, deletion of frills from the core product, etc. but this reduction I price should not be at the cost of quality and service. Organizations have to focus on price performance of the product.

Promotional Innovation

Changing demographicsand spending profilespresent count- less opportunities for a creative response by the corporate sector. Promotional campaigns have to be innovative to target theyoungsters by engaging them through combination of media options. Promotional campaign such as video mounted on trucks traveling low cost theatrical need to be undertaken for any serious promotional effort in rural areas.

Rural Retailing

Retailing in India is slowly on the rise with changing consumerpreferences and tastes and evolution of a global structure. Rural markets are relatively virgin markets, which have evolved on their own with very little direct contact withthem bythe corporate world, but their size is compelling and attractive. Retail sector offers opportunities for exploration and investment in rural areas, with Corporate and Entrepreneurs havingmade a foray in the past. India's largely rural population has caught the eye of retailers looking for new areas of growth. Market structure in India is dichotomous having rural and urban markets.

One of the principal reasons behind the explosion of retail outlets and its fragmented nature isthat retailing is probably the primary form of disguised unemployment/ underemployment. The overcrowded agricultural sector, stagnating manufacturing sector, the hard nature of jobs and low wages in both virtually force many Indians to the service sector. So, it is almost a natural decision to open a small shop or store depending on the available means and capitals due tothe lack of opportunities. This phenomenon explains the million of kirana shops and small stores.

ITC launched the country's first rural mall 'Chaupal Sagar', offering a diverse product range from FMCG to electronics appliance to automobiles, attempting to provide farmers a one-stop destination for all of their needs. There has been yetanother initiative by the DCM Sriram Group called the 'Hariyali Bazaar' that has initially started off by providing farm related inputs and services but plans to introduce thecomplete shopping basket in due course. Other corporate bodies include Escorts and Tata Chemicals (with Tata Kisan Sansar) setting up agri-stores to provide products/ser- vices targeted at the farmer in order to tap the vast rural market.

Innovation and Creativity in Rural Retailing

To tap the vast potential of rural India, the models of marketing which is for the urban markets will not fruitful. Innovative models are required to tap the potential of the rural India. Some proved innovative models which are being used by the corporate in the rural India are discussed below.

GODREJ'S Aadhaar and Manthan

Manthan focuses on supplying animal feeds for dairy and poultry. On the other hand Aadhar is a supermarket, which retails agricultural inputs such as fertilizers, pesticides, animal feeds and small implements along with FMCGs, appliance and also services, like valuable technical guidance, soil & water testing services.

HUL Project Shakti

Hindustan Unilever Limited (HUL) and its constituent companies have been in India since 1931. HUL has been proactively engaged in rural development since 1976. The principal issue in rural development is to create income-generating opportunities for the rural population. Such initiatives are successful and sustainable when linked with the company's core business and is mutually beneficial to both the population for whom the programme is intended and forthe company. Based on these insights, HUL launched Project Shakti in the year 2001, in keeping with the purpose of integrating business interests with national interests there are more than 40,000 Shakti Entrepreneurs across India.

HUL: Shaktimaans-To-Power-Rural-Reach

The company is utilising the skills of the likes of Sudharkar to distribute its products in remote villages which have a de- mand for such products, but don't have a distribution network. Every day, Sudharkar sets out on a bicycle which has been pro- vided by HUL for him to commute to villages to distribute prod- ucts and sachets of popular brands like Wheel, Lifebuoy, Pond's, Brooke Bond, among others, to aspiring consumers. Earlier, these consumers had to satisfytheir needs bypurchasing products from nearby villages where the company has direct distribution. Now, their needs will be met in their own villages by the visiting shaktimaans.

A shaktimaan is a male member of a shakti entrepreneur family. In 2000, HUL collaborated with self-help groups to expand its ru- ral reach under project shakti. It partnered women entrepreneurs called shakti ammas from rural areas of Andhra Pradesh and 14 other states by offering them opportunities for business. HUL soon figured out that the menfolk from shakti households, who would by now be familiar with the operations and product range of the company, could be used for the next leap — to reach vil- lages with a population of less than 2,000. They have been chris- tened shaktimaans.

DSCL HARYALI Stores

DCM Shriram Consolidated Ltd. (DSCL), capitalizing its over 35 years of experience in the agri-input markets & firsthand knowledge of Indian farmers, is setting up a chain of centers aimed at providing end-to-end ground level support to the Indian farmer and thereby improving his "profitability" & "productivity"."Hariyali Kisaan Bazaar" is a pioneering micro level effort, which is creating a far-reaching positiveimpact inbringing a qualitative change and revolutionizing the farming sector in India. It is also an example of how well meaning corporate can contribute to development of agriculture by building sustainable business models. It seeks to empower the farmer by setting up centers, which provide all encompassing solutions to the farmers under one roof. Each "Hariyali Kisaan Bazaar" centre operates in a catchment of about 20 Kms. A typical centre caters to agricultural land of about 50000-70000 acres and impacts the life of approx. 15000 farmers

ITC E-CHOUPAL

ITC's trailblazing answer to these problems is the e-Choupal initiative; the single-largest information technology-based intervention by a corporate entity in rural India, enriching the farmer with knowledge; elevating him to a new order of empowerment. E-Choupal delivers real-time information and customized knowledge to improve the farmer's decision-making ability, thereby better aligning farm output to market demands; securing better quality, productivity and improved price discovery. The model helps aggregate demand in the nature of a virtual producers' co- operative, in the process facilitating access to higher quality farm inputs at lower costs for the farmer. The e-Choupal initiative also creates a direct marketing channel, eliminating wasteful intermediation and multiple handling, thus reducing transaction costs and making logistics efficient. The e-Choupal project is already benefiting over 3.5 million farmers. By 2012, the e-Choupal net- work will cover over 100,000 villages, representing 1/6th of rural India, and create more than 10 million e-farmers

M & M Shubh Labh Stores

The venture willleverage on Mahindra's agribusiness division, which is involved in contract farming, contract services, exports and agri retailing. MSSL is involved in contract farming across100,000 acres in eight states, covering 30,000 farmers. The company had set up agri centers in various districts of the country either under the Mahindra Krishi Vihar franchisee model or directly as Mahindra Agribusiness. These centres are one-stop shop for agro service, retailing of agro inputs and procurement of produce.

Bharat Petroleum

Bharat Petroleum is planning totarget cluster of smaller villages with a population of about 200 to 250 households. It is planning to set up the pumps for these small villages will be smaller in size and therefore will be low priced units in terms of the cost of the infrastructure to establish these outlets. These retail outlets will serve a radius of seven to eight such villages.

Reliance Rural Hub

It is piloting a rural-business-hub (RBH) model in a Gujarat village, which if successful and implemented could rival that of DSCL's Hariyali Kisan Bazaar and Future Group's Aadhar. RBH would offer farm input, food, grocery, consumer durables, and financial and health services. It will also provide farmers a platform to sell their produce, an equivalent of village haat.

Warna Bazaar

Warna Bazaar is the name of two superstores in Kolhapur and Sangli in Maharashtra, which are set up in the area of 10,000 sq. ft. Along with that they have 30 stores of 500-1,000 sq. ft at the village level. These stores retail products like apparel, food, grocery, agri-inputs, vehicles, consumer durables and hardware.

TATA Kisan Kendras

The Tata Kisan Sansar network reflects the Tata Group's belief that technology can and must be harnessedto solve India's social and economic problems. The concept of precision farming be- ing implemented by the TKS has the potential tocatapult rural India from the bullock-cart age into the new era of satellites and IT. TKSs stock seeds, pesticides and fertilizers that farmers can buy at affordable prices, and they lease out farm equipment and im- plements to farmers who cannot afford to buy expensive modern machinery. One of the biggest worries for small farmers in India is finance. The sansars take care of this need too. Farmers can get credit, insure their crops against natural disasters, and even avail of buyback facilities.

Kasti Society

Kasti society has established organized co-operative rural retailing near Ahmednagar in Maharashtra, which serves 70 surrounding villages. Numbers of shops are arranged in the shopping center for- mat with three supermarkets with total retail space of 5,000 sq.ft.

Indian Oil

Indian oil is setting up large number of retail outlets to sell non fuel products like soaps, and fertilizers along with petrol and die- sel in the rural market.

CONCLUSION

The world is changing like never before. This change is more attributed to the way business practices and strategies are changing across the globe. Business organizations are rapidly changing their domain and strategy to take benefit of this emerging global order. Rural marketing endeavours have to be seen and implemented as investment for better tomorrow. Thus successful company will be one which meets consumer's expectations through products offered at affordable cost and still is in position to earn a decent return on investment because of its strong, efficient and intelligent distribution channel.

REFERENCES

1. Badi, R.V and N.V Badi (2006), Rural Marketing, Himalaya Publishing House, New Delhi
2. Dogra, Balaram and Shuman, Karminder, (2008) Rural Marketing: Concepts and Practices Tata McGraw-Hill Publishing House, New Delhi.
3. FICCI – Nielsen Report, Bridging the Urban Rural Divide.
4. Gopalaswamy, T.P, (2008) Rural Marketing Environment, Problems and Strategies Vikas Publishing house, New Delhi.

5. Kashyap, Pradeep and Raut, Siddartha (2007), The Rural Marketing Book, Biztantra, New Delhi.
6. Krishnamoorty, R (2008), Introduction to Rural Marketing, Himalaya Publishing House.
7. Lamba , A.J, The Art of Retailing, Tata McGraw Hill Publishing Companies, New Delhi .
8. Singh, Sukhpal(2008) Rural Marketing: Focus on Agricultural Inputs, Vikas Publishing House, New Delhi.
9. Velayudhan, Sanal Kumar (2007), Rural Marketing: Targeting the Noon Urban Consumer, Sage Publication, New Delhi.

CHAPTER - 40

Advances in Food Technologies in Relation to the Youth and Enterprise

H Umesh Hebbar
Principal Scientist, Department of Food Engineering,
Central Food Technological Research Institute
Mysore - 570 020

The Indian food industry is poised for huge growth, increasing its contribution in world food trade every year. In India, food processing is gaining importance as now it is being considered as profit industry. The increased demand for processed food in the domestic market and Indian foods getting popularized in the global market has put food processing industry on high priority list by the entrepreneurs. Accounting for about 32 percent of the country's total food market, the food processing industry is one of the largest industries in India and is ranked fifth in terms of production, consumption and export. The total food production in India is likely to double in the next ten years with the country's domestic food market estimated to reach US$ 258 billion by 2015. The food industry in India has been attracting a lot of attention from foreign investors in the area of processing, development of infrastructure for storage and transportation and retail marketing and scenario is changing rapidly.

The government of India is making all efforts to encourage investments in the sector, mainly to improve infrastructures for cold chain development & processing and also grant-in-aid for setting up laboratories that are equipped for testing food products.The Vision Document-2015 prepared by Ministry of Food Processing Industries (MoFPI), envisages trebling the size of investment in the processed food sector by increasing the level of processing of perishables from 6 to 20 percent, value addition from 20 to 35 percent and share in global food trade from 1.5 percent to 3 percent by 2015. To achieve these targets, an investment of Rs 100,000 crore (US$ 16.24 billion) is required by the year 2015.As most of the food processing industries are largely unorganized, and rural based there exists a good potential for rural youth to exploit this situation.

The need of the hour for food processing sector is i) infrastructure for storage and transportation, particularly for perishable foods and processing units at strategic locations,ii) adopting technologies that are simple, cost effective and environmental friendly and can be used by small, medium and large-scale processing industries and iii) technologies that can be employed at the farm level to convert perishable foods into intermediate product, if not the end product.With the increased awareness among the customers about health and nutrition, the onus is on the processor to adopt technologies that helps in retaining nutritional quality as well as sensorial characteristics of the product. The increasing competition in the market is forcing the industry to look for cost effective processing technologies.

Food processing involves many unit operations; right from cleaning of raw materials to packaging the product, and the operations involved in any process depends mainly on the end product. Food processing is broadly classified as primary and secondary processing based on the level of processing. The technology developed depends on the scale of operation i.e. rural technologies that are simple and rely on manual operation or semi-auto systems or state-of-the-art technologies for large scale production wherein automation comes to the forefront. Innovations are being broughtin terms of equipment as well as technologies used.Food packaging is another important area and having great potential for growth, as shelf-life of the product greatly depends on the packaging material used. The change in life-style of customers has increased the demand for product packaging that is easy to handle & store and convenient to use. The environment concerns have led to the development of biodegradable packaging materials. Packaging materials that are light in weight but durable, resistant to harsh conditions and recyclable will be sought after in the future. The intelligent packaging materials that can indicate the quality of food are gaining popularity. Nanotechnology also is expected to play a major role in this segment. Undoubtedly food packaging is going to play a major role in the growth of food processing industry.

Advanced processing technologies have been developed primarily for large scale processing. These technologies help to produce food products consistently at superior quality. However, when we deal with small scale operation and at farm level, the techniques to be used are entirely different and are normally labor intensive. The challenge here is to ensure the safety and qualityof theproduct, as increased human intervention enhances the chances of contamination. Some of the initiatives that could be taken up for small scale processing or farmers for value addition are

- Integration of pre-harvest stage to produce desired quality products
- Use of Appropriate packaging to extending shelf-life of produces
- Minimal processing technology for extending the shelf-life of vegetables and fruits
- Conversion of highly perishable food products to intermediate products for further processing
- Value addition at farm level using simple machineries e.g. small mill to handle pulsesfor de-husking and milling
- Use of mobile units for processing
- Solar energy based dehydration units to convert into intermediate moisture products/end products
- By-product utilization and value addition to waste by recovering bioactive compounds.

In summary, we can say, the food processing sector can provide ample opportunities,and youth can play a major role in the growth of the industry. Use of advanced communication systems to access information, integrating processing with pre-harvest activities, adaptation of 'site-specific' processing techniques, networking with market could boost the prospects rural or farm level processing.

CHAPTER - 41

Sweet Corn Cultivation: An Economic Opportunity for Rural Youth in Bihar

D.K. Baranwal[1] and S.S. Solankey[2]
[1]*Department of Plant Breeding and Genetics, Bihar Agricultural College, B.A.U., Sabour (Bhagalpur) – 813 210, Bihar*
[2]*Department of Horticulture (Vegetables & Floriculture), Bihar Agricultural College, B.A.U., Sabour (Bhagalpur) – 813 210, Bihar*

ABSTRACT

Sweet Corn (*Zea mays var. saccharata var. rugosa*) also called Indian corn, sugar corn, and pole corn is a kind of maize having high sugar content (12 – 15%). It is a very delicious and rich source of energy, vitamin A and vitamin C. It is consumed as raw, boiled or steamed green cobs/ grain. Cooked/ processed sweet corn has significant antioxidant activity which has been suggested to minimize the chance of heart disease and cancer. In addition to this, it is also utilized in preparation of soup, salad and other famous recipes. Besides green cobs, the green fodder is also available to the farmers for their cattle. It requires deep fertile and good drainage loam or sandy loam soil with neutral pH. It germinates best at the temperature range from 20°C - 25°C. Green cobs are harvested after 20 days of pollination during kharif but the days may accorded by season to season. It is ready to harvest in 70-75 days during kharif season and 105 – 110 days during rabi season. After harvesting, green cobs should be immediately transported to the cold storage in refrigerated trucks to avoid the conversion of sugar to starch because loses flavor at high temperature. Being a short duration crop (70-80 days), farmer can earn 2.0 to 2.5 lac rupees per hectare because it require less input i.e. water, manpower and other crop management practices. Therefore, it can generate a good employment for rural youthdue to high demand in hotels for preparation of various dishes.

Key words: Sweet corn, employment, fodder, protein, sugar and vitamins

INTRODUCTION

Maize (*Zea mays* L.) is a miracle crop emerging as the third most important cereal crop next to rice and wheat. It is grown for food, feed and as a source for numerous industrial products. Its array of diversity of uses and large hidden potential for exploitation led the renowned Nobel laureate, Norman E Borlaug to say that " last two decades saw the revolution in rice and wheat; the next few decades will become the maize era". India is the seventh largest producer of maize in an area of 8.11 million hectares with a production of 18.9 million tones and 2335 kg ha^{-1} productivity. Whereas, in Andhra Pradesh, it is cultivated in an area of 0.85 million hectares s with a production of 4.22 million tones and 4965 kg ha^{-1} productivity (CMIE, 2010). Sweet corn (*Zea mays var. saccharata var. rugosa*) is one type of maize and contains 12 to 15% sugar in immature grains (Bhatt, 2012). Sweet corn is generally dull yellow and white but dull yellow color is preferred.It is one of the most popular vegetables in the Europe, USA and other developed countries of the world. For obtaining of best and uniform kernels the fields must be isolated over 100 meters from field corns or other sweet corn with different kernels color or planted in different period in order to avoid the same flowering stage. Sweet corn is consumed at the soft dough stage with succulent grains, emerges as an alternative dish of urbanites namely, vegetable, roasted ears, soups, corn syrup and sweeteners etc. It also found a special niche in the preparation of native beer. Therefore, it can generate a good employment for rural youth and its marketing is not difficult because it has high demand in the metro cities and hotels for preparation of various dishes with hi-level of nutraceutical values.

Presently, greater emphasis is given to the cultivation of sweet corn due to increasing demand. There is an increasing tendency to produce sweet corn at commercial level to augment the income of the farming community dwelling in the outskirts of big cities and metropolis. Bihar has a very good climatic conditions and soils for sweet corn cultivation. As well as it can generate theemployments among rural youth. It is a new crop for Bihar and has an ample scope to get more profit and generate employments.

Soil and Climate

Sweet corn grows best at a soil pH between 6.0 and 6.8.The ideal soil for sweet corn is a loamy sand or sandy loam. Generally, it is known as occurrence of early cob in light soil (sand or loam) and late cob in heavy soil (silt, clay). Light soils are warm up faster than heavy soils, so seed germinates earlier. For hot and dry summer conditions, heavier soils have the advantage of good water holding than light soil. Sweet corn is light loving crop and do not grow under shade. It is susceptible to frost. The optimum temperature range for its proper growth and yield is 60 to 85 ^{0}F and if temperature goes below 50 ^{0}F it cannot be grow.

Irrigation

Sweet corn is a heavy water user. However, sweet corn requires ample water from germination to harvest stage. The most critical period for irrigation is about 2 weeks before silking. Water deficit during this period will seriously impair cob quality. Generally, one inch water per week is desired to supplement natural rainfall, as per need.

Varieties

Sweet corn hybrid, Sugar-75 matures in 80 to 85 days and suitable for both *kharif*and *rabi*cultivation in Andhra Pradesh, India. One famous Sweet corn composite variety Madhuri is suitable for Rabi and Spring cultivation and cob is ready to harvest after 50 -55 days of sowing.Priya, orange corn and Almoda are also popular varieties for Indian climate (Singh *et al.*, 2003).

Seed Rate

In general 12- 15 kg seeds are enough for sowing in one hectare land. The seeds would be treated with fungicides like Thiram @ 3 gm / kg of seed or Apron 35 S.D. @ 4 gm/ kg (Singh *et al.*, 2003).

Sowing Time

Sweet corn can be grown in all three seasons*viz.*,Rabi, Kharif and Spring in Biharcondition. Mostappropriate time for Rabi maize is 15 October – 15 November; for Kharif maize, last week May to last week June and for spring, mid February – mid March.

Sowing Method

Sweet corn should be sown in rows always. Row to row distance should be 60 – 75 cm and plant to plant distance should be 20 – 25 cm. The appropriate plant population for one hectare should be 65,000 to 80,000.

Fertilizer Requirement

In general, 150 kg Nitrogen (N_2), 70 Kg Phosphorous (P_2O_5) and 50 – 60 Kg Potassium (K_2O) along with 20 Kg Zinc sulfate ($ZnSO_4$) will be sufficient for one hectare area. At the time of sowing, whole amount of phosphorous, potassium, Zinc sulfate and one third amount of nitrogen would be dressed in 10 – 15 cm depth of rows as basal dose (Singh *et al.*, 2003). Remaining amount of

Nitrogen should be dressed in three equal split after 20, 40, and 60 days after sowing near the root zone.

Interculture and Weed Control

The sweet corn crop should be keep weed free for 30 – 45days after planting because this duration is most critical for crop weed competition point. Two – three manual weeding would be sufficient for this purpose. Ploughing with country plough between maize rows is frequently used by people (Singh *et al.*, 2003). However, hand hoe, khurpi and spades are also used for this purpose.

The herbicides Atrazine is commonly used in maize for effective weed control at the rate of 1.0 – 1.5 litre per hectare and mixed in 1,000 litres of water sprayed on soil surface just after sowing.

Major Diseases

1.Stalk rot

It is bacterial and fungal borne disease both and survives on diseased debris. The bacterium infects entire plant particularly collar region while fungi infects lower internodes at entire crop growth period. The infected plant parts become soft, twisted and toppled over the ground. For effective control, use Captan at the rate of 150 gm per 100 litres of water in poorly drained field pockets.

2. Sugarcane downy mildew

Long yellowish and whitish stripes appear on upper plant leaves and plant became stunted. For effective control the disease, spray Mancozeb 75WP 1.5 kg and 0.25 kg Zinc sulphate in 500 litres of water per hectare.

Major Insect Pests

1.Stem borer

The hatched caterpillars migrate towards the central shoot followed by bore in top internode and growing point damaged. It is control by spray of 2 kg Carbaryl 50 WP in 1000 litre of water per hectare.

2.Cut worm

The grey colour larvae of the pest cut seedlings at ground level. Dusting of 2 per cent Folidoldust at the base at the rate of 20 -25 kg per hectare is effective for controlling the pest.

Yield

Sweet corn can be harvested within 80 to 90 days after sowing. They are harvested earlier by 35 to 45 days compared to normal grain corn.The varying plant densities and nitrogen levels significantly affected the green cob and fodder yields (Bhatt, 2012 and Sahoo and Maha, 2007). There was significant improvement to an extent of 35.8% in cob yield due to increasing plant density from 66,666 to 80,000 ha^{-1}. Lower yields at 66,666 plants ha^{-1} were due to production of number of cobs. When planting density was further increased from 80,000 to 100,000 ha^{-1} the cob yield reduced considerably to the tune of 53.1% as compared to 80, 000 ha^{-1}. At higher plant density of1,00,000 ha^{-1} more competition for resources occurred and reduced the values of different yield attributes.The individual plants growth at higher plant density was affected more due to mutual shading and greater interplant competition. The internal competition within the individual plant between vegetative and reproductive parts had more dominantcompetition in comparison with between plants increased with higher plant densities (Bhatt, 2012 and Sahoo and Maha, 2007).

Grain Protein Quality

The protein content ofkernel was influenced by plant densities and nitrogen levels. There was reduction in protein content of grain with increased plant density which might be due to inadequate availability of nitrogen as a result of competition(Bhatt, 2012 and Singh *et al.*, 1997). The cob and fodder yields and protein content were significantly affected by nitrogen levels. Each successive increase in nitrogen levels enhanced the green cob yield to the tune of 36, 13.2 and 12.3% with the application of 160, 200 and 240 kg N ha^{-1} respectively over 120 kg N ha^{-1}(Bhatt, 2012). Increased plant height and dry matter production witheach successive increase in nitrogen levels enhanced thegreen fodder yield to an extent of 15, 20.9 and 30.7%with application of 160, 200 and 240 kg N ha^{-1} over 120kg N ha^{-1} respectively.Application of 240 kg N ha^{-1} not only enhanced the cobyield but also improved the quality of grain as evidencedin higher protein content. The increased supply ofnitrogen increased the protein content of grain sincenitrogen is the constituent of protein molecule (Bhatt, 2012; Krishna *et al.*, 1998 and Raja, 2001).

Economics

The maximum net returns and net returns, per rupee invested (B:C ratio) were found with plant density of 50 × 25 cm (80,000 ha^{-1}) both lower and higher levels over 80,000 ha^{-1} reduced the net returns and benefit cost ratio (Bhatt, 2012; Sahoo and Maha, 2007and Ashok, 2009). There was a marked improvement in net returns and benefit cost ratio with each successive increase in nitrogen level from 120 to 240 kg ha^{-1}. The maximum net returns were noticed with 240 kg N ha^{-1}. The benefit cost ratio was also enhanced with higher nitrogen levels. Higher yields of green cobs and fodder directly contributed to the returns at higher nitrogen levels (Bhatt, 2012 and Ashok, 2009). As per the findings of Bhatt (2012) the treatment combination of 80,000 plants ha^{-1} with application of 240 kg N ha^{-1} due to interaction gave maximum net returns and B:C ratio (4.25).

CONCLUSION

In Bihar, there is suitable climate for corn cultivation and an intense demand to produce sweet corn at commercial level to generate income for farming community. It has potential to generate highest income day basis due to short crop duration. Protein content and kernel quality is depending on plant density and nitrogen application. Significant improvement in cob yield and B: C ratio was found at increased plant density (about 80,000 per ha). Highest kernel protein content and highest cob yield can be achieved by judicious application of Nitrogen (240 Kg per ha). Obviously, sweet corn cultivation will be excellent opportunity for rural youth to generate employments and coping malnutrition severity in Bihar.

REFERENCES

Ashok, KR (2009). Production potential and nitrogen use efficiency of sweet corn (Zea mays) as influenced by different planting densities and nitrogen levels. Indian Journal of Agricultural Sciences, 79(5): 351 – 355.

Bhatt, PS (2012). Response of sweet corn hybrid to varying plant densities and nitrogen levels. African Journal of Agricultural Research. 7(46): 6158 – 6166.

CMIE (2010). Centre of Monitoring India Economy Pvt.Limited, Apple Apetile, Mumbai, India.

Krishna, A; Raj, KSVand Sambasiva, RA (1998). Effect of planting pattern and nitrogen on fodder maize (Zea mays) inter cropped with cowpea (Vignaunguiculata). Indian Journal Agronomy. 43(2): 237 – 240.

Raja, V (2001). Effect of nitrogen and plant population on yield and quality of super sweet corn (Zea mays). Indian Journal Agronomy. 46(2): 246 – 249.

Sahoo, SC and Mahapatra, PK (2007). Yield and economics of sweet corn (Zea mays) as affected by plant population and fertility levels. Indian Journal of Agronomy. 52(3): 239 – 242.

Singh, AK; Singh, GR and Dixi, TRS (1997). Influence of plant population and moisture regimes on nutrient uptake and quality of winter maize (Zea mays). Indian Journal Agronomy. 42(1): 107 – 111.

Singh, C; Singh, P and Singh R (2003). Modern techniques of raising field crops (2nd Ed.) Oxford and IBH Publishing Co. Pvt.Limited, New Delhi. pp. 84-111

CHAPTER - 42

Economics of Production and Marketing of Desiccated Coconut Powder in Karnataka

Arjuman Banu, B.M. Shashidhara, T.N. Venkata Reddy G.N. Nagaraj and K.B. Umesh
Department of Agricultural Marketing Co-operation and Business Management UAS, GKVK, Bangalore

ABSTRACT

India is one of the leading coconuts producing country in the world. Yet, it does not figure in the list of leading exporting countries of desiccated coconut. The study makes an attempt to examine the Economics of production and marketing of desiccated coconut. The study was conducted in Tiptur taluk of Tumkur district. The primary data was collected from the 10 desiccated coconut entrepreneurs from Tipturtaluk of Tumkur district of Karnataka.The study reveals that desiccated coconut units were utilizing 78.9 per cent of their installed capacity. The net return per tonne of coconut processed by desiccated coconut units was lakh Rs. 32.1, lakh Rs.81.1, and lakh Rs. 114.5 in case of small, medium, and large sized units respectively. On an average Rs. 144.01 lakhs was needed for the establishment of desiccated coconut unit. Out of which the investment on land (53.64%) was a major cost followed by machinery (26.15%), building (15.11%), and power installation (5.11%). The break even analysis showed that all categories of units had processed more than breakeven level. The investment made on desiccated coconut units was realized back within 2 to 3 years. The major constraints faced by desiccated coconut units were lack of skilled labor, limited market for desiccated coconut, higher requirement of working capital, and use of obsolete machines. Motivating the entrepreneurs to improve the quality standards, and modernization of desiccated coconut units need to be considered to improve the performance of desiccated coconut sector.

Key words: Coconut, production, economics, marketing, investment

INTRODUCTION

The coconut palm, Cocos nucifera is a member of the family Arecaceae (palm family). The term coconut refers to the entire coconut palm, the seed, or the fruit, which, botanically, is a drupe, not a nut. The term is derived from 16th century Portuguese and Spanish cocos, meaning "grinning face", from the three small holes on the coconut shell that resemble human facial features. Coconut is produced in 92 countries worldwide on about 11.80 million hectares (29.5 ac) of land. The world production is estimated to be 61.72 million tons (FAO, 2010) with an average yield of 5.26 tons / ha.India ranks third in the world coconut map and became the largest producer of coconut with the production of 16.9 billion nuts from about 1.89 million hectares.

In Karnataka coconut is cultivated in all the 30 districts of the state. It is grown in coastal belt as well as in interior parts. However, Tumkur, Hassan, and Chitradurga and Dakshina Kannada districts are the major districts cultivating coconut, together accounting for over 75 per cent of the state's area. In Karnataka, Tumkur district ranked first both in terms of area (1,10,937 hectares) and production (5474.91 lakh nuts) of coconut.

Desiccated coconut is coconut meat which has been shredded or flaked and dried to remove as much moisture as possible. The desiccated coconut is generally kept in a cool, dry place out of the light. Dried coconut can be stored in the freezer or fridge to keep it fresh. Desiccated coconut can also be stored at room temperature up to a year safely.

MATERIAL AND METHODS

The present study was conducted in Tiptur taluk of Tumkur district of Karnataka. A large number of desiccated coconut units are located in Tiptur taluk. Hence the present study is taken up in this district. In the district nearly 70 desiccated coconut production units are operating. Out of seventy, ten desiccated coconut units in Tiptur taluk which are currently operating are selected for the study.The descriptive statisticswas used for estimating the cost and returns of desiccated coconut production. Averages, percentages and ratios were worked out using the data collected and are presented in tabular form to facilitate easy comparison. Breakeven analysis was employed to know the minimum level of production required to recover the fixed capital employed in desiccated coconut production units the concept is important in any business as it indicates minimum amount of business necessary for operating the enterprise in the short run without loss.

RESULTS AND DISCUSSION

The general characteristics of desiccated coconut units reveals that, 30 per cent are small sized (3 units), 40 per cent (4units), are medium sized and 30 per cent (3 units) are large sized units. The presence of more number of medium sized units is mainly because majority is partnership firms. Among the small sized units majority are sole proprietorship firms because small sized units require less capital investment.Out of the total number of desiccated coconut entrepreneurs 40 per cent (4 entrepreneurs) have taken it up has an independent venture and 50 per cent of entrepreneurs have ventured into desiccated coconut production in the form of vertical integration to reduce the number of intermediaries and thereby reducing marketing cost. Most of the desiccated coconut units are located in towns like Tiptur, because of the advantages like better transportation facilities, availability of skilled labor, better power supply compare to villages and ease of locating customers like bakers and commission agents.

Total Cost of Desiccated Coconut Production in Desiccated Coconut Unit of Different Sizes

The structure of the variable cost clearly indicated that cost of raw material (Rs. 1, 09,040/tonne) was the most significant component (77.50 %) of the total production cost.Marketing cost involves transportation cost, commission charges, sales tax and packing material. Packing material and Sales tax (Rs. 3850 and 2932.8 per tonne) constitutes a major component in marketing cost. Among the different categories of desiccated coconut units large units incurred higher marketing cost per tonne (Rs.11, 756) because they are selling large proportion of their total output to other states which adds to marketing cost in the form of additional tax as well as transportation cost. Interest on fixed capital (Rs.674 per day) forms next major component of total cost due to the fact that higher investment on establishment of units (Rs. 144.01 lakhs, Rs. 298.74 lakhs and Rs. 234.58 lakhs in case of small, medium, and large units respectively). Large sized units total cost is comparatively less than medium sized units because large sized units are very old (30 years). When compared to large and small sized units, medium sized units paid higher interest amount of fixed capital per tonne (Rs.1131.10 as against Rs.475.46, Rs.263.06 in case of small and large units respectively). Fixed cost per tonne was marginally higher in medium sized units followed by small sized units compared to large sized units due to higher interest amount on fixed capital. Depreciation cost (Rs. 1737.7, Rs. 2750.8 and 930 respectively), salaries to permanent employees (Rs. 200, Rs. 350 and Rs. 450 respectively).License fee (Rs. 15.68, Rs. 12.44and Rs.8.92 respectively) and other expenditure like insurance premium. Though establishment cost is high in case of large sized units, fixed cost per tonne is low (Rs. 5657.93). This is because of economies of scale production and better utilization of processing capacity (76.19 % of installed capacity). There is need to increase the utilization capacity by small sized units.

Returns from Desiccated Coconut Units of Different Sizes

Desiccated coconut is the main product while processing coconut, and hence constitutes the major component of gross returns. Byproducts obtained in desiccated coconut production are pairing and shells. Even though the investment was higher in larger sized desiccated coconut units, the total cost per tonne of coconut processed was (Rs 77,694.15, Rs 1,49,735.5 and Rs.1,91,652.4 in case of small, medium and larger sized desiccated coconut units respectively), less compared to other sized units. The gross returns realized per tonne is more in large sized units (Lakh Rs. 265.2) compared to medium (Lakh Rs. 530.4) and small sized (Lakh Rs. 689.52) units because small sized units sell their product whenever price of product is high whereas in case of medium and larger sized units selling of product is on contract basis at a particular price. Net returns realized per tonne are high in large processing units (Rs. 114.56, Rs. 81.19 and Rs. 32.12 in case of large, medium and small sized desiccated coconut units respectively). The higher net returns among large sized desiccated coconut units were mainly due to the economies of large scale production.Thus, it is clear that the large sized desiccated coconut units earn more profit by more efficient as well as large scale production of desiccated coconut and byproducts. Though desiccated coconut units of all sizes are net gainers (ranging from lakh Rs. 32.12 in case of small sized units and lakh Rs.81.19 in large sized units lakh Rs.114.56), the results indicate profitability and the size of the unit are directly proportional. The net returns in all the units indicate positive value addition.

CONCLUSION

- The investment on desiccated coconut units is economically viable and economic viability improves with increase in the size of the DC units.
- To modernize the processing units can increase the production efficiency and improve the economies of scale to make desiccated coconut price competitive.
- An uninterrupted quality power supply will enhance the efficiency of desiccated coconut production units.
- Popularizing the machinery for deshelling of coconut may address the labor scarcity problem to some extent.

Table 1 : Total cost of desiccated coconut production in desiccated coconut units of different sizes

(Amount in rupees)

Sl. No.	Particulars	Size of desiccated coconut unit							
		Small		Medium		Large		Pooled	
		Amount	Percentage to the total	Amount	Percentage to the total	Amount	Percentage to the total	Amount	Percentage to the total
A	**Variable Cost**								
1	Operational Cost								
A	Cost of raw material	58000	74.65	116000	77.47	150800	78.684	109040	77.50
B	Utilities (Gas, Generator)	60	0.08	168.24	0.11	149.74	0.078	130.21	0.09
C	Repairs and maintenance	20	0.03	22	0.01	28	0.015	23.2	0.02
D	Office maintenance	10	0.01	12	0.01	18	0.009	13.2	0.01
E	Labor (150/-)	6000	7.72	7500	5.01	8250	4.305	7275	5.17
F	Telephone charges	6	0.01	9	0.01	15	0.008	9.9	0.01
G	Power	850	1.09	1050.33	0.70	1200.33	0.626	1035.23	0.74
	Total operational Cost	64946	83.59	124761.6	83.32	160461.1	83.725	117526.8	83.53
2	Marketing Cost								
A	Packaging material cost	2000	2.57	4000	2.67	5500	2.87	3850	2.736
B	Sales tax (2%)	1560	2.01	3120	2.08	4056	2.12	2932.8	2.084
C	Commission charges	600	0.77	12	0.01	1500	0.78	634.8	0.451
D	Transportation cost	300	0.386	500	0.33	700	0.37	500	0.355
	Total marketing Cost	4460	0.06	7632	5.10	11756	6.13	7917.6	5.63

Contd...

3	Total working capital (1+2)	69406	89.33	132393.6	88.42	172217.1	89.86	125444.4	89.16
4	Interest on working capital @ 8% pa	5552.48	7.15	10591.49	7.07	13777.37	7.19	10035.5	7.13
	Total variable cost (3+4)	74958.48	96.48	142985.1	95.49	185994.4	97.05	135479.9	96.29
B	Fixed cost								
A	Depreciation	1737.78	2.24	2750.8	1.84	930	0.485	1900.6	1.35
B	Rental value of land(15,000/month)	300	0.39	2500	1.67	4000	2.087	2290	1.63
C	Salaries to permanent laborers	200	0.26	350	0.23	450	0.235	335	0.238
D	Insurance premium	6.75	0.009	6.06	0.00	5.95	0.003	6.234	0.004
E	License fee	15.68	0.020	12.44	0.01	8.92	0.005	12.356	0.009
F	Interest on fixed capital @ 12%pa	475.47	0.61	1131.1	0.76	263.0667	0.137	674	0.479
	Total Fixed Cost	2735.67	3.52	6750.433	4.51	5657.937	2.95	5218.257	3.71
C	**Total Cost (A+B)**	**77694.15**	**100**	**149735.5**	100.00	**191652.4**	100.00	140698.2	100.00

Table2 : Return from desiccated coconut units of different sizes

Sl. No	Particulars	Size of DC unit			Pooled
		Small	Medium	Large	
1.	Quantity of coconut processed (tonnes/year)	1200	2400	3120	2256
2.	Quantity of desiccated coconut obtained (tonnes)	300	600	780	564
3.	Quantity of byproduct obtained (tonnes)				
	a) Parings	60	120	156	112.8
	b) Shell	240	480	624	451.2
	Total (tonnes)	300	600	780	564
4	Price of DC realized (Rs./tonnes)	78000	156000	202800	146640
5	Price of parings realized (Rs./tonnes)	30000	60000	78000	56400
6	Price of shell realized (Rs./tonnes)	5500	11000	14300	10340
7	Value of main product (lakh Rs.)	234	468	608.4	439.9
8	Value of by-product				
	a) Parings	18	36	46.8	33.8
	b) Shell	13.2	26.4	34.3	24.8
	Total (lakh Rs.)	31.2	62.4	81.1	58.6
9	Gross returns (lakh Rs.)	265.2	530.4	689.5	498.5
10	Total cost / day (lakh Rs.)	77694.1	149735.5	191652.4	149294.6
11	Total cost (lakh Rs.)	233.0	449.2	574.9	447.8
12	Net returns (lakh Rs.)	32.1	81.1	114.5	50.6

Note

1. Paring is nothing but outer brown portion (testa) of coconut kernel. Shell is hard covering of the kernel. Paring and shells are obtained as byproduct during desiccated coconut production.
2. On an average 25 kg parings, 100 kg shells were recovered from 800 to 1000 coconuts.
3. One tonne of coconut comprises of 1800 to 2000 number of coconuts.

REFERENCES

1. George, M.V. and Pillali (1999). Marketing strategy for stabilization of coconut prices. Indian Coconut Journal, 30(3): 202-207
2. Sudhahari, B. and Munkundan, K. (1992). Economics of desiccated coconut industry with special reference to Kozhikode and Kannur districts in Kerala. Indian Coconut Journal, 23(6):12-16

CHAPTER - 43

Multitier Cropping System for Profitability and Scalability in Vegetable Production

S.S. Solankey[1], Shirin Akhtar[2] and Aditya[3]

[1,2] *Department of Horticulture (Vegetable & Floriculture), Bihar Agricultural College, Bihar Agricultural University, Sabour (Bhagalpur), Bihar – 813 210*

[3] *Department of Extension Education, Bihar Agricultural College, Bihar Agricultural University, Sabour (Bhagalpur), Bihar – 813 210*

ABSTRACT

India is the second largest producer of vegetables in the world after China with an annual production of 162.19 million tonnes from an area of 9.2 m ha. The requirement of vegetables in the country has increased a lot to meet the nutritional requirement of an estimated 1200 million population expected by 2020-21. Although the productivity levels of vegetables have increased manifold, it would not be sufficient to feed the ever increasing population. Production per unit area of land, time and inputs can be increased by improving efficiency of the rate and extent to which solar energy is harvested for conversion to economic produce. In multiple cropping systems, the possibility of more efficient use of resources like sunlight, nutrient and water per unit area is higher leading to higher production stability. It can be very much successful in high imput requiring crops like vegetables. Intensive cropping involving multi-tiered vegetable crops like pointed gourd + elephant foot yam + cowpea grown on the single piece of land at the same time can generate profits up to Rs 1,30,000 per acre. India ranks first in the youth population. If the energy of these dynamic youth is properly utilized in the field of better agricultural practices, especially vegetable cultivation, it will help improve the national productivity of vegetable crops and generate employment as a whole. Thus, intercropping involving multi-storied/ multi-tier cropping system can serve as well-established practice for sustained supply of vegetables throughout the year through which rural youth can generate high income for sustainable livelihoods and increased per capita income.

Key words: Multi-tier cropping, youth, income, vegetable, employment.

INTRODUCTION

Mono-cropping is exception, while mixture (of species)is the rule of nature. In multi-tier systems, the possibility of more efficient use of resources like sunlight, nutrients andwater leads to increased production stability. Besides, the root systems of the component crops are also located at distinct zones so as toexplore the soil for moisture and nutrients. Intercrops serve as an insurance against the menace of pestand disease, vagaries of weather, market fluctuation and helpto increase the net profit to growers (Reddy and Mohammad, 2009). India is the second largest producer of vegetables in the world after China with an annual production of 162.19 million tonnes from an area of 9.2 million hectares (Anonymous, 2013). Although the productivity levels of vegetables have increased manifolds, it won't be sufficient to feed the ever increasing population. This will complicate the issue of price rise further leading to inflated price of vegetables (Anonymous, 2009). India ranks first in the youth population. If the energy of these dynamic youth is properly utilized in the field of better agricultural practices, especially vegetable cultivation, it will help improve the national productivity of vegetable crops and generate employment as a whole.

Multistoried /multitier cropping system: As the name indicates, under this system two or more crops of different heights are grown simultaneously on a certain piece of land in any certain period of time.Thus land, water, and space are utilized in most efficient and economic way. Multistoried/ multitier cropping system aims at better use of the production components such as soil, water, air, space, radiation and other inputs on sustainable basis (Singh *et al.*, 2014).

Some of the examples of multistoried cropping system are:

- Sugarcane + Okra + Onion
- Cotton + Radish + Cluster bean + Beet root
- Cotton + Radish + Beetroot + Coriander
- Sugarcane + Mustard + Onion.
- Coconut + Pepper + Pineapple.

Advantages of Multistoried Cropping System

- Better use of growth resources including sunlight, nutrient and water.
- Suppression of weeds, reduced plant and disease incidence and yield stability.

- Ecological stability i.e. improved soil health and agro-ecosystem.
- Flow of income during cropping periods.
- Other miscellaneous benefits like distribution of labour, physical support of one crop to another and homegardening leading to more food supply

Disadvantages of Multistoried Cropping

- Labour intensive.
- Control of pest and disease or chemical weed control may be difficult.
- Difficulty in mechanization.
- Adverse competitive effect or by allelopothy.

Vegetable Based Multistoried Intercropping

Different workers have studied the benefits of multistoried vegetable cropping system. In one such study by Kumar *et al.* (2012), in summer season tomato was intercropped with okra, potato, corn, chilliand eggplant whereas among winter vegetables, pea in combination with cauliflower, potato, lettuce, radish, turnip, coriander and spinach was grown. The yield was generally more when each vegetable was grown as a single crop. But the focus of this study was to evaluate them in combinations. Pea + radish intercropping was found to be the most economical. This combination gave significantly maximum net income of Rs. 1,81,648 ha^{-1} due to highest land equivalent ratio (1.54), followed by yield of pea and radish (6.5 and 68.2 t ha^{-1} respectively) resulting in a net incomeof Rs. 80,800 ha^{-1}from pea +lettuce combination, which yielded 7.0 and 5.5 tha^{-1}, respectively. Pea intercropped with spinach and potato remained least economical and resulted in the minimum net income of Rs. 43,215 and Rs. 44,949ha^{-1}, respectively. Similarly, summer vegetables produced more yields when grown as single crops, than their respective yields in intercropping except tomato grown in okra and corn. The best companion of tomato was okra and their combination gave maximum net income of Rs. 1,34,241 ha^{-1} due tomaximum yield of tomato (27.61 t ha^{-1}) as compared to that in sole crop (20.34 tha^{-1}) and high market price of both vegetables. Tomato + corn combination resulted highest net income of Rs. 12,579 due to high yields of tomato in intercropping and stability of corn yield in both years. The maximum net incomewas also due to high market price of tomato and corn. However, heavy reduction in yields of all vegetables (grown alone or incombination) in hot summer season compared to those in mild summer season has been found in all cases.

Three-tiered Cropping in Coconut Orchard

Rahim and Islam (1998) produced a good indication of multi-storied cropping system in threelayered canopy configuration where the first layer consists of aroids (*Colocasia*) or gingeror pineapple, the second layer banana and the third layer was coconut. The orchard plants (coconut) also get benefit from thecultivation under multistoried system.

Cotton Based Multitier Intercropping System

This multitier system was developed through rigorous testingof eight cotton based systems in comparison to sole cotton under irrigated condition of south zonewith *Bt* cotton (Sankaranarayanan *et al.*, 2012).

- Cotton + Radish + Cluster bean+ Beet root
- Cotton + Radish + Beetroot + Coriander

Selection of varieties: The recommended cultivars of different vegetables under this system are as follows:

Cotton (*GossypiumhirsutumL.*)	:	RCH–20 Bt.
Coriander (C*oriandrumsativum*L.)	:	Surabhi.
Radish (*Raphanussativus*L.)	:	PusaChetki.
Beet root (*Betavulgaris* L.)	:	Detroit Dark Red.
Cluster bean (C*yamopsisteragonolaba*L.)	:	PusaNavbahar.

Methods of planting followed: The field was ploughed once with tractor drawn mould board plough and then harrowedtwice. *Bt* hybrid was planted at 120 x 45 cm, where two ridges at 60 cm apart were formed, and various intercrops (three crops) were planted on 4 sides of the 2 ridges in sequence. Radish,coriander and others vegetables (beet root and cluster bean) were planted at intra-row spacing of 10,15 and 20 cm respectively. No additional irrigation was provided for the intercropping systems(similar to sole cotton crop).Yield of cotton under sole crop is 2,615 kg ha^{-1} however under intercropping a yield of 2,545 kg ha^{-1}has been obtained. In the multi-tier cropping system comprising of cotton +radish + cluster bean + beet root, the recorded yield of radish cluster bean and beet root has been found to be 6,660 kg ha^{-1}, 4,536 kgha^{-1} and 5,671 kg ha^{-1}respectively. The other prominent multi-tier cropping system was radish + beet root + coriander where yield of radish,beet root and coriander has been recorded as 5,091 kg ha^{-1}, 5,347 kg ha^{-1} and 3,536 kg ha^{-1} respectively.Mono-crop of radish, beet root, cluster bean, and coriander produced yields of 21,475 ha^{-1} kg,17,123 kg ha^{-1},7,585 kg ha^{-1} and 15,142 kg ha^{-1} respectively.

Elephant Foot Yam - A Profitable Crop for Vegetable-based Multi-tier Cropping System

Under changing climate and unpredictable rainfall it is a necessity for farmers to secure production by adopting multitier cropping system. In this method farmers can grow vegetable crops i.e. sweet corn, bottle gourd, bitter gourd, French bean etc. with elephant foot yam to get more benefit. By adoption of this system they can grow two to three vegetable crops simultaneously in a single piece of land at thesame time and finally improve their lifestyle and economic status. Intensive multi-tiered cropping can involve crops like:

Pointed gourd + elephant foot yam + cowpea: This multitier system is very popular in northern plains of India especially in Bihar where elephant foot yam is grown on large scale.

Selection of varieties	:	Elephant foot yam : Gajendra, Sree Padma, Kusum
Pointed gourd	:	Rajendra Parwal-1, Rajendra Parwal-2
Cowpea	:	KashiKanchan, KashiGauri, ArkaGarima.

In this system farmers planted pointed gourd vines in the first fortnight of November and when the vines sprout (20 – 25 days after planting) they trained it on 7 ft bamboo poles in the square system with the help of nylon threads and make net like structure for proper spreading. After that transplanting of elephant foot yam is done in the furrows, followed by cowpea that is sown in the rest of embedded land. In general cowpea crop becomes ready for harvesting in 40 – 45 days after sowing and 5 – 7 pickings can be taken. However, the vines of pointed gourd take longer time to fruit and fruiting starts in the month of May – June as the vines cover the whole net over the bamboo *machan*. Moreover, the crop of elephant foot yam will beready for harvest in October –November month. If the economics of this multitier cropping system is analyzed in per acre of land then farmers can get 30 quintals of cowpea green pod, 70 – 80 quintals of pointed gourd fruits and 50 – 60 quintals of elephant foot yamcorm from a single acre. Finally after selling of these vegetables farmers can generate a profit of approximate Rs. 1,30,000 per acre of land.

Structure of the Multi-tier Home Garden

Generally home gardens consist of herbaceous layer near the ground, a tree layer at the upper level and anintermediate layer between the two crops. The bottom area can be split in two with the lowest sectionbeing less than one meter in height and dominated by different vegetables and medicinal plants. The secondlayer is from one to three meters in height and composed of food plants (such as cassava and banana). Theupper section can also be divided in two, consisting of fully-grown timber and fruit trees in the uppermost layer(over 25 meters) and below it, medium-sized fruit and timber trees between 10 – 20 meters. This type of modelcan provide an array of products for consumption and sale. Thus in a very small area, a plethora ofspecies can be found which provide a variety of uses and purposes.

- Coconut + Banana + Taro
- Coconut + Banana + Ginger
- Coconut + Banana + Turmeric

Dulbergiasissoo based multistoried cropping system

Rahim*et al.* (2007) studied impact of growing vegetables in multistoried cropping system in *Dulbergiasissoo* and coconut orchards. The salient findings of their research are as follows:

1. Brinjal (*Solanummelongena*):The highest yield (52.4t/ha) was recorded under open condition which was followed by *D. sissoo*+ guava based cropping system (46.1t/ha). Significantly the lowest yield (37.7 t/ha) was found under the *D. sissoo*+ lemon based cropping system. The economic analyses showed that *D. sissoo*+ guava based cropping system is clearly preferred production system (B: C ratio 5.12).

2. Elephant foot yam (*Amorphophalluscampanulantus*):Yield of *Amorphophallus*was influenced significantly by different production systems. The highest yield of *Amorphophallus*(24.1t/ha) was recorded under *D. sissoo*+ guava based cropping system. The benefit cost ratio was also higher in *D. sissoo*+ guava based cropping system and it was 4.6 times higher than the sole cultivation.

3. Arvi (*Colocasiaesculenta*):The lowest yield of *Colocasia*(13.6t/ha) was found under open condition and was significantly followed by *D. sissoo*+ lemon based cropping system. The highest yield of *Colocasia*was noted under *D. sissoo*+ guava based cropping system. The economic analyses show that cropping systems gave better B: C ratio than that of sole cropping. The highest yield of *Colocasia*(31.8t/ha) was recorded under *D. sissoo*+ guava based cropping system. Yields obtained from *D. sissoo*+ lemon based cropping system and open condition were statistically similar. Again, *D. sissoo*+ guava based cropping system gave highest B: C ratio which was 3.1 times higher than the sole cropping of *Colocasia*i.e. production under open condition.

Coconut based Multistoried Cropping System

1. Brinjal (*Solanummelongena*):The highest yield (52.4t/ha) was recorded under open condition followed by coconut + guava based cropping system (38.4t/ha). Significantly the lowest yield was found under the coconut + lemon based cropping system. The economic analyses show that coconut + guava based cropping system is clearly preferred production system (B: C ratio 6.50).

Vegetable Based Multitier System for Soil Health Improvement and Profitability

In a study conducted by Jakhar*et al.*, 2012, it has been found that among different treatments of multitier cropping of ginger: pigeonpea yielded minimum runoff of 45.2 mm. Square plantation of papaya in sole ragi and intercropped ragi: pigeonpea (6:2) plots with boundary plantation of *Gliricidia* on bunds reduced net soil loss by 36 and 45%. Cultivated fallow reported maximum nutrient loss to the tune of 109.9, 0.31 and 10.70 kg ha"[1] for organic carbon (OC), Phosphorus (P) and Potassium (K), respectively. Sole ginger and ginger intercropped with pigeonpea in papaya + *Gliricidia* system recorded 58 and 41% lower OC loss than the farmers' practice of ragi broadcasting. Multitier cropping of papaya + *Gliricidia* withginger: pigeonpea intercropping recorded highest values of enrichment ratio for OC and P to the tune of 2.25 and 4.56, respectively. Ginger, pigeonpea and ragi recorded an increase of 23, 14.4 and 25.8% in cation exchange capacity, when raised under multitier cropping system. In comparison to sole ragi cultivation, its intercropping with pigeonpea under multitier system showed an increase of 14.7 and 34.9% for infiltration and porosity, respectively. In comparison to sole cultivation,ginger under papaya intercropping gave 15 qha^{-1} higher yield and in ginger + pigeonpea intercropping system the yield enhanced by 21 qha"[1], respectively. Among all treatments, papaya+ *Gliricidia* based multitier cropping system of ginger intercropped with pigeon pea (8:2) gave maximum net returns of Rs. 2.47 lakhs with B: C ratio of 2.55 followed by sole ginger.

SUMMARY AND CONCLUSION

In high density population areas, a multi cropping system is imperative to provide households with enough food for consumption as well as income for other essential goods. Multi-tier cropping systems are appropriately such dynamic interactive practices aimed at better use of the production components such as soil, water, air space, solar radiation and all other inputs on sustainable basis. The introduction of non competitive, short duration, multi intercrops into fruit orchards or heighted plants, salvaged the risk perturbed by mono-cropping. Higher production, economic return and resource utilization realised with vegetable based multi-tier system is as well as advantageous in more than one ways. Hence, channelizing the dynamic energy of our youth into such cultivation system would definitely change their outlook towards conventional farming and make a pathway for higher income generation. Thus, adaptation of vegetable based multi-tier cropping system can become a means of livelihood security by the rural youth.

REFERENCES

Anonymous (2009).The Survey of Indian Agriculture.The Hindu Kasturi Building, Anna Salai, Chennai.

Anonymous (2013).National Horticulture Board, Department of Agriculture and cooperation, Government of India pp.4.Retrieved from www.nhb.gov.in.

Jakhar, P., Barman, D., Gowda, H.C.H. and Madhu, M. (2012). Multitier cropping system for profitable resource conservation and sustainable management in sloping lands of eastern India. Indian Journal of Agricultural Research. 46(4): 309 – 316.

Kumar, Pawan, Nanwal, R.K., Dhindwal, A.S. and Yadav, S.K. (2012). Competition relations; multi-storied cropping and yield stability in Intercropping (Chapter7). pp. 77-88. Manual on cropping systems and sustainable agriculture.Department of Agronomy, CCS Haryana Agricultural University, Hisar, Haryana, India.

Raghu Rami Reddy, P. and Shaik, Mohammad (2009). Evaluation of cotton (Gossypiumhirsutum)-based intercropping system through different approaches under rainfed conditions. Indian Journal of Agricultural Sciences 79 (3): 210–4.

Rahim MA, Islam KS (1998) Multi storied tree cropping system: a sustainable land use system and vertical yield achievements in a land scarce community forestry conditions. In: Proceedings of the international seminar on cultivating forests: alternative forest management practices and techniques for community forestry, Bangkok.

Rahim, M.A., Bari, M.S., Islam,Ferdousi and Naher, Nazmun (2007).Multistoried Cropping System - A New Room for Indigenous Vegetables and Medicinal Plants.Acta Hort. 752: 493 – 497.

Sankaranarayanan K., Nalayini P. and Praharaj C.S. (2012). Multi-tier cropping system to enhance resource utilization, profitablity and sustainablity of Btcotton (Gossypiumhirsutum) production system. Indian Journal of Agricultural Sciences 82 (12): 1044–50.

Singh, R.P., Pande, P., Solankey, S.S. and Chatterjee, A. (2014).Cropping Systems in Vegetables. In: Olericulture Fundamentals of Vegetable Production. Vol. I, (Editor: Singh, K.P. and Bahadur, Anant). Kalyani Publishers, New Delhi. pp. 348-373.

CHAPTER - 44

Prospective of Conservation Agriculture for Sustainable Integrated Farming System – A Case Study

Gurnam Singh[1], Sardul Mann[2], and Satyajeet Yadav[3]
CCS Haryana Agricultural University, Hisar, Krishi Vigyan Kendra,Fatehabad Haryana 125050
1DES (Agricultural Economics), KVK, Fatehabad

ABSTRACT

Agriculture is an important component of Indian economy contributing over 14% to the National GDP and employment to about 52% of the workforce. The small farm families form the backbone of Indian agriculture. The sustainability of small farms needs to be continuously supported with appropriate technologies and also with promoting off farm employment opportunities in the rural sector. The role of KVKs' in preparing the next generation farmers' through imparting training to update their knowledge and skills in modern agricultural technologies related to technology assessment; refinement, demonstration and training of extension personnel to orient them in the frontier areas of technology development. Sh. Narender Singh s/o Sh. Jagdish Rai, Graduate, 42 years old youth from Dharania Village came in contact with KVK, Fatehabad in 2008 and got training from this centre. He was motivated to diversify his traditional system of agriculture with integrated farming system of Dairy Farming, Vegetable, Seed production, inter cropping of Guar in cotton. He also used neem cake and bio control agents in poly houses for management of root knot nematode. He is re-cycling cow dung from the dairy farm and producing bio gas for domestic consumption with the diversification of farming system. The total expenditure was Rs. 5,55,645 and gross return was Rs. 13,63,100. The farmer benefit cost ratio is 2:45.

Keywords: Poly house, Employment, Skill, Rainfall, Seedling, Nursery Bio-gas

INTRODUCTION

Agriculture is an important component of Indian economy contributing over 14% to the National GDP and employment to about 52% of the workforce. The small farm families form the backbone of Indian agriculture and a recent National Sample Survey Data reveals that small holdings in Indian agriculture exhibit higher productivity than larger holding. The sustainability of small farms needs to be continuously supported with appropriate technologies and also with promoting off farm employment opportunities in the rural sector. Nowadays, Indian Agriculture has multi-facet challenges in the form declining productivity of land, labour and water, other natural resources. In spite of this situation technologies continue to play a major role in achieving sustainable production in agriculture and allied sector. The growing concerns for sustainable agriculture have been sent as a positive response to limits of both low inputs, traditional agriculture and intensive modern agriculture relying on high levels of inputs for crop production.

Role of Krishi Vigyan Kendra

In India there are 639 Krishi Vigyan Kendras. The role of KVKs' in preparing the next generation farmers' through imparting training to update their knowledge and skills in modern agricultural technologies related to technology assessment; refinement, demonstration and training of extension personnel to orient them in the frontier areas of technology development. KVKs' also work as resource and knowledge centre of agricultural technology for supporting initiatives of public, private and voluntary sector for improving the agricultural economy of district through production and supply of good quality seeds, planting materials and various bio products to the farming community.

General Description of the District

Fatehabad is situated in western Haryana, Punjab in North, Hisar district in east and Rajasthan in South and Sirsa district in West surround the Fatehabad district. Fatehabad district came into existence on July 15, 1997. It was carved out of Sirsa and Hisar districts. The district has three sub divisions namely Fatehabad, Tohana, Ratia and six development blocks namely Fatehabad, Tohana, Ratia, Bhuna, Bhattu-Kalan and Jakhal. Fatehabad is located at 29.30°N, 75.27°E. It has an average elevation of 224 meters (734 feet). This city is situated on National Highway No. 10 connecting the Sirsa to Delhi. The total geographical area of 252304 ha. As of 2011 census and having an population of 1883044 male constitute 989668 and female 893376. The climate of district is of very pronounced character i.e. very hot in summer and very cold during winter. Temperature ranges from -1 to 48 degrees Celsius. Annual rainfall is 267 mm. Topography of the district is plain and sand dunes soils are sandy, sandy loam and clay with pH ranging from 8 to 9. In Fatehabad 56, 36 and 8 percent water poor, marginal and good quality respectively.

Cropping pattern: Paddy-wheat and cotton-wheat is the main crop rotation followed in the district. The average yield of paddy, cotton and wheat are 5809, 768 lint and 46.81 quintals per hectare respectively. The other crops grown are Guar, Bajra, Oilseeds and Pulses.

Case Study

Sh. Narender Singh s/o Sh. Jagdish Rai is a small progressive farmer cultivating 3.6 ha land. He started agriculture with traditional crops of cotton, bajra, guar and wheat after doing his graduation in the year 2000-01. He realized very soon that his farm income was insufficient to meet out the family requirements.

Fortunately, he came in contact with KrishiVigyan Kendra, Fatehabad on 15.12.08. After acquiring trainings in diversification in Agriculture, he started vegetable, horticulture and seed production programme in the crop of cereal and vegetables under the technical guidance of KVK Scientists. He also visited CCS HAU, Hisar for establishing polyhouse for vegetable production. His net earnings from vegetable production from polyhouse are Rs. 232000/- and Rs. 285000/- during the year 2012-13 and 2013-14 respectively. He is very innovative and established drip irrigation system in cotton to save water. He is very regular in laser land leveling. For the last two years he is producing vegetable seeds of potato, garlic, onion and carrot. He is re-cycling cow dung from the dairy farm and producing bio-gas for domestic consumption. He is also innovative by doing inter-cropping of guar in cotton, coriander in papaya and onion in papaya. He is using neem cake and bio control agents in poly houses for management of root knot nematodes. He has four improved cattle (2 buffalos and 2 cows).

Brief Bio-Data

Name	:	Narender Singh
Father's Name	:	Jagdish Rai
VPO	:	Dharania
Teh.&Distt.	:	Fatehabad
Mob. No.	:	97280-71270
Age	:	42 years

Education : Graduation (B.A.)

Land Holding : 3.6 ha

Members

- District Progressive Kisan Club, Fatehabad
- Scientific Advisory Committee (SAC) of KVK Fatehabad
- ATMA (Deptt. of Agriculture) Fatehabad
- Life Membership of Haryana Kheti.

Awards

- Ch. Devi Lal Kisan Puraskar in 2002-2003.
- Sir Chhotu Ram Kisan Puraskar on Chhotu Ram Jyanti in 2008-09.
- District Level Best KisanPuraskar on KisanDiwas, 23.12.2013.
- Honoured by Dainik Jagaran Newspaper.

Technology Used

- He is very regular in laser leveling of his entire fields by which he saves a lot of water and energy.
- He used Bed Planter for growing vegetables and vegetable seed production programme (Potato, Garlic, Onion and Carrot)
- He has established drip irrigation system in cotton.
- He has installed a bio-gas plant with the help of this he is maintaining soil health by applying through FYM and Vermi-composting.
- He is using neem cake and bio-control agent in poly-houses for root knot nematode management.
- He has adopted crop diversification (cereals/pulses/vegetables/oil seed/polyhouses).

Salient Achievements

- He is producing vegetable seed of carrot and onion and sold onion nursery also.
- He is also growing molya resistant variety Raj MR-I and growing mustard crop in nematode prone area.
- He is doing inter-cropping of guar in cotton, papaya + coriander, papaya + onion.

- He is producing hybrid seed of Desi cotton AAH-I, after taking parent line from CCS HAU, Hisar.
- He is a good seed producer of vegetable crop in-collaboration with NSC and other seed agencies.

Sh. Narender Singh is a Small but progressive and innovative farmer of District and has increased his farm income through diversification and adoption of improved technologies:-

Economic analysis of Vegetable Production in Poly House; Area = 1960 m²

Seasons	Crop	Variety	Area	Expenditure* (Rs.)	Gross Returns (Rs.)	Net Returns (Rs.)	BC Ratio
Nov.12 to April 13	Cucumber	Hilton	1960 m²	1,28,000	3,60,000	232000	2.81
Aug.13 to Nov. 13	Cucumber	Hilton and Multistar	1960 m²	40,000	1,60,000	120000	4.0
Nov. 13 to April 14	Tomato	Savellia and Himsikhar	1960 m²	45,000	2,10,000	165000	4.66

Economic Analysis of Crops grown during Kharif 2013

Crop	Area (ha)	Variety	Yield (Qtl)	Expenditure* (Rs.)	Gross Returns (Rs.)	Net Returns (Rs.)	BC Ratio
Cotton	0.2	AAH-I	3	17200	54000	36800	3.13
Cotton	1.6	Rashi-134, Mahyco 6304, Bio seed 6488	32	136445	176000	39555	1.29
Cotton+ Guar	1.2	Bio seed 6488, HG-563	21 (Cotton) 9 (Guar)	105000	115500 45000 **160500**	55500	1.52
Moong	0.2	Basanti	3	4000	18600	14600	4.65
Onion nursery (seedling)	0.4	Selection	40	80000	224000	144000	2.8

**Included rental value of land.*

Images ofCase study of Sh. Narender Singh

Produced and sold Onion Seedlings (Own Selection)

Tomato Crop (Variety Savellia and Himsikhar) (Nov. 13 – May 14)

Potato Crop (Variety KufriPuskar)

Use of Drip in Vegetable Ridge Guard (Hybrid US-11)

Onion Bulb Production (Own Selection)

Carrot Seed Production (Variety PUSA Rudira)

Onion Seed Production
(Own Selection)

Carrot Seed Production
(Own Selection)

Moong Crop
(Variety Basanti)

Cucumber Crop (Variety Hilton)
(Nov. 12 – April 13)

Cucumber Crop
(Variety Multistar)
(Aug. 13 – Nov. 13)

Molya Resistant Variety Raj MR-1 and
Mustard Variety RH-30 in
Nematode Prone Area

CONCLUSION

As an occupation, agriculture seems to be considered flow esteem. Thus, unemployed youths are forced to choose another career. There are not enough initiatives from the government to provide opportunities to learn vocational skills.

This skill can be used in their daily life but also helps in choosing alternative careers. Halfhearted support to policies that encourage rural youth to take up agriculture farming will not yield expected results. The focus on formal education alienated these youth from their traditional occupations forcing them to migrate to urban areas. The discussion with young farmers/ rural youth revealed that they are innovative and can be helpful if vocational trainings/ skills are imparted to them. This case study indicated that these rural youths can see their future in agriculture related entrepreneurs if proper guidance, skills and opportunities are provided at young age.

CHAPTER - 45

Constraints in the Way of Mushroom Entrepreneurship

Binita Rani[1], Rakesh Kumar[2] and R. K. Sohane[3]
KVK, Patna1, KVK, Arwal[2] and Directorate of Extension Education[3]
Bihar Agricultural University, Sabour, Bhagalpur

ABSTRACT

Human resources are one of the most strategic and critical determinants of growth. A country cannot make rapid social and economic advancement unless it has enterprising people with necessary knowledge, skill and attitudes. Entrepreneurship development not only helps individual to become self reliant but also adds to wealth creation process, thereby paving the way to past growth and development of the nation. Rural entrepreneurship is required to capitalize on the opportunities available and to create job, mainly for rural youth and the farm family. In order to achieve this we are promoting mushroom production and processing as an enterprise for rural youths and farm family through their capacity building activities like training, consultancy, forward and backward linkages and to some extent by making available the much required inputs in the initial stages as the scenario of mushroom production in state is far from satisfactory. With the vast opportunities both at the international and national levels lying at the door steps of India, nothing less than a mushroom revolution in the coming decade will help to achieve the targets of nutritional security of people. During last year we have trained 215 farmwomen and youth in mushroom production and processing out of which 110 trainees started producing it at small scale and selling in the local market. Three villages of Patna district have been developed as "mushroom village" namely Sadikpur, Chaknawada and Mahmadpur. A sample of 100 mushroom growers of Patna district was taken for present investigation. All of them agreed that growing mushroom is a beneficial enterprise as it is capable of recycling agricultural wastes and generating employment opportunities. But, they revealed that important constraints in recommended practices of mushroom cultivation are spawn availability, perishable nature and marketing of mushroom in local villages because of poor infrastructural facility and lack of institutional finance available to them. So, in order to encourage mushroom Production at household level in villages, it is required to uplift farmers, initiating new applications, providing them with adequate marketing information and sufficient agri-business finance knowledge. Two of the most important problems in India, unemployment and malnutrition, will be addressed to a larger extent by promoting the cause of mushroom growing and its utilization as health food.

Key words: Mushroom, entrepreneurship, villages, constraints, production

INTRODUCTION

Entrepreneurship development is an essential factor, which plays a great role in agricultural sector in the era of liberalization, privatization and globalization, open competition and free market economy. Agro-based entrepreneurship activities among farmers, farm women and rural youths are vital in the contest of generating gainful employment, minimizing the regional disparities and development of socioeconomic conditions. Now there is a need to mobilize the rural youth to develop themselves as an entrepreneur based on the rural resources and market demand. The major problem identified among the rural youths involved in this sector was lack of knowledge, skill, financial support, availability of cold chain and marketing of produce.

Mushroom cultivation is considered as one of the promising enterprises capable of recycling agriculture wastes and generating employment opportunities. Therefore, growing mushroom is a beneficial enterprise for small or even marginal farmers as the vast opportunities both at the international and national levels are lying at the door steps of India and mushroom revolution in the coming decades will help to achieve the targets. To tackle this issue we are organizing training for farmers with the help of lectures, practical and skill demonstrations on mushroom cultivation floored by the support of critical inputs like spawn and polybags to popularize this commodity and to promote mushroom growing as an enterprise for rural development. Regular follow up was also arranged to help them achieve competency but the adoption of this technology is not satisfactory due to some constraints related to mushroom cultivation faced by the farmers.

The present study was therefore undertaken with specific objectives to investigate the constraints coming in the way of adopting mushroom as an enterprise at larger level.

METHODOLOGY

The study was carried out in different villages of Patna district. A sample of 100 mushroom growers out of 215 persons who participated in various training programmes organized by KVK, Patna during 2013-14 was drawn by resorting to disproportion random sampling method. Keeping in view the objectives of the study, a well structured interview schedule was prepared. The schedule was designed with relevant questions or statements and for data collection; trainees were interviewed personally in their respective villages.

RESULTS AND DISCUSSION

During the year 2013-14, 215 farmwomen and youth were trained for mushroom production and processing at KVK, Patna. Out of which 110 trainees started practicing mushroom at small scale and selling in the local market. Three villages of Patna district have been developed as "mushroom village" namely Sadikpur, Chaknawada and Mahmadpur (Table 1). A sample of 100 mushroom growers of Patna district was taken for present investigation. All of them agreed that growing mushroom is a beneficial enterprise as it is capable of recycling agricultural wastes and generating employment opportunities.

Table 1: Distribution of trained farmer on mushroom enterprise from different village of Patna

Criteria	Village			
	Sadikpur	Chaknawada	Mahmadpur	Other
No. of Household	280	315	212	-
No. of farmers trained	61	64	51	39
No. of farmwomen	16	12	30	4
No. of rural youth	45	52	21	35
No. of farmers adopting mushroom cultivation	35	28	31	16

Personal attributes like age, education and occupation were selected for this study. The study indicated that 78 % of the respondents belonged to age group between 25 to 40 years where as only 26 % of respondents had education up to college level. All of them were practicing farming as their occupation. Thus, it is inferred that even illiterate farmers and youths were interested in participation in the training programme. It might be due to the fact that they were interested in acquiring skills and knowledge gain in training and to do it in actual field situation to generate additional income. After one week of training, all of them took spawn from our KVK, and started its production at their house hold level and tried to market in the local area. After practicing mushroom production for one year they were asked about various constraints faced by them in continuing this practice. The most perceived constraints faced by the farmers in recommended cultivation practices of mushroom were complex technology for spawn production and timely unavailability of quality spawn (100 %), marketing of produce (90 %), storage of mushroom (88%) followed by value addition of mushroom (76%). Other constraints related to mushroom cultivation are chemical sterilization of straw and contamination of polybags (Table 2).

Table 2: Nature and extent of constraints faced by mushroom growers

Sr No	Nature of constraints faced	No. of responded (%)
1	Quality supply of spawn	100
2	Input requirement (polybags and formalin)	65
3	Chemical sterilization of straw	02
4	Bagging	12
5	Yield of mushroom	40
6	Storage of mushroom	86
7	Marketing of produce	90
8	Contamination of polybags	16
9	Value addition to mushroom	76
10	Insurance & loan facility	100

The major marketing constraint in adopting recommended cultivation practices of mushroom were "perishable nature of the commodity" which needs storage facility and cold chain for increasing shelf life. It is not consumed in the local market at very high level as poor ones are not consuming it regularly because of its cost i.e Rs. 100 per Kg, even value added products in the form of papad and pickles are not purchased by local people. The main financial constrains in adoption of recommended cultivation practices was "difficulty in borrowing loans" and lack of insurance facility for this enterprise. Overall the main constraints faced by the farmers were input supply, poor infrastructural facility and lack of institutional finance available to them.

CONCLUSION

Mushroom, being rich source of protein, vitamins and minerals is capable of taking care of nutritional needs of individual taking care of better health. On the basis of these findings, it may be concluded that in order to encourage mushroom production as an enterprise by rural youth and farm women, it is required to ensure availability of quality spawn at their door step and enhance their capability of producing value added mushroom products with long shelf life. They must be provided sufficient agri-business finance knowledge to avail loan facility and subsidy. The most important considerations are market net working which may be improved by formation of SHG or Kisan club in turn improving investment in mushroom cultivation and real empowerment of the rural India with production of nutrient rich fungus without farming land requirements.

CHAPTER - 46

Traditional Knowledge of Dairy Farmers about Their Animal Husbandry Practices of Nalanda District of Bihar

Sanjeev Kumar[1], Anand Kumar[2], Suryamani Kumar[3] and Jitendra Kumar[4]

[1]Programme Co-ordinator, KVK, Nalanda; [2]SMS, KVK, Nalanda; [3]Assistant Professor, SGIDT, Patna; [4]Dairy Field Officer (Govt.of Bihar) Nalanda

ABSTRACT

Documenting the traditional knowledge system for technology development considers as the first step towards a sustainable agriculture. A study was undertaken in some villages of operational area of KVK, Nalanda district to investigate and document the traditional knowledge system of dairy farmers in the area of Breeding, Feeding, Health care and Management. The findings revealed the materials and practices related to animal husbandry from the study area. The majors among them were like use of Eucalyptus leaves in controlling Haemorrhagic septicemia (Gala ghotu), allowing animal to remain stand in mud for few hours, spread out tobacco leaf power on the infected part of FMD disease, applying brine solution on the animal's buttock for the treatment of mastitis, feeding banana stem juice in order to controlling diarrohea in dairy animals. Papaya fruits fed to the cattle increases fertility in them, spraying of chhoa for controlling termites, incorporation of tobacco leaf, betel leaf and curry leaf enhances the shelf life of ghee at least by 2 months, use of Eucalyptus leaves in controlling Pneumonia in cattle.

Key words: Traditional knowledge, dairy, breeding, feeding, health care, management

INTRODUCTION

The documentation of traditional knowledge system (TKS) in agricultural education, research and extension pave the practical approach towards the sustainable development. The documentation of TKS is become necessary before the valuable information is lost forever. It may have scientific basis and its techniques could be utilized to other similar agro-climatic situations. It may be an alternative or a complement and easier to secure adoption than recent technology. It may also generate ideas for future research (Rao *et al.,* 1995). Rural communities have reservoir of expertise in management of agro ecologies along with associated agricultural livestock and aquatic systems (Prain, 1994; Kumar *et al*., 2010). In the present technology generation system, there is often no role of farmer's except a nominal or contractual participation by way of conducting adaptive trails on their field (Rai, 2005). Despite the excellent technological achievements during last two decades in the area of dairying, socio-economic and technological constraints in transfer of technology are causing hurdles in the promotion and adoption of dairy production (Sah *et al*., 1998; Kumar *et al*., 2010). Farmers belongs to Nalanda district, take care of their dairy animals with their local traditional knowledge system available in particular production environment and socio cultural conditions. To improve the productivity of dairy animals, TKS of dairy farmers of Nalanda must consider and that should form the basis for technology integration process.

The present study is undertaken to ascertain the dairy farming knowledge in the study area of Nalanda district of Bihar. Based on findings from above study, will further supports the documents available on Traditional Technical Knowledge relevant to dairy farming, which may be further considered either alone or its integration with modern technologies in order to adapt them to local condition. Keeping in view of these facts, a study has been undertaken to identify and document the traditional knowledge of farmers about dairy farming practices in few villages of Nalanda district of Bihar.

MATERIAL AND METHODS

The study was conducted in eight villages of operational area of Krishi Vigyan Kendra (KVK), Nalanda of Bihar Agricultural University, Sabour. Villages namely Dwarikabigha, Chatiyana, Belchi, Lachhubigha, Katari, Bechubigha, Lakhichak and Mahddipur of Nalanda district were selected randomly. Fifteen dairy farmers from each of the eight villages were selected randomly. Only those dairy farmers were selected, who were having at least one dairy animal at the time of survey. Before the data collection, a Base line survey was undertaken in the study area keeping in view of the objective. Personal interview of dairy farmers with the help of semi-structured interview schedule were conducted in the above mentioned villages to collect data. The information collected from the dairy farmers were compiled as objectively as possible and list of traditional practices followed in various aspect of dairy farming like breeding, feeding, health care and management were prepared and analyzed.

RESULT AND DISCUSSION

Traditional knowledge related to breeding

The traditional breeding practices generally observed by dairy farmers in the operational area of KVK, Nalanda are shown in Table - 1. Although, some of breeding practices of dairy farmers in the study area were found to agree with scientific breeding practices, but most of them were practiced for long-long time. A considerable percentage of farmers i.e. respondent (83.90%) identified the conceived animals by stop crying or bellowing. However, farmers (74.65%) recognized the conceiveness of dairy animals by not come in heat after service upto 22 days. A small proportion of farmer (15.40%) also identified the same by keeping tail of animal far from vagina. Besides that, in pregnancy diagnosis (PD), good percentage of farmers (33.12%) reported to put a small stone on the animals back and if it does not fall they confirm that animal is pregnant, which a very crude method of judgment.

Table 1 : Traditional knowledge related to Breeding

S. No.	Practices	Symptoms/ Methods	%Dairy Farmers
1.	Detection of Heat period	- Bellowing	100
		- Mounting on other animals	86.54
		- To approach near bull(s)	43.68
		- Swelling of Udder and teats	35.44
2.	Drying of lactating dairy animal	- After 3-5 months of pregnancy	65.15
		- After 2-3 months of pregnancy	59.60
		- Stop milking when milk become salty taste	48.40
3.	Identification of conceiveness of dairy animals	- Stop crying or bellowing	83.90
		- After service do not come in heat upto 22 days.	74.65
		- Vagina becomes a red colour	45.50
		- Keeps tail far from vagina	15.40
4.	Improvement in fertility of dairy animals	- Feeding of Papaya fruits to dairy animals	25.90

Contd...

		- Feeding of mixture of leaves of Banyan tree (*Ficus bengaleansis*), Peepul tree (*Ficus religiosa)*, Mango and Neem	21.50
5.	Indication of heifer maturity for service	- Heifer age (3-4 years)	100
		- Teeth numbers	81.14
		- Well developed body	71.15
		- Glazy appearance	55.90
		- Chest length (12 inches)	45.51
6.	Pregnancy confirmation of dairy animal	- Well developed body	91.00
		- Discharge of mucus from vagina	89.90
		- Swelling of udder	73.15
		- Whitish colour of urine	59.60
		- Put stone on animal's back & doesn't fall	33.12
		- Movement of calf in flank region	31.90

Our finding is also supported by Srivastava (1982) who reported the same method being followed in trial area of Bihar for PD. Similar types of findings were found by Sah *et al*., (2003) and Kumar *et al.,* (2010) in Banka district of Bihar. The method for animal pregnancy confirmation as revealed by the farmers (31.90%) in study area was to movement in the flank region, if cold water is given to animal. Similar trends were also reported by Sah *et al*., (2003) and Kumar *et al.,* (2010) towards the practice to dry lactating animal, farmers (65.15%) had the presumption that after 3-5 months of pregnancy, dairy animal automatically ceases to provide milk. However, a good percentage of respondent i.e. farmers (59.60%) had the opinion that only after 2-3 months of pregnancy animals automatically become dry. Few proportion of farmer (48.40%) felt that the milking of the lactating animal should be stopped when milk becomes salty taste. Our finding was close conformity with the study reported by Sah *et al*., (2003) and Kumar *et al*., (2010). It is also evident from the Table - 1 that some of practices to judge the maturity of heifer for service on the basis of no of teeth, well developed body glazy appearance of body etc. were found which were found to match with the scientifically recommended method. A small proportion of farmers had also opinion that feeding of papaya fruits or mixtures of leaves of Banyan tree (*Ficus bengaleansis*), Peepul tree (*Ficus religiosa)*, Mango and Neem to dairy animal improves the fertility of dairy animals.

Traditional knowledge related to feeding aspect

The traditional knowledge related to feeding practices of dairy animals observed in the operational area of KVK, Nalanda are shown in Table - 2. It is observed that only 50.20 percent of farmers allow their newly born calves to take milk within one hour of birth. However, rest of farmers allows feeding of milk to calves later on. It was found for the study area of Nalanda that most of the farmer did not allow newly born calves to take colostrums. They were not only unfamiliar of the advantages of colostrums feeding to calves, but also had the misconception that colostrums feeding encourage the diarrhoea in neonate calves. This finding coincides with the results of Sah *et al.* (2003). Most of farmers did not have facility for providing water at stall feeding and the dairy animals generally depended on different sources of water available in the study area like bore well (42.62%), open well (39.17%) and village pond (18.21%). To enhance the milk production of dairy animals, farmer were providing feed to their animals, like oil seed cake, maize and wheat grain, carrot feeding, calcium powder etc. which was followed by majority of farmers in the study area. About 39.72% of farmers were also providing Gur as a feed. Similar observation also reported by Srivastava (1982), who reported that feeding of Gur, Methi, rice bran and gundali grass to enhance milk production in the dairy animals. Among other feeds used to increase milk production were basan, dried fruit of mahawa, banana and calcium powder, subabul seeds etc. It has been observed during study that Paddy and wheat straws bran, and bhusa were major component of dry fodder, while green fodder included different types of grasses tree leaves and sudan grass and lentil shoots. Sah *et al.*, (2003), also reported the use of oil seed cake, maize, wheat grain, Gur, Basan, dried mahawa fruit and banana for enhancing the production of milk.

Table 2 : Traditional Feeding practices of Dairy Farming

S. No.	Particulars	Practices	% Farmers
1.	Clostrum Feeding of neonate calf	- After one hour	50.20
		- Within one hour	40.37
		- After 6 hour	9.43
2.	Feeds used to enhance milk yield	- Oil seed cake	89.90
		- Maize grain	73.00
		- Gur	39.72
		- Wheat grain	37.40
		- Dried fruits of mahwa	22.42
		- Carrot	25.73
		- Banana	15.23
		- Berseem grass	13.40
		- Seeds of subabul (*Leuceana leucocephala*)	9.74
3.	Source of water for dairy animals	- Bore well	42.62
		- Open well	39.17
		- Village pond and canal water	18.21

Among the tree leaves, leaves of mango, neem, kohai, rayon etc. fed to the dairy animals in central Gujarat as green fodder, was reported by Gupta and Patel (1991). Similarly, Kokate and Tyagi (1991) also reported in their study regarding the feeding of grasses like fuli (Theunda spp.) and dhuli (Pscheoamium nigosum) grasses. It was also found that dry fodder were given to pregnant dairy animals at their animal houses, green fodder receive by them during grazing and additional amount of approximately 0.5 to 0.6 kg/animal of concentrate mixture were also receive by pregnant dairy animals. For lactating dairy animals dry and green fodder were provided as stall feeding. In addition, generally 1 -2 kg concentrate mixture per dairy animal per day was also provided according to their capacity of milk production. Dry dairy animals were given with dry fodder as stall fed and green fodder during grazing. Our findings were also supported by Sah *et al,* (2003).

Traditional knowledge related to health care aspects

The traditional systems still prevalent for treatment of various diseases in dairy animals are being practiced in the study area of Nalanda district, since generations despite of recent development in the medicines. The Table-3 shows about the various traditional treatment followed by farmers of Nalanda district for different diseases.

Table 3 : Traditional practices for treatment of disease in dairy animals

S. No.	Diseases	Traditional Treatments
1.	Anoestrus Condition	- Feeding of sufficient quantity of mustard cake. - Feeding of mixture made from Jayphal and mustard oil or mustard oil and black pepper. - Sprouted wheat grain/or boiled kulthi grain given to dairy animals.
2.	Anthrax	- Providing mixture of garlic, onion and turmeric power to dairy animals for feeding.
3.	Constipation	- Use ground caster seed with water - Feeding of gur with water - Feeding of peeper, gur and betel leaf (*Piper betle*)
4.	Cough and cold	- Use blanket for covering of dairy animals body. - Feeding of black Pepper and Mustard Oil. - Use egg and desi ghee (1 egg in 100 g deshi ghee). - Massaging of animal body with paste of Jayphal and mustard oil.
5.	Diarrhoea	- Seeds of Datura (*Datura stramonium*) is burnt and left

Contd...

		in open overnight and then given to dairy animals
		- Feeding banana stem juice to dairy animals
		- Providing paste of Bakul (mimusopselengi) leaves and Desi Ghee.
6.	Eye infection	- Cleaning of eye with muslin cloth soaked either in saline water or luke water.
7.	F.M.D.	- Allowing dairy animal to remain stand in mud for sometime.
		- Spreading Tobacco leaf powder on the infected part
		- Feeding mix of sesame oil, desi ghee and banana
		- Washing of infected part (Foot and Mouth) body with Alum
		- Washing of infected part (Foot and Mouth) body with neem oil
8.	H.S. (Galaghotu)	- Use of Eucalyptus leaves
		- Fresh dung of buffalo on the neck of cow is rubbed, if cow is suffering and vice-verse, if buffalo is suffering.
9.	Internal worms	- Feeding of neem oil
		- Feeding of castor oil
		- Feeding of leaves and bark of neem along with cucumber (*Cucumis sativus*)
10.	Jaundice	- Feeding of turmeric paste
11.	Mastitis	- Use of brine solution on the buttock of affected animals.
		- Create smoke near udder
		- Use of tobacco power in the infected part of disease
		- Washing of teats of udder with dettol water.
		- Application of $Ca(OH)_2$ Paste.
12.	Milk Fever	- Feeding of green fodder and mixed concentrate mixture
		- Washing with luke water followed by massaging the dairy animals with mustard oil for sometime.
13.	Repeat breeding	- Dry fodder only given to animals upto 22 days after service.
14.	Tympany	- Feeding of paste made of ginger, black pepper, sindha salt and black salt to dairy animal.
		- Providing Hukka water mixed with decoction of ajwain to dairy animals.
15.	Urinary problem	- Leaf of Datura (*Datura stramonium*) mixed with honey
		- Feeding of coriander and pepper mixed with sesame oil
16.	Wound	- Use of paste made from onion and ghee, chewing tobacco nishwhar (Memecylon edule) leaves after cleaning of wound with turpentine oil and luke water.

Therefore, there is urgent need to document the various traditional systems followed by farmers and systematically probe which may be worthy for inducting into recent veterinary medicine system. Our observation also gets support with the reports of Sah *et al.*, (2003).

Traditional knowledge related to management

The traditional Knowledge with respect to management practices of dairy animals are shown in Table–4. It is apparent from Table-4 that majority of farmers (63.30%) keeps the dairy animals in Kuchcha house and only 27.30 percent farmers keeps cemented shelter for their dairy animals.

Table 4 : Traditional management practices of Dairy Farming

S. No.	Particulars	Practices	Farmers (%)
1.	Calf suckling before milking	- 3-5 minutes suckling	61.10
		- 2 minutes suckling	28.22
		- 5-10 minutes suckling	10.68
2.	Castration	- Crushing of testes with wooden hammer	74.21
		- Burdizzo Castrator	25.79
3.	Clean milk production	- Cleaning of milk utensil with water	100
		- Cleaning hand with water	100
		- Washing of udder and teat with cloth soaked in water	100
		- Washing of teat just after milking	76.00
		- Cutting of nail regularly	72.00
		- Good habit during milking	60.00
		- Using solution of idophor / potassium permagnate as disinfectant	23.50
		- Using milking machine	0
4.	Housing of dairy animals	- Kuchcha house	63.30
		- Cemented house	31.9
		- Open area near home premises or under tree	4.80
5.	Method of milking	- Knuckling	63.10
		- Full hand method	20.20
		- Stripping	16.70
6.	Naval Separation	- Naturally fall off	100.00

7.	Placenta Expulsion	- Feeding grain of paddy	97.50
		- Feeding of soanf decoction	29.60
		- Feeding of syrup containing *Ajwain, Black pepper, Soanf, dried ginger, gur etc*	25.10
		- Use of lime like $Ca(OH)_2$ paste on the horn	17.10
		- Feeding jamoon (*Sizygium jambos*) bark decoction	14.50
		- Walking of dairy animal allowed	13.70
		- Manually	9.20

Similarly, large (63.10%) of farmers adopting Knuckling method of milking. However, about 16.70% of farmers adopt stripping method of milking which is not scientifically recommended for milking of dairy animals. Only 20.20% of farmers follow the full hand method of milking actually recommended for milking. Generally, farmers castrated their male calves at the age of 3 to 4 year. The majority of farmers (74.21%) got their male calves castrated by crushing the testes by using wooden hammer which seems to be unscientific. A small proportion (25.79%) of farmers castrated their animals by using Burdizzo castrator. It was observed that farmers in the study area believed that animals lose its health if castrated at early age which is unscientific. Farmers of study area did not follow naval cutting and it was left till fall off naturally. The majority of farmers (97.50%) were adopting the feeding of paddy grain to dairy animals for placenta expulsion. One of the interesting finding for placenta expulsion is use of lime like calcium hydroxide paste on the horn of dairy animals and it was being followed by 17.10% of respondents. The study also reveals and as presented in table-4 is that majority (61.10%) farmers adopted 3-5 minutes dam's suckling before milking. It is also apparent from table-4 that 10.68% of farmers, allow 5-10 minutes dam's suckling before milking. However, generally farmers of study area not allowed their calf remain attached with dam. Besides that, it was also found that farmers did not allow for weaning the calf just after calving. In others words, farmers were unaware about the advantages of the colostrums feeding with respect to health of their calves. General practice for producing clean milk was followed by the farmers in the study area. However, farmers were not following any effective measures for hygiene milk production, as evident from Table-4. The main reason was unawareness among the farmers with respect to clean milk production and they also lacked with required facility. Our finding was in almost conformity with the findings of Sah *et al.* 2003 and Kumar *et al.* 2010. An effort with respect to clean milk production has been made by the K.V.K., Animal husbandry department, Dairy Development Office of the district and other NGOs engaged in dairy development so that clean milk production could be done in the study area.

CONCLUSION

Traditional knowledge plays a major role in finding location specific solutions for problems. Most of this knowledge is not based on formal research, but on careful observations and experience from the farm family and milk producers.

The findings of the present study revealed that both traditional as well as recent scientific practices were prevailing among the farmers in the study area of Nalanda district, but the traditional practices were more dominant as found from the study. However, some of practices, though unscientific in nature, are supported by the farmers in the study area. So, there is need for systematic research from the scientist in the related area to understand the existing dairy farming practices and traditional system of dairy farming very carefully and find out its compatibility with the condition and resources available with farmers in the study area. The TKS merged with recent scientific aspect of the system will make the first step of developing and disseminating the sustainable dairy technologies for sustainable dairy development.

REFERENCES

Edison, S. (2004). Coastal Agro-ecosystem Research Perspectives: NATP Strategies. J. Indian Society of Coastal Agricultural Research, 22,(1&2): 9-12.

Gupta,A. K. and Patel, K.K.(1991). Experimenting Farmers, Pastorilists and Artisans. Report of a survey of local innovations in dry region of Gujarat. Honeybee Newsletter 2 (1):21.

Kokate, K. D and Tyagi, K. C (1991). Dairy Farming practices of Tribal Cattle Owners. Indian J. Extn. Edu. 27(3&4): 70-75

Kumar, S; Kumar, B; Mandal, B. K. and Sankhala, G (2010). Indigenous Knowledge System of Farmers about their Dairy Husbandry Practices of Banka District of Bihar. Research J Agri Sci. 1(4): 370-374.

Mangala Rai (2004). Coastal Ecosystem in India- An overview. J. Indian Society of Coastal Agricultural Research. 22(1&2):13-18.

Prain, Gordon D. (1994). Local Knowledge and global Science: The need for partnership in plant genetic resources research. In: Local Knowledge, Global Science and Plant Genetic Resources: Towards a Partnership. Gordon D. Pralo and Bagalanon (eds.). Proceedings International Workshop Genetic Resources. UPWARD, Los Banos, Laguna.

Rao, S: V.N.; Van Den Ban, A.W.; Rangnekar, D.V. and Ranganathan, K. (1995). Indigenous technical knowledge and livestock. In: Handbook for Straw Feeding Systems. K Singh and J.B. Schiere (eds.)pp 119-128, ICAJR, New Delhi, India.

Sah, A. K. (1998). Dairy Scenario: Rising Hope. Yojana. November 1998; 42-44

Sah, A. K; Chand, R and Kumar, S (2003). Traditional Knowledge System of Dairy farmers: An Empirical study. Indian J. Dairy Sci. 56(2): 107-111.

Srivastava, R. M. (1982). Cattle in Culture and Economy of Tribal Munda of Bihar. Ph.D. Thesis submitted to Kurukshetra University.

CHAPTER - 47

Challenges Being Faced by Educated Rural Youth Involved in Dairy Farming - A Case Study in Muktsar, Punjab

Madhu Shelly
Assistant Professor (Animal Science), Krishi Vigyan Kendra, Muktsar (under Punjab Agricultural University, Ludhiana), Punjab

ABSTRACT

Dairy farming plays a significant role in rural economy. It provides more avenues for increasing rural incomes and sustaining livelihood. Milk and other dairy products are a major source of protein and calcium for vast vegetarian population of India and demand is likely to double within a decade. Many of the educated youth in rural areas are going back to dairy farming as a natural choice with passion, getting involved deeply, continuously learning and innovating in the process. However these young and intuitive farmers face many challenges especially with regard to technical input and availability of green fodder round the year.In order to address these problems, this study was conducted in Muktsar district of Punjab. The young farmers selected belong to age group ranging from 24-35 years. Seventy percent of the respondents were graduates, among them 25 percent being technocrats. About 65 percent had medium family size (5 members). Eighty five percent of respondents had small land holding ranging from 5-8 acres. For majority of the respondents' dairy was the primary occupation. Sixty percent of the respondents had obtained 45 day artificial insemination (AI) training from Dairy Development Board, Punjab and themselves did AI on their farm. Many (75 percent) had herd size of 35-40 animals. Majority of the farmers had hired labour for milking and daily care of animals with family members involved in management of the farm. Most of them (65 percent) believed in preparing their own silage and breed improvement. The study revealed that many educated youth in dairy farming were of progressive and scientific temper with entrepreneurship potential but were facing many impediments like low milk yield (4.5 – 6.5 kg/ animal/day), poor breed up gradation, green fodder shortage, a large number of services per conception (3 in majority cases) with artificial insemination (AI), low competence of veterinary service providers in remote rural areas in addition to high veterinary costs, difficulty in marketing milk from crossbred dairy cows due to low preference. It was felt that if sincere and sustained efforts were made by all the stakeholders involved, dairy farming had the potential for all round socio-economic development of rural areas.

Key words : Dairy farming, sustaining livelihood, increased rural income, scientific temper, entrepreneurship potential, socio-economic development

INTRODUCTION

India is the world's largest single milk producer, with a total of 132.4 million tones of liquid milk produced in 2012-2013 (NDDB, Milk production in India 2014). Livestock production is the most important agricultural activity in the country, contributing about 24.8% to the agricultural GDP (Gross domestic product). The value of India's milk output during 2012-2013 has been estimated at over Rs.2900 billion, higher than the combined value of other major agricultural crops like paddy, wheat and sugarcane (NDDB, Facts at a Glance 2014).

Dairy farming dominates livestock production, providing 18 million people 70% of them women, with employment. The dairy sector is also the major source of income for an estimated 27.6 million people (NIRD 2012). The majority of milk production is still carried out by small scale often landless farmers, who get a large share of the total price paid by consumers. This makes milk production very attractive offering many of country's farmers a way out of poverty. In addition to rearing livestock, collection of milk and its processing is a source of stable income for many. For example Milkfed in Punjab has a workforce of nearly 5000 employees with another 10000 involved in collection of milk, technical inputs and other related requirements. Milkfed has a network of more than 5000 village milk producers cooperative societies supporting lakhs of small holder dairy farmers (website of Milkfed).

Farm sector in India represents a perfect combination of crop and cattle population wherein both complement and supplement each other. The milch animals are fed on crop residues and milch animal dung is used as manure for crops. Therefore milk is a cash crop for small holders, using family labour as a value added market commodity.

Dairy farming is the most important economic activity for small and marginal farmers in the rural areas. With multiple benefits of nutritive food, supplementary income and productive employment for family labour, dairy is more often a way of life rather than an allied occupation. In order to provide sustainable income for rural farmers, government is actively encouraging diversification in agriculture with emphasis on allied animal husbandry sectors. Government policies aimed at enhancing technical inputs in dairy, has provided new tools like improved semen straws for breed upgradation of low yielding indigenous animals at the doorstep of farmers. Dairying with crossbred cows and high yielding buffaloes is fast emerging as a lucrative business with unlimited opportunities at innovation and value addition.

Milk and milk products are a major source of animal protein and calcium for vast vegetarian population of India and demand is likely to increase further. Based on the estimates by National Dairy Development Board (NDDB) the demand for milk is likely to reach 180 million tonnes by 2022. This leads to mismatch between demand and supply, hence creates opportunities for new entrants in to this industry. Important minerals and vitamins like calcium, choline, potassium, vitamin D, magnesium, phosphorous, vitamin A, riboflavin, vitamin B6 and vitamin B12 are also present in required quantity. Milk is good for bone health (lowering risk of osteoporosis, Flynn 2003) for blood pressure and heart health (Elwood 2005) in addition to muscle building. It also lowers risk of colorectal cancer and depression. Besides milk is a source of 'high biological value' protein which means that it provides us with all the essential amino acids that the body cannot make itself. Studies have also identified other fats in milk such as conjugated linoleic acid which may potentially protect against several major chronic illnesses (Larsson et al, 2005).

However all these benefits, facts and figures notwithstanding, a large proportion of farmers in the villages are on their way out of dairy farming. Economic difficulties faced by farmers as a result of low farm incomes (vis-a-visnon-farm) and great instability and variability in the income from farming. The common response of farmers specially youth to the problem is to migrate to urban areas or to non-farm occupations which provide higher returns to per unit of labour applied NSSO, 2003 also reports of increasing migration from rural to urban areas.

In this bleak scenario, a silver lining is still evident. Many educated youth in rural areas do believe in the strength of dairy farming and its future prospects. They are turning to dairy as a natural choice but with an entrepreneurial bent of mind. It is a fact that dairy is closely interwoven with the socio economic fabric of rural people in Punjab. Moreover dairy provides a support system to milk producers without disturbing their agro economic systems. Future of dairy shall depend on the learning ability, technological skills, innovating capacity and entrepreneurship potential of the educated rural youth for whom dairy is a vocation of choice rather than chance. However problems and shortcomings that arise in their way need to be tackled immediately by all the stakeholders for the overall prosperity and socioeconomic development of our villages. The present study was carried out to identify the challenges being faced by these motivated youth.

MATERIALS AND METHODS

In Muktsar district, twenty four educated young farmers were identified from among the trainees of "Commercial dairy farming" training being conducted by KrishiVigyan Kendra, Muktsar. Data were collected personally through a structural interview schedule that were pre-tested and adjusted prior to the objectives of the study. Data were collected from January 2011 to March 2012. Challenges faced by these youth related to feeding and breeding of dairy animals, milk production and marketing as well as herd management and veterinary care has been further analyzed. Simple frequency analyses have been used to interpret the data.

RESULTS AND DISCUSSION

Socio-economic profile of selected dairy farmers

Favorable socio-economic profile of selected dairy farmers affect the profitability of dairy enterprise to a great extent. A perusal of Table 1 reveals that the selected farmers belonged to age group of 24-35 years. The young farmers have the will to learn and experiment whereas the old are resistant to change. The entrepreneurial potential and scientific temper of youth is also more. Seventy percent of the respondents were graduates, among them 25 percent being technocrats, with a bachelors degree in engineering. The love for technology which they get through education reaps benefits in farming also as majority of them are internet savvy and had good contact with extension functionaries. Sixty percent of the respondents had obtained 45 day artificial insemination (AI) training from Dairy Development Board, Punjab and themselves did AI on their farm. This helped them tackle the problem of quacks and in turn shorten the inter-calving interval in order to make the animal profitable for the owner. About 65 percent had medium family size (upto 5 members) and it was observed that family was a big support for them as females looked after daily management of animals and the elders looked after the children. This helped the young farmers concentrate and innovate. Since 85 percent of the respondents had land holding between 5-8 acre and a herd size of 35-40 animals, therefore dairy was their primary occupation. The farmers considered crossbred cows as more profitable for dairy enterprise although 8 percent of them kept buffaloes also.

Table 1 : Socio economic characteristics of selected rural youth (n=24)

Variable	Categories	No. of farmers	percentage of farmers
Age status	young (24-35 years)	24	100
Educational qualification	upto high school Graduate Post graduate	5 17 2	20.83 70.83 8.33
Occupation status	Dairy as main Dairy as subsidiary	16 8	66.67 33.33
Annual income of farmers	Low (<1 lakh) Medium (1-5 lakh) High (>5 lakh)	2 20 2	8.33 83.33 8.33
Family size	upto 5 members> 5 members	16 8	66.67 33.33
Land holding size	Landless (no land) small (1-5 acres) medium (5.1-10 acres) large (>10 acres)	0 12 8 4	0 50.00 33.33
No. of animals (herd size)	Small (1-35 animals) Medium (35-50 animals) Large (>50 animals)	4 18 2	16.67 75.00 8.33
Type of herd	crossbred cows	18	75.00
	buffaloes	24 16.67	8.33
Contact with extension functionaries	low medium high	8 14 2	33.33 58.33 8.33
Internet savvy	low medium high	2 10 12	8.33 41.67 50.00
Training status	No training 2 week general training 45 day AI training (from DDB, Punjab)	4 6 14	16.67 25.00 58.33
Economic motivation	low medium high	— 16 8	 66.67 33.33
Scientific temper	low medium high	4 18 2	16.67 75.00 8.33
Entrepreneurial potential	low medium high	2 18 4	8.33 75.00 16.67
Adoption of new technology	low medium high	2 16 6	8.33 66.67 25.00

The production potential and performance of dairy animals depends on the efficient care and management suitable for their special needs which includes balanced feeding, efficient breeding, proper marketing and management with timely health care. The entrepreneurial rural youth are facing many challenges with regard to these parameters and degree varies from person to person.

Challenges with Regard to Feeding

Rural young farmers in many cases considering dairy as a family tradition continue to practice what their forefathers or their neighbors practised. About 66.67 percent of them lacked proper knowledge about balanced feeding of animals (Table 2). High cost of compounded cattle feeds available in the market was a major concern (91.67 percent). Kumawat *et al.*, 2014 also reported similar findings. Poor availability of green fodder (58.33 percent) along with unavailability of fodder seeds were other impediments. There was low awareness among farmers regarding the importance of mineral mixture although many were concerned about its quality. University made mineral mixture still enjoyed good reputation with farmers. Majority of rural youth did believe in preparing silage on farm but 50 percent reported its spoilage due to unavailability of proper technical inputs.

Table 2 : Challenges regarding feeding of dairy animals

Sr. no.	Challenge	No. of farmers	Percentage of farmers
1.	High cost of compounded cattle feed in market	22	91.67
2.	Non availability of green and dry fodder	14	58.33
3.	Lack of proper knowledge about balanced ration	16	66.67
4.	Low awareness about mineral mixture	14	58.33
5.	Non availability of quality mineral mixture	18	75.00
6.	Lack of proper technical inputs in preparation of silage	12	50.00
7.	Unavailability of fodder seeds	10	41.67

Challenges with Regard to Breeding

Ill defined breeding policy with respect to different regions in the country is a major problem. Table 3 reveals that farmers reported lack of expert services at AI centers (91.67 percent) and unequipped AI centers (41.67 percent). Some farmers reported that they had to depend on quacks as AI centers were not available near their living places (33.33 percent). Saini (1975) also reported distant location of various units as a major constraint. General disregard with respect to importance of proper breeding resulted in stray bulls being used for natural service in villages. A large number of services per conception (three in majority cases) with AI is another concern.

Table 3 : Challenges regarding breeding practices

Sr. no.	Challenge	No. of farmers	Percentage of farmers
1	Ill equipped AI centres	10	41.67
2	Lack of expert services at AI centres	22	91.67
3	Lack of AI centres in the nearby area	8	33.34
4	Lack of pedigree bulls for natural service	8	33.34
5	Right breed identification	6	25.00

Challenges with Regard to Proper Management

The scientific management practices recommended by experts (after a lot of research)are essential for efficient milk production. About 83.33 percent showed a lack of knowledge about these while 41.67 percent expressed their in ability to provide the required grazing place and space for shed (Table 4). Another major problem is retention of labour in dairy farming (66.67 percent). Therefore people with a tradition of keeping animals and having an attachment with regard to their welfare (like dairy farmers) can reap far great profits. The need of the hour is to blend tradition with science. This calls for a managerial upgradation of farm operations. Clean milk production through the use of teat dips and personal hygiene of the milkers needs to be given due attention.

Table 4 : Challenges regarding proper management of dairy animals

Sr. no.	Challenge	No. of farmers	Percentage of farmers
1	Lack of knowledge regarding scientific animal management	20	83.33
2	Lack of grazing place and space for shed	10	41.67
3	Problem of labour for taking care of animals	16	66.67
4	Poor knowledge of clean milk production practices (eg.teat dips)	14	58.33
5	Careless attitude regarding summer and winter care of animals	12	50.00

Challenges with Regard to Health Care

Most of the rural veterinary hospitals are poorly equipped lacking facilities for cold storage of vaccines. Along with this low awareness and sometimes low preference for vaccination (41.67 percent farmers) seriously affected the disease prevention work of veterinary doctors (Table 5). Problem of quacks (50 percent), high cost of medicines (41.67 percent), low supply of medicines (Sharma 1984) in civil veterinary hospitals are the factors posing a big challenge for rural young dairy farmers. Nearly half of the farmers expressed low awareness regarding deworming.

Table 55 : Challenges regarding health care of dairy animals

Sr. no.	Challenge	No. of farmers	Percentage of farmers
1	Low awareness and sometimes low preference for vaccination	10	41.67
2	Lack of adoption of scientific practices for health care	8	33.33
3	No deworming of milch animals	12	50.00
4	Unavailability of adequate veterinary service	14	58.33
5	High cost of quality veterinary service	18	75.00
6	Problem of quacks in villages	12	50.00
7	High cost and unavailability of medicines and disinfectants	10	41.67
8	No veterinary hospitals in villages	4	16.67

Challenges with Regard to Milk Production

Table 6 reveals that majority of farmers (66.67 percent) felt that low milk yield (4.5 – 6.5 kg/ animal/day) is a great hindrance in sustainability of dairy farming. About 58.33 percent also reported low fat content in milk due to which they were unable to get good milk prices at collection centres.

Table 6 : Challenges regarding milk production in dairy animals

Sr. no.	Challenge	No. of farmers	Percentage of farmers
1	Low milk yield	16	66.67
2	Low fat content in milk	14	58.33

Challenges with Regard to Milk Marketing

Due to more popularity of buffalo milk in Punjab, farmers reported (58.33 percent) that milk of crossbred cows had less takers (Table 7). Farmers were unable to get remunerative prices of milk in all the cases. Value addition of milk and manufacture of milk products hold a great potential. However many had low motivation for this activity. Farmers need to remember that shelf life of the product along with its quality hold the key to a successful marketing strategy. As in any other enterprise, there should be proper planning about demand forecast, inventory of resources, need for external inputs, skill level of the available human resources and their training needs as well as infrastructure and services needed for carrying out various operations and marketing.

Table 7 : Challenges regarding milk marketing

Sr. no.	Challenge	No. of farmers	Percentage of farmers
1	Low preference for milk of crossbred animals	14	58.33
2	Lack of remunerative prices of milk	24	100.00
3	Lack of cold storage facility at collection centers	8	33.33
4	Low motivation in making value added dairy products	12	50.00

CONCLUSION

Rural youth because of their energy and change potential need to realize their responsibility with regard to their surroundings. It is a fact that the problems are locality specific and resources to tackle them must be generated locally for sustainable development. Migration of educated rural youth from villages to towns and cities is the most serious form of brain drain adversely affecting rural India's development, sustainability and prosperity. Therefore if the educated youth choose to live in villages and launch a new white revolution based on the integrated use of resources (agriculture residues, animal husbandry and wisdom of older generation), our demographic dividend has the potential to make India a developed nation self sufficient in every way. Dairy has profound potential to bring about socio-economic change (as demonstrated by Operation Flood) but the need today is to reorient and channelize the efforts of all the involved stakeholders at various levels for the noble purpose of rejuvenating this family farming once again to achieve higher targets.

REFERENCES

1. Elwood PC (2005) Milk consumption, stroke, and heart attack risk: evidence from the Caerphilly cohort of older men. Journal of Epidemiology and Community Health 59; 502-505.
2. Flynn A (2003) The role of dietary calcium in bone health. Proceedings of the Nutrition Society 62; 851-858.
3. Kumawat Raju, Singh N.K., Meena Chirangi Lal (2014) Economic analysis of constraints faced in adoption on sample dairy farms in Bikaner district of Rajasthan. Global Journal of Science Frontier Research: Agriculture and Veterinary 14 (6): 35-43.
4. Larsson SC, Bergkvist L, Wolk A (2005) High-fat dairy food and conjugated linoleic acid intakes in relation to colorectal cancer incidence in the Swedish Mammography Cohort. Am J Clin Nutr 82; 894-900.
5. Milkfed(2005) Report of the Punjab State Cooperative Milk Producers Federation.
6. National Dairy Development Board, Milk Production in India, http://www.nddb.org/English /Statistics/Pages/Milk-Production.aspx, accessed in May 2014

7. National Dairy Development Board, Facts at a glance, http://www.nddb.org/English / Pages/Facts-at-Glance.aspx
8. National Institute of Rural Development, Study on the improvement in rural livelihoods through dairy farming, retrieved from http://www.nird.org.in/nird_docs/ven_finrepo.pdf on 12 October 2012
9. NSSO, 55th round (2003), Migration in India (July 1999 to June 2000). National Sample Survey Organization, Ministry of Statistics & Programme Implementation, Government of India.
10. Saini, S.P.S. (1975)A study of factors affecting adoption of selected recommended practices. M.Sc. Thesis, Punjab Agri. Univ., Ludhiana.
11. Sharma, K N S (1984). An economic study of cross bred cattle in Bangalore. Ph.D. Thesis, Kurukshetra Univ., Kurukshetra.

CHAPTER - 48

A Study of Knowledge about Dairy Entrepreneurship and Relationship of Variables Towards the Training Need of Milk Producers in Patna District

Pankaj Kumar[1], J. Oraon[2], Rajesh Kumar[3] and S.K. Rajak[4]
[1,4]*Asstt. Professor (Extension Education), BVC, Patna*
[2]*University Professor & Chairman (Extension Education) RVC, Kanke, Ranchi*
[3]*Asstt. Professor (A.H) BAU, Sabour, Bhagalpur*

ABSTRACT

The present study was carried out to assess the knowledge level of dairy Entrepreneurs of Patna District and to find the co-relation between the socio-economic variables and the training needs. This study was conducted by personally interviewing 100 dairy farmers, sent for training at B.V.C. Patna by different Agencies. Majority of the respondents (58.20%) had medium level of knowledge flowed by 26% respondents with high level of knowledge, while some of the respondents (15.80%) had low level of knowledge about dairy entrepreneurship. The study of co-relation between socio-economic variables and training need of the respondents showed that variables like education (0.469), social participation (0.500) and the knowledge (0.734) were significantly and positively co-related with the training need. Whereas variables like age (0.063) family size (0.193) and land holding (0.072) were found to be positively but non-significantly co-related with the training needs while, the caste (-0.093) was found to be negatively and non-significantly co-related with the training needs at 0.05% level of probability. Therefore it is imperative on the part of change agency to utilize the variables gainfully so as to give needed push to knowledge level in the area of Dairy entrepreneurship development.

Key words: Knowledge level, relationship, training needs, dairy entrepreneurship

INTRODUCTION

India is predominantly an agrarian economy, with more than 75% of rural households are still depending upon agriculture and allied activities for their livelihood.But it apparent from the present scenario that our land-man ratio is decreasing day by day. Hence one should must explore the possibilities of rural employment generation through the allied sectors lie dairying, fisheries, poultry farming etc. Among many livestock enterprises, dairying is the most ancient occupation established in the rural setting of our country. Nowadays, dairy sector contributes significantly in generating employment opportunities and supplementing the income of small and marginal farmers and landless labourers of rural India, besides providing food security.

But the milchanimal's productivity is not satisfactory and the rapid growth of milk production in India is mainly because of the increase in the number of animals rather than that of improved productivity.

The low productivity of dairy animals is of great concern and average productivity of Indian cow is only 987 Kg/lactation as against the world average of 2038 Kg/lactation. The gradual breed deterioration generally occurs from negligence over centuries and consequent rise in the population of non-descript cows (80%) and buffaloes (50%) along with the chronic shortage of feed and fodder coupled with their nutritive values and low fertility of our dairy animals has resulted in the low productivity. In India, low animal productivity results due to climatic, social and economical factors. India possesses enormous bovine wealth, but their per capita production is far lower than the other countries of the world due to reasons that the farmers do not adopt improved dairy management practices at the desired level.

Keeping the above problem in view, the present study was taken up to assess the knowledge level of dairy farmers in Patna district and finding the co-relation between socio-economic variables with their training needs to make the training programme need based and problem oriented.

METHODOLOGY

1. **Locale of the study:**The present study was conducted at Bihar Veterinary College, Patna (Bihar).

2. **Selection of the respondents:** For the present study 100 dairy farmers were purposively interviewed; with the help of interview schedule at BVC Patna. The respondents were selected among the

trainees group sent by different agencies like ATMA, BAMETI, and NGO's for training on improved dairy farming. only those respondents were selected for interview who have two or more milch animals and selling milk/ products for economic gain.

3. **Statistical methods used:**Frequency and percentage was used for the statistical assessment of the knowledge level as per the responses recorded by the dairy farmers. Statistical test such as Pearson's coefficient of co-relation was used to find the co-relation between the variables.

RESULTS AND DISCUSSION

It was found that majority of the respondents (58.20%) had medium level of knowledge followed by the respondents with high level of knowledge about dairy entrepreneurshipactivities (26.00%) , while 15.80% of the respondents showed low level of knowledge. Further, it was found that majority of the respondents (56.71%) were having medium level of knowledge because of illiteracy, poor educational facilities and weakness of extension agency in imparting technical knowledge by organizing training programmes and animal health camps in order to bring awareness among them.

It was also observed that majority of the respondents (90.69%) did not have any knowledge about the brucellosis followed by 85.42% respondents who were ignorant about the different vaccination schedule. However, 84.60% respondents did not know about economic losses caused due to infectious diseases. Symptoms of diseases such as H.S., B.Q., FMD and mastitis were not known to 68.60%, 62.69%, 60.00 % and 65.00% respondents respectively.Important managemental operations such as use of insecticides and de-worming at regular interval were not known to 64.20% and 60.00% respondents respectively.

The study revealed that majority of the respondents (70.57%) did not know the necessity for the separate housing of pregnant animal before 2-3 weeks prior to parturition. It might be because of lack of technical knowledge about the importance of housing and monetary constraints for constructing the animal house. Again most of the respondents (70.60%) did not know the drying off animals before parturition. Also, 49.56% of the respondents were unknown about steaming-up i.e. allowing extra concentrate ration in the last trimester of pregnancy.

The study shows that majority of the respondents (70.80%) were unaware about cutting the naval cord and application of Tincture iodine to it. However only 40.50% respondents did not know about colostrums feeding and its importance for the new born calf.

Majority of the respondents (90.88%) did not know the urea treatment of dry fodder, while 60.89% of the respondents did not know about the methods of feed preservation whereas 60.09% of the respondents were unaware of providing balanced ration to the milking animals. Cultivation of fodder was not know to 37.33% of the dairy farmers interviewed. This might be due to lack of knowledge about the methods of conservation of feed, urea treatment over the feed to enhance its nutritive value and likewise feeding management practices particularly during summer and in scarcity.

As regards the knowledge level about clean milk production, majority of the respondents (80.98%) did not know that the milker should be healthy and free from bad habits. This might be because of their ignorance about their personal health and spread of zoonotic diseases. 76.60% and 63.20% respondents expressed their ignorance about the value addition of milk and functional dairy cooperative societies in their area respectively.

Relationship of personal and socio-economic characteristics of dairy farmers with training needs

It was assumed that, training need is the function of personal and socio-economic characteristics. So, correlation coefficient was administered to test this assumption. The results are presented as:

Sl.No.	Independent variable	'r' value
1.	Age	0.063^{NS}
2.	Caste	-0.093^{NS}
3.	Education	0.469**
4.	Family size	0.193^{NS}
5.	Land holding	0.072^{NS}
6.	Herd size	0.433**
7.	Annual income	0.458**
8.	Daily milk production	0.578**
9.	Daily milk sale	0.486**
10.	Social participation	0.500**
11.	Socio-economic status	0.391**
12.	Knowledge	0.734**

NS – Non Significant

* Significant at 0.05 level of probability

** Significant at 0.01 level of probability

Age of the respondent was not found to be significantly related with training needs of dairy farmers. The possible reason might be that, training need could be felt important depending upon the situation, need and knowledge of the respondent and not mostly upon the age. It was observed that, there is negative and non-significant relationship between caste and training need of the dairy farmers. It was noticed that, there is positive and significant relationship between education and training need of the daily farmers. It might be because education imparts knowledge and creates awareness and curiosity of learning skills and newer things. These observations are in line with those of Jondhale and Chole (1989). With respect to land holding, it was observed that there is a positive and non-significant association between the training needs.

Herd size, Annual income, daily milk production and daily milk sale were found to have a positive and significant relationship with training needs. Social participation was found to be positively and significantly associated with training needs of the farmers. Probable reason may be that, social participation brings awareness among the farmers. These findings are in line with those of Ingole (1990) and Gaikwad (2003). Socio-economic status was found to be positively and significantly associated with training needs of the dairy farmers. Higher socio-economic status implies more social participation and annual income which tends them to be attentive for training programmes. Knowledge was found to have a positive and significant association with training needs of the dairy farmers. It suggests that, knowledge creates awareness and change in the attitude to sharpen the skills through training programmes.

CONCLUSION

- As regards to knowledge, majority of the respondents were having medium level of knowledge.
- Training needs were significantly and positively correlated with education, herd size, annual income, daily milk production, daily milk sale, social participation, socio-economic status and knowledge at 0.05% level of probability.
- However, age, caste, family size and land holding were found to be positively but non-significantly associated with training needs at 0.05% level of probability.
- Thus the study is of immense importance to training institute like K.V.Ks, S.A.Us. and other related organisations. Keeping in view the facts, appropriate and suitable training programme for dairy entrepreneurs can be formulated based on their training need and knowledge level with respect to different A.H. practices and also the study has much value to the extension agencies engaged in transfer in technology mainly in the area of dairy entrepreneurship development.

REFERENCES

Gaikwad, S.P. (2003): Training needs of goat keepers. M.Sc. (Agri.) thesis submitted to Mahathwada Agril. Univ., Parbhani.

Gurav, K.V. and L.P. Kamble (1995): Training needs of rural women. Mah. J. Extn. Edu., 14: 93-95.

Ingole, P.G. (1990): A study of training needs of dairy farmers. M.Sc. (Agri) thesis submitted to Marathwada Agril. Univ., Parbhani.

Jondhale, S.G. and R.R. Chole (1989): Training needs of dairy farmers. Mah. J. Extn. Edu., 8 : 157-160.

Kumar, P.(1997): Adoption of improved A.H practices among milk producers of Samastipur Distt. M.V.Sc. Thesis, Deptt. Of extn. Edn, R.A.U. Pusa, Bihar.

Reeja, G.P. and M.R. Subhadra (2002): Knowledge of farm women about scientific slection and housing of calves: An impact study. Indian Vet. J., 79: 286-288.

CHAPTER - 49

New Opportunities for Sustainable Family Dairy Farming in Bihar

Dharmendra Kumar[1], Asit Chakravarti[2] and Rajesh Kumar[3]
[1]SMS (Animal Science), Krishi Vigyan Kendra, Banka, Bihar -813102
[2]Senior Scientist (LPM), ICAR Research Complex for Eastern Region, Patna
[3]Assistant Professor-cum-Junior Scientist (AH), Dept. of Agronomy, BAU, Sabour

ABSTRACT

Indian agriculture is an economic symbiosis of crop and cattle population.Millions of family dairy farming dominates India's dairy industry, contributing 62 percent of total milk produced in the country. So, encouraging family dairy farming in rural areas is necessary to increase the milk production at national level and increase the socioeconomic status of landless and marginal farmer. In rural areas there is a major problem of milk marketing. In urban scarcity of quality milk, other side in villages especially faraway from city there is surplus milk and farmers are not interested to rear more dairy animals. Marketing of surplus milk is main problem because milk is a highly perishable item and farmers not having time and also due to less quantity of surplus milk (3-5 liter) they are not interested to go market. For this we can work in group and collect milk at one place from all members and one person go to market for selling of milk. Extra milk can be used to manufacture other milk products like ghee, peda, paneeretc by other member. Some members also engaged in other work like treatment of animals, collect information of different schemes from related offices. While engaging with rural communities, we recognised that village women are the primary caretakers of cattle and play a significant role in dairy farming. As a result, the Village Woman Dairy DevelopmentProgramme - an initiative focusing on empowering village women engaged in dairy farming.Benefits in dairy farming depend on cost of feeding, medicines, labour and income from sale of milk. Milk price is fixed and not in farmers control so we increase the income through reducing the cost of milk production. Approx 70% cost of milk production on feeding of animals. To reduce cost of feeding and medicine we fed balanced ration to dairy animals. Also reducing cost of production by feeding locally available unconventional feeds through total mixed ration feeding and urea mineral molasses block supplementation.

Key words: Family dairy farming, Woman Dairy Development Programme, Balanced ration

INTRODUCTION

It is well established fact that India is basically an agriculture country. Majority of its population live in village whose primary occupation is agriculture. Indian agriculture is an economic symbiosis of crop and cattle population. For a farmer dairying is a subsidiary source of income. It plays an important role in the agriculture economy. Millions of rural smallholder milk producers (family dairy farming) dominate India's dairy industry, contributing 62 percent of total milk produced in the country. i) The dairy sector today provides self-employment to 80 million farm householdswith the triple benefits of nutritive food, supplementary income and productive employment for family labour, mainly for women. ii) Dairying fits in well in diversified farming programs. This diversification will help to use the farm labor efficiently and dairying will be one such diversified activity on a farm. iii) Dairying helps in having an alternative source of income to the farmer.iv) Milch animals are efficient Consumers of roughages:The different types of roughages such as Paddy straw. Wheat straw etc. sometimes goes waste on some farms. If the farmer has dairying these roughages can efficiently be converted into most nutrition's milkand manure. These roughages are bulky and it is not economical to transport them over long distances. v) Dairying provides a stable income: The prices of most of the agriculture produce show great fluctuation. The prices of dairy products do not show such wide range of fluctuation in their prices.vi)Income is distributed through out the year: If the farmer depends only on agriculture he would be receiving income from his farm only on harvesting the crop. This would mean that the income of the farmer would only be seasonal. Since the produce from a dairy animal is distributed almost throughout the year vii) Dairy production improves the family diet and reduces the food cost.Milk is a basic food and an important item in the family food budget viii) Dairying aids in the maintenance of soil fertility. The manure provided is distributed on the land and returns plant food nutrients to the soil ix) Supplies animals required for draft purpose.Moreover milk is a cash crop for smallholders, converting low-value agriculture byproducts and crop residues, and using family labour as a value-added market commodity.

Livestock Scenario in Bihar

Livestock sector plays an important role in socio-economic development of rural households. During last 6 years milk production increased by two folds from 2.63 million tones in 2001-02 to 5.45 million tones in 2006-07 against the milk production at national level increasing from 84 M.T. to 100 M.T. during the period (BBAHS – 2012). During past three years, increase in milk production in Bihar was much higher than the increase in milk production at National level. Among the different categories of households, the increase in number of bovine in milk was much higher on medium farms (from 64 to 116 per 100 households) followed by small (from 46 to 70 per 100 households) and marginal (from 37 to 56 per 100 households). The importance of livestock sector in state's economy can be understood by the fact that it contributed more than 45 percent to state agricultural GDP which is much higher than the contribution of this sector at national level (Gross National Agriculture Production) which stands at 27 percent. Agriculture in Bihar is dominated by small land holders as the average size of land holdings is small (0.37 ha.) and about 78 percent farm holdings are of less than 0.50 ha with an average size of land holdings being 0.10 ha only.So, encouraging family dairy farming in rural areas is necessary to increase the milk production at national level and increase the socioeconomic status of landless and marginal farmer.The factors which can improve the dairy production scenario of Bihar state depicted below.

I. Marketing

In urban areas there is no problem of milk marketing but in rural areas there is a major problem of milk marketing. In urban scarcity of quality milk other side in villages especially faraway from city there is surplus milk and farmers are not interested to rear more dairy animals. Marketing of surplus milk is main problem because milk is a highly perishable item and farmers not having time and also due to less quantity of surplus milk (3-5 liter) they are not interested to go market. For this we can work in group and collect milk at one place from all members and one person go to market for selling of milk. Extra milk can be used to manufacture other milk products like ghee, peda, paneeretc by other member. Some members also engaged in other work like treatment of animals, collect information of different schemes from related offices. While engaging with rural communities, we recognised that village women are the primary caretakers of cattle and play a significant role in dairy farming. As a result, the Village Woman Dairy DevelopmentProgramme - an initiative focusing on empowering village women engaged in dairy farming.

So, small dairy farmers group (Village Dairy Cooperative Societies) is essential for upliftment of family dairy farming. Benefits to work in group:

i) We can make own cattle feed from locally available raw materials.

ii) We install combined pumpset and fodder cutter in village.

iii) Improved breed of animals purchasing from other members, good breed calf sell in members only.

iv) Vaccination and treatment of animals done by member who trained for this. Purchasing of medicines on wholesale rate.

v) Natural insemination by rearing improved breed bull.

vi) Interested in increase the milk production with other members.

vii) Solve the problem of other members.

viii) One member involves in making ghee, khoa, paneer, peda etc. from surplus milk and other persons involve in marketing.

ix) We can install dairy machinery like pasteurization and others large milk processing machine in groups.

x) One member can go to block and district offices to take knowledge of schemes and benefited to all members.

xi) Make a milk parlor in market for selling of milk and other products where 1-2 members are involves.

xii) Run gobar gas plant in group from which get electricity, light, compost etc. produced.

xiii) If one family shaving one liter milk/ day can get approx 1000 Rupees in a month if society in village, other wise he consume this extra milk due to problem of marketing. In this way from only 100 families we shave 1 lakh rupees per month in a village.

Dairy co-operative (Sudha) is only successful organization in Bihar but covers less than 15 percent of villages during 24 years. The dairy co-operative system has lost its stream in Bihar. Moreover, it should not be allowed to monopolize the milk marketing system. During survey, several farmers showed concern about low prices paid by co-operative (Singh K. M. *et al.*, 2010). Milk marketing needs more emphasis but private milk processing and marketing organizations are not getting institutional support in Bihar which could be done by promoting private entrepreneurs through institutional financing and government support.

Government is actively supporting the dairy sector by implementing various schemes. It all started with the White Revolution under the title Operation Flood (OF) Programme launched in 1970. By promoting Anand Pattern of dairy cooperatives, OF envisaged sustained increase in resource productivity culminating in improved quality of life of milk producers and assured supply of quality of milk and other dairy products to consumers at reasonable price in a free market environment. The objectives of Operation Flood included: Increased milk production ("a flood of milk") and augmenting rural incomes

Village Dairy Cooperative Societies

The basic unit in the Anand Pattern is the milk producers' cooperative society, a voluntary association of milk producers in a village who wish to market their milk collectively. Every milk producer who has at least one milch animal that is owned can become a member of the cooperative by paying a nominal entrance fee and purchasing at least one share. At a general meeting of all the members, the representatives are elected to form a managing committee, which runs the day-today affairs of the cooperative society by setting the policies and appointing necessary staff. Out of nine managing committee members, one member is elected as chair person. Every morning and evening the society buys the surplus milk from its producer-members. The producer is paid for milk usually within 12 hours (for morning milk, in the same evening and for evening milk, the next morning). The payment is made on the basis of fat and SNF (solids-not-fat) content of the milk supplied by the individual producer. The district milk union organizes the transport of collected milk twice daily (whenever the dairy society has no bulk coolers) or according to the schedule based on the quantity of the milk collected and the capacity of bulk coolers from all its affiliated member cooperatives. The producers in this system are not only ensured regular and remunerative payment for their milk, but also have access to the milk production enhancement inputs, support services, to further improve their productivity and income. The micro-level support services such as veterinary first aid are organized by village societies. One of the staff members of the society is trained to carry out these functions. Societies also market the liquid milk locally at the village level. In addition, supply of balanced cattle feed and fodder seeds and milk products are also channeled through societies. In addition to regular payment and support services, members also receive the price difference, dividend and bonus deriving from the net profit of the society's business as per the provisions of the society's by-laws.

Selecting the animal to farm with – Cows v/s. Buffaloes

Cows	Buffalo
• Good quality cows are available in the market and it cost around Rs.2500 to Rs.3000 per liter of milk production per day. (e.g. Cost of a cow producing 10 liter of Milk per day will be between Rs.25,000 to Rs.30,000). • If proper care is given, cows breed regularly giving one calf every 13-14 month interval. • They are more docile and can be handled easily. Good milk yielding cross breeds (Holstein and Jersey crosses) has well adapted to Indian climate. • The fat percentage of cow's milk varies from 3-5.5% and is lower then Buffaloes.	• In India, we have good buffalo breeds like Murrah and Mehsana, which are suitable for commercial dairy farm. • Buffalo milk has more demand for making butter and butter oil (Ghee), as fat percentage in milk is higher then cow. Buffalo milk is also preferred for making tea, a welcoming drink in common Indian household. • Buffaloes can be maintained on more fibrous crop residues, hence scope for reducing feed cost. • Buffaloes largely mature late and give birth to calves at 16 to 18 months interval. Male calves fetch little value. • Buffaloes need cooling facility e.g. Wallowing tank or showers / foggers with fan.

A suggestion to help you in deciding the animal to farm with

Middle class health-conscious Indian families prefer low fat milk for consumption as liquid milk. We suggest you to go for a commercial farm of mixed type. (Cross breed, cows and buffaloes kept in separate rows under one shed). Conduct a through study of the immediate market where you are planning to market your milk . You can mix milk from both type of animals and sold as per need of the market. Hotels and some general customers (can be around 30%) prefer pure buffalo milk. Hospitals, sanitariums prefer cow's milk. When

What are the various breeds? What is the economic life of animals?

Popular buffalo milch breeds are Murrah, Surti, Mehasani, Jaffrabadi, and Nali – Ravi and Badhawari. The indigenous milch breeds of cattle are Gir, Sahiwal, Red Sindhi and Tharparkar. The exotic breeds of cattle are Holstein Friesian, Jersey and Brown Swiss.

Economic life of buffaloes is 5-6 lactation and that of Crossbreed cows is 6-7 lactation.

II. Feeding

Benefits in dairy farming depend on cost of feeding, medicines, labour and income from sale of milk. Milk price is fixed and not in farmers control so we increase the income through reducing the cost of milk production. Approx 70% cost of milk production on feeding of animals. To reduce cost of feeding and medicine we fed balanced ration to dairy animals.Ration balancing is the process to balance the level of various nutrients of an animal, from the available feed resources, to meet its nutrient requirements for maintenance and production.

i) Ration balancing of dairy animals

The feed given to animals usually comprises one or two locally available concentrate feed ingredients, grasses and crop residues. This leads to imbalanced feeding which means that proteins, energy, minerals and vitamins in the ration are either more or less. While imbalanced feeding adversely affects the health and productivity of animals in various ways, it also reduces the net daily income to milk producers from dairying because the potential of milk production of animals is not fully exploited. At times, when the animals are overfed, this can also raise the cost of milk production.Therefore, milk producers need to understand the implications of imbalanced feeding and recognize the importance of giving their animals' balanced ration.

Disadvantages of imbalanced feeding

- Low milk production, poor growth and reproduction.
- Milk production of animals lower than their genetic potential.
- Shorter lactation length and increased inter-calving period.
- Animals more prone to metabolic diseases such as milk fever and ketosis.
- Slow growth of young animals delaying the age of first calving.
- Low productivity and shorter duration of productive life.

Types of dietary feed ingredients

- **Compound cattle feed**: This is considered to be a balanced source of nutrients for growth and milk production. However, only 10 to 12 per cent of the total feed ingredients are used to produce compound cattle feed.
- **Other feeds**: Feed ingredients like mustard cake/meal, groundnut cake/meal, sunflower meal, cottonseed cake/meal, soya bean meal, guar meal, linseed cake, safflower meal, deoiled rice bran, rice polish, wheat bran, maize bran, maize grain, sorghum grain, wheat, broken rice, millets and chunnies are fed as such, depending on availability and cost.

- **Crops residues & grasses**: Wheat straw, paddy straw, sorghum straw, maize stovers, straw of pearl millet and locally available grasses are fed as basal feed.
- **Green fodder**: Maize, sorghum, oats, hybrid napier,bajra, lucerne, cowpea and berseem are available seasonally and fed in a limited quantity.
- **Mineralmixture**: This is a source of macro and micro minerals, usually lacking in the animal's ration.

Benefits of RBP

- Uses locally available feed resources to balance the ration of animals at least cost.
- Increases milk production with more fat and solids-not-fat.
- Helps increase the net daily income.
- Improves reproduction efficiency.
- Helps reduce inter-calving period, thereby increasing the productive life of animals.
- Improves the general health of animals.
- Improves the growth rate in calves leading to early maturity.

ii) Total mixed ration

TMR can be defined as, "the balancing and combining of all feeds into one complete ration" in a form that precludes selection.This complete ration provides all essential nutrients that are required to meet the needs of the animal.It is designed to be the sole feed source given over a 24 hour period and fed *ad lib.* for optimum results.

Advantages

i) Increased DMI, Peak DMI is achieved 4 to 8 weeks earlier than conventional systems.

ii) Optimal microbial protein synthesis: (maximised at pH of 6.3 to 7.4).

iii) Increased milk production by approximately 5 to 8 percent.

iv) Improved fat percentage:Improved rumen fermentation, and optimal pH (6.2 to 6.8) achieved, resulting in both maximal rumen fermentation and cellulose digestion leading to the formation of acetic acid, the precursor of butter fat production.

v) Reduced digestive upsets: due to each bite of feed having the same composition, minimizing pH fluctuations in the rumen.

vi) No need to feed mineral/vitamin supplements as all the requirements included in the TMR.

vii) Eliminate concentrate feeding at milking.

viii) Variety of less palatable feeds can be utilized in the ration, as these are masked by the other ingredients by the elimination of selection.

ix) Better control of the cow's diet.

iii) Urea mineral molasses block

The basis for Indian milk production is the millions of nondescript cows and buffaloes in rural areas, fed mainly on crop residues and agro-industrial by-products.– that are deficient in nitrogen, minerals and vitamins. As protein supplements such as oil cakes are only availability at a very high price in developing countries, if available at all, this has led to the use of non-protein-nitrogen sources, such as urea, to compensate for the nitrogen deficiency in fibrous feeds, thus enhancing their digestibility, intake and nutrient availability through optimization of rumen fermentation. The use of solid feed supplementation blocks, i.e. urea-molasses blocks or multi-nutrient blocks, to provide the nitrogen, minerals and vitamins lacking in fibrous feeds offers several advantages: ease of transport, storage and use, and reduced risks compared with other approaches, such as giving a small amount of urea in drinking water, sprinkling of urea solution on fibrous feeds before feeding, or urea-ammonization of crop residues. The problem of feeding urea to animals has been overcome in India by the introduction of feed supplementation blocks in the form of urea-molasses mineral block (UMMB) licks developed at the National Dairy Development Board (NDDB), Anand, India. The slow ingestion of urea provided through such licks ensures its efficient, non-toxic utilization.

Advantages of feeding UMMB licks to animals (Makkar *et al.*, 2007)

- Stimulates rumen fermentation, thereby, increases straw intake by animals.
- Increases microbial protein synthesis and supply at abomasum level, giving higher productivity.
- Improves daily milk (by 0.5–1.0 kg) and fat (by 0.3–0.5 percent) yields.

- Increases lactation length.
- Maintains health and reproductive functions.
- Improves growth rate of animals on straw-based diets.
- Decreases in inter-calving days and in the age at first calving are additional beneficial effects of feeding the blocks.
- Feeding of crop residues with UMMBs can sustain a milk yield of up to 4 or 5 liters per day in cattle
- No risk of urea toxicity.

Medicated urea-molasses block: To develop a self-medication device for prolonged low-level administration of anthelmintics, using UMMBs as the carrier. Technology for the production of UMMBs incorporating flukicide was also developed to control liver flukes in dairy animals. Fenbendazole(0.5 mg/kg body weight) has been found most suitable for incorporation in UMMBs as the drug is stable in blocks, does not require a milk withholding period and has a safety index of 67 in cattle and 1000 in small ruminants (Sanyal, 1994; Sanyal and Singh, 1992; Sanyal, 1993). The effective level was reached between days 4 and 6 of drug administration and a plateau was maintained thereafter. The daily low level dosing, the parasites would be exposed to the toxic concentration of the drug for much longer periods without additional use of the drug.

CONCLUSION

To encourage the farmers for dairy farming increase the profit of farmers by feeding balanced ration and reducing the cost of feeding hence decrease the per liter cost of production. Feeding of total mixed ration increase the utilization of non conventional feeds. Apart from cooperatives, the dairy sector is still characterized by small-scale, scattered and unorganized milch animal holders; low productivity; inadequate andinappropriate animal feeding and health care; lack of basic infrastructure for procurement, transportation, processing andmarketing of milk and lack of professional management.To encouraging dairy farming by rural farmers through dairy farmer group (Village Dairy Cooperative Societies) by solving milk marketing problems and also raising competitive nature in farmers.

REFERENCES

Govt. of Bihar. (2012). Bihar Basic Animal Husbandry Statistics 2012.based on BAHS Series-2012 Govt. of India, UNDP 2011, FAO Stat 2011

Harinder, P.S. Makkar, Manuel Sánchez and Andrew, W. Speedy (2007).Urea- molasses multinutrient blocks: simple and effective feed supplement technology for ruminant agriculture. FAO Animal Production and Health.

Sanyal, P.K. (1993). Plasma levels of fenbendazole metabolites in buffalo and cattle after long-term intraruminal administration. Veterinary Quarterly, 15: 157–159

Sanyal, P.K. (1994). The pharmacokinetic behaviour of fenbendazole in buffalo and cattle.Journal of Veterinary Pharmacology and Therapeutics, 17: 1–4.

Sanyal, P.K. & Singh, D.K. (1992). Plasma oxfendazole profile in buffalo and cattle: a marker of pharmacokinetics of fenbendazole. Buffalo Journal, 8: 157–161.

Singh, K. M., Singh R. K. P., Jha, A. K. and Meena, M. S. (2010). Dynamics of Livestock Sector in Bihar: A Temporal Analysis Agricultural Situation in India, 13:687-702.

CHAPTER - 50

Transforming Youths for Better Future Through Dairy Faming

Ramesh Kumar[1], Dahiya D.S.[2], Sheokand, R.S.[3] Sheokand, B.S.[4] and Dhanda S.[5]

[1]*Scientist (Animal Genetics & Breeding), Disease Investigation Lab, Krishi Vigyan Kendra, PanduPindara,Distt. Jind (Haryana)*

[2,3,4,5] *Directorate of Extension Education, Lala LajpatRai University of Veterinary and Animal Sciences, Hisar (Haryana)*

INTRODUCTION

Livestock is an integral part of agriculture in our country. In the decade, India's population has grown drastically by 13% to 1.2 billion people. Meeting the nutritional needs of this swiftly growing is a gargantuan challenge. Researchers and young farmers have an important role to play in the India Government strategy to relieve pressure on food resources. The dairy industry is a large and important sector in India. We are the highest milk producers in world and milk is an integral source of nutrition and household income. But per animal production is very low as compared to developed countries. The challenge is to improve production and reproduction of dairy animals. Indian farmers feed their animals with imbalanced ration. Supplementation of mineral and vitamins in feed not only improves health status but also in turn increase the income of dairy farmers. The economic contribution of dairy sector seems to be quite substantial in the agricultural economy as well as in the national economy. Keeping in view the growing importance of dairy sector in the changing national and global economic scenario, the dairy farming vocational training programs are organized in the Extension Center of LUVAS at Jind. To undertake this work,856 young dairy farmers of the area who have successfully completed dairy farming vocational training programmes during the year 2010 to 2013 were selected for the collection of data. The information collected from the trained youths by pre and post evaluation of training, visiting and discussing with them personally was analyzed for their adoption percentage of different technologies to dairy entrepreneurship. It was observed that overall average adoption rate was 36% and unemployed rural youths(78%) with education (10+2 and above) showed keen interest in dairy farming. The knowledge level of the respondent before and after training for breeding, feeding, health care and management was 28.0, 30.0, 38.0, 42.0 and 49.0, 60.0, 76.0, and 86.0%, respectively. Cleaning of sheds (95%), Vaccination against contagious diseases (95%) and ectoparasitic control (85%) were highly adopted by farmers followed by deworming of animals(69%), washing of udder and teats (65%) and colostrums feeding to calves in time etc. It was inferred that balanced feeding fetched more income to farmers (Rs.450-590 per animal /month) with 56% anestrus animals showing estrus symptoms. Mineral mixture feeding increased milk yield to the tune of 2-9 % with 59% animals coming in heat.Colostrum feeding and deworming led to better growth and decreased calf mortality. It was concluded that such type of trainings motivated theyoung farmers for scientific management of herds leading to maximum profits from this vocation.

MATERIALS AND METHODS

Sample Selection: In the present paper an attempt has been made to analyze the various training courses organized by the KVK, Jind on dairy farming. This in depth survey of respondents was conducted on the trainees who acquired specialized vocational training on dairy farming during the years 2009-10, 2010-11,2011-12 and 2012-13 Farmers were interviewed in order to assess the impact of training courses organized by the KVK Jind. A total of 856 dairy farmers imparted vocational training of 11 days duration during these years. These trainees were interviewed for the study purpose

Method of data collection: The method used wasface to face interviews with the trainees. The semi-structured interview schedule was developed to know change in knowledge about different aspects of dairy farming before and after the training. To assess the gain in knowledge, knowledge score of the trainees by pre training test and post training test were compared.

Training programs conducted for farmers : Atotal number of 856 farmers of Jind district of Haryana were given practical training on dairy farming from 2009-2010 to 2012-2013, which was conducted by the KrishiVigyan Kendra, Jind.The trainings were imparted on skill development regarding breeds and their characteristics, breeding management, care and management of the newlyborn calves, better housing and balanced feeding,prevention and control of internal and external parasites and diseases, selection of male calves for breeding and clean milk production, value addition of milk, development of linkages with local veterinary dispensary etc. The farmers were trained to work for vertical expansion of the scientific dairy production system taking into account of the limited resources of the farmers to keep the operational cost to minimum.

RESULTS AND DISCUSSION

Status of farmers: The study was conducted in Jind district of Haryana state. A survey proforma was developed comprising general information, background of the farmers such as their age, education, occupation, land holding, sex etc. Dairy farming practices like status of housing, feeding and watering deworming, vaccination schedule of animals and mineral mixture feeding to animals were also included. Various sources of information were also studied.A total of 856 farmers, rural youths and farm women were given practical training on scientific dairy farming from the year 2010 to 2013. Out of 856 farmers, rural youths and farm women, only 517 farmers (60.4%) had adopted the scientific farming practices. Rate of adoption was highest in Narwana tehsil on percent basis (70.8%), whereas the rate of adoption was lowest in the Safidon (53.4%).Average rate of adoption from the year 2010 to 2013 was 60.4% (Table 1).

Table 1: Selection of trained dairy farmers for collection of data in Jind district of Haryana.

Name of the Tehsil	Total no. of trainees	No. of trainees adopted dairy entrepreneurship	Adoption (per cent)
Jind	352	212	60.2
Safidon	298	159	53.4
Narwana	206	146	70.8
Total	856	517	60.4

Data revealed that unemployed rural youths (52.2%) showed keen interest in dairy farming. However, when respondents were classified on the basis of sex, it seemed to be predomenated by males (Table2). However,females performed more work in rural areas of this district as far as dairy operations were concerned. These findings are in agreement with the observations of Rangnekar (1992). The herdsize ranged from 1 to 20 animals. The respondents opinioned that increasing the herd size also increased the income but finance and labour problem were the main constraints. It was very difficult to maintain a large herd using family labour and resources. Most of the participants were below 30 years age and educated. Since the land holding of most of the participants is less than 5 acres and they have to grow crops for their livelihood, therefore it was very difficult for them to rear more number of animals.

Table 2 : Socio-personal profiles of trainees.

Variables	Number of trainees	Percent of trainees
1 Age (years)		
Less than 25	498	58.2
25-30 years	217	25.4
30 years and above	141	16.5
2. Education		
Up to 10th class	196	22.9
10+2	428	50.0
Graduates	232	27.1
3.Occupation		
Housewife	209	24.4
Unemployed youth	447	52.2
Agriculture	97	11.3
Laborers	103	12.0
4. Herd Size		
More then 20	48	5.6
11-20 animals	102	11.9
6-10 animals	397	46.4
1-5 animals	309	36.1
5. Land Holding		
>5 acre	282	32.9
2-5 acre	361	42.2
<2acre	96	11.2
Landless	117	13.7
6.Sex		
Male	647	75.6
Female	209	24.4

It was found in the periodically conducted surveys in the district Jind that a large number of dairy farmers have developed the elite herds of Murrah buffaloes and cross-bred by adopting improved technologies. The knowledge level of the respondents in areas of dairying, viz., breeding, feeding, health care and management was 28.0,30.0, 38.0 and 42.09 per cent, respectively whereas, the extent of knowledge level after dairy farming training was 49.0, 60.0, 76.0 and 86.0 per cent in these areas, respectively. The overall increase in knowledge level for dairy farming was79.07 per cent (Table 3).

Table 3 : Impact of the training programs in the area of Dairy Production

Practice	Knowledge Level Before Training	Knowledge Level After Training
Breeding	28.0	49.0
Feeding	30.0	60.0
Health Care	38.0	76.0
Management	42.0	86.0

Data in Table 4 depicts that vaccination against F.M.D. and H.S.and cleaning of sheds was maximum adopted practices leading to minimum losses due to diseases and ectoparasitic control. Similarly washing of udder and teats with clean floors resulted in mastitis control which in turn led to maximum profit from this vocation.These findings are in agreement with Singh *et al.* (2011)Data in Table 5 presents the economic returns to farmers by technological interventions.Milk yield increased significantly with decrease in anoestrus cases which in turn fetched more profits to farmers by adoption of balanced feeding, mineral mixture supplementation and deworming etc. Decreased mortality and better growth in calves were other factors to increase income to farmers. These findings were in agreement with Ghaffar (2007) and Singh *et al.* (2011)).

Table 4: Impact of trainings on dairy farming.

Sr. no.	Practice	Farmer's practice	Recommended	Adoption rate(%)
1	Mineral mixture/ balance feeding	Use of cotton seed cake, mustard cake, choker, greenfodder,wheat,straw, paddy straw	Balance ration with mineral @ 2%including proper ratio of cakes, cereals & by products.	50
2	Washing of udder &teats	Before milking only	Before and after milking	65
3	Cleaning of sheds	Only cleaning, no washing	Cleaning & washing of pucca floors	95
4	Deworming of calves & adults	Deworm the calves &adults only when recommended by veterinarian.	Deworming according to schedule	69
5	Ectoparasitic control	Use kerosene oil or pick up by hand	Spraying with recommended medicine with proper dose & schedule	85
6	Artificial insemination	Mating with non-descript draft bulls	A.I with semen of elite proven bulls	46

Contd....

7	Anoestrus/repeat breeders	Use methi, bajra etc. for anoestrus. Go on for natural mating in repeaters	Rectal examination of reproductive organs by V.S. proper treatment & balance feeding	45
8	Pregnancy diagnosis	Only guess	Rectal examination after 3 months of insemination/ mating	35
9	Vaccination against contagious disease	Only FMD	Vaccination against FMD and HS at recommended schedules	95
10	Colostrums feeding to new born calf	After removal of placenta	Within one hour of birth of calf	35

Table 5 : Economic returns by technology/intervention in dairy farming

Technology/ intervention	No. of farmers	Milk yield increase per day/animal	Additional returns/ animal per month	Remarks
Balance feeding	429	3-10%	Rs. 450-590	56% anoestrus animals showed oestrus symptoms
Mineral mixture feeding	856	2-9%	Rs.350-400	59% anoestrus animal came to heat.
Deworming in adult animals	390	3-8%	Rs.300-550	Good health
Deworming in calves	795	Appreciable growth over control	Increased body weight gain but difficult to quantify	As a result of better growth, age at first calving was reduced. Mortality due to endoparasites decreased
Colostrums feeding within one hour of calving	856	Improved body resistance	Decreased infections	Better growth and decreased mortality

CONCLUSION

It was found that after attending the training programs there was gain in knowledge level of the beneficiaries. After adoption of the newer technologies learnt by the entrepreneurs, they gotmore income and employment. It is recommended that various KVK should tailor and organize need based vocational training programs for entrepreneurship development so that the rural people are benefited It is imperative that the dairying being an integral component under the mixed farming system need be strengthened in the shuttle cities of different states and in blocks around the major towns of India.This will help in achieving the sustainable production and in turn will increase the income and employment in the rural areas. It is also recommended that through such training programs organized by KVK's, socioeconomic conditions of the rural farm families could further be improved through scientific dairy farming.

REFERENCES

Singh S.,Kumar,R.,Sharma C.S., Kumar U. and Kumar A. (2011) Improved Dairy Production and Processing Through Training Programmes. J. Dairying, Foods & H.S., 30 (3) : 165 - 169,

Ghaffar A. (2007). Integrated Approach for Improving Small Scale Market Oriented Dairy Systems in Pakistan: Economic Impact of Interventions. Italian J. Anim. Sci. 6(2):1240-47.

Kadian, K. S. and Kumar, Ram. (2000). Multivariate Analysis of Information Seeking Pattern of Diary Farmers of Himachal Pradesh.Indian J. Dairy Sci.53 (5): 375-380.

Rangnekar, S. D. (1992). Women in Livestock Production in Rural India. Proceedings, 6th AAAP Animal Science Congress held on 23-28 November, Bangkok, Thailand.

CHAPTER - 51

Backyard Poultry Farming: A Profitable Enterprise for Rural Youth Farming

Adhiti Bhanotra[1], Minu Singh[2] and Jancy Gupta[3]
[1&2] Ph.D Scholar at SKUAST-J, & NDRI Karnal, Veterinary Extension Education
[3]Ex-Head & Principal Scientist, Dairy Extension Divison, Karnal, Haryana

INTRODUCTION

Backyard poultry production is an age old practice in rural India. Rural poultry farming in India contributes to about 40% of the national egg production. In ancient time poultry birds were reared for different purposes such as for its fighting ability. In old civilization, the cocks are known for their morning alarm and the selection is also done for these traits. Their fighting ability is more responsible for their spread than there nutritive value of the chickens. Backyard poultry farming is now gaining considerable momentum in the country and enables rural community to participate, in the economic, political and social sustainable development of the rural communities.

Backyard poultry farming is being recognized as important component of socio- economic improvement among the weaker section of society; specially rural youth, landless labour, small and marginal farmers. Backyard poultry generates self-employment, provides supplementary income with protein rich food at relatively low cost. Backyard poultry farming is the one among different enterprises in the present situation which gives some assured income to rural youths of the country.Before 1950, poultry keeping in India was largely a backyard venture. During last 3-4 decades Indian poultry industry has shown excellent growth. Poultry farming is a profitable venture as a means of livelihood and income prosperity for rural youth and farmers. The small and marginal farmers may choose backyard poultry farming alternative method for generating good income with their limited land and water use. Keeping in view of demand of poultry eggs and meat in the country and its potential for income and employment generation among rural youth, there is a need to increase the poultry production in the country.

Backyard poultry farming can play very important role in the diversification of agriculture, particularly in view of the fact that with ever increasing population, there is tremendous pressure on land. If poultry is taken up on sound basis, it can help to harness triple benefitsi.e.additional employment income with least investment in backyard production.

Most of the backyard poultry production comprises rearing of indigenous birds with poor production performances. The potentiality of indigenous birds in terms of egg production is only 70 to 80 eggs/ bird/ year and meat production is also very less. However, the backyard poultry production can be easily boost up with improved varieties of chicken and can promise a better production of meat and egg. To improve rural youth farming, backyard poultry is a handy enterprise with low-cost initial investment, but high economic return along with guarantee for improving protein deficiency among the poor.

Importance of Poultry Farming

Poultry are efficient converter of food into egg and meat and feed items unsuitable for human consumption and bye products can be used successfully as feed for poultry. Egg contains 12.0% protein, 11.3% fat and 1% carbohydrate white broiler meat contains 18.2% protein and 6. 2% fat of higher biological value.

Poultry farming can be done on small as well as large scale. When it is done as small scale venture, the investment is also minimal. Hence it is an important enterprise for self-employment of rural youth.

1. Due to small generation interval the returns from this venture is quite quick compared to other livestock species. In layer farm income starts giving after 6 months while in meat production it starts after 2 months.
2. Beside the use of poultry as food its manure is a boon to agriculture farming. Forty layer birds produce one ton of poultry manure in 18 months on deep litters which sufficient for 1 acre of land. Poultry manure is a rich source of organic matter nitrogen; phosphorus average poultry manure contains 2.6% nitrogen, 2.0% phosphorus and 1.5 % potash and trace elements.
3. The biggest advantage with the poultry is that it can be done as family business where all the members whether male or female, adult or children, old or young can contribute substantially for successfully running of this enterprise. Thus it provides sufficient job opportunities for all the members of family.

4. Poultry farming can be done in the back yard and due to this reason it is playing an important role in the social and economic uplifting of weaker section of the society.
5. Backyard poultry farming acts as an 'ATM', because as per family need the birds and eggs can be sold at any time anywhere with cash in hand.
6. The growth rate of poultry (Broilers) is very fast. If reared in organized way, a house hold can earn Rs. 15 -20 per bird net profit at the market age (4-5th weeks) of the bird.
7. Feed cost is negligible due to better utilization of agricultural by-products and leftover feed and grains.
8. Boost up in family income for better utilization of family labourwho are not able to perform other agricultural works like old family member or children.
9. Quality of chicken and egg is better in terms of organic farming as the birds are raised in stress less environment with natural input.
10. Poultry needs small land and resources than needed for other livestock species.

Improved Strain of Poultry Birds for Backyard Farming

Desi or indigenous birds are generally poor performers in egg and meat production. To obtain maximum profit from backyard poultry farming there is a urgent need in the country to improve the status of backyard poultry farming with improved strain of poultry birds, who performs an excellent result when raised in backyard with low inputs. Improved chicken that can be introduced in backyard poultry farming should have following characters-

- Adaptability in village condition
- Self-propagation
- Good brooding capacity
- Mothering ability
- Well body conformation
- Hardy in nature
- Good scavengers
- Attractive and coloured plumage
- Escaping capacity from predators
- Disease resistance

Considering the above characteristics the varieties that can be recommended for backyard poultry farming are Gramapriya, Vanaraja, Giriraja, Girirani, Krishna-J etc. The annual egg production capacity of Gramapriya and Vanaraja birds are 200 to 220 and 200 to 230 egg/ bird/ year, respectively. First egg lays at 200 to 230 days depending on plane of nutrition and other management parameters. Egg weight varies from 55 to 60 gram. Mature body weight 2.5 to 3.5 kg.

Housing Management for Backyard Poultry Farming

No elaborate housing is required for backyard poultry farming but, it should protect the birds from sun, rain and predators. If free range system is practiced the birds are let loose in day time for foraging and at night sheltered in shed. For better production performances certain criteria that can be considered are –

1. The poultry house should be in east-west orientation to protect from summer wind and cold stress and also for direct sunlight in winter months.
2. During summer direct sun light should be avoided to reduce the summer stress in birds.
3. Low cost housing material like wood, bamboo, grass, thatch etc can be utilized.
4. The poultry house should be free from water seepage or moisture.
5. Floor should be in elevated land or above ground level (minimum 2ft) and free from water crack, easily cleaned, rat proof and durable.
6. There should be free air movement in upper part of the shed to reduce gas formation inside the shed.
7. Height of the side wall in poultry house is generally 7 ft to 8 ft. The centre height is 9 ft to 12 ft with slope in either side.
8. Roofing material like thatch, tiles, asbestos etc can be used.
9. Brooder house should have easy ventilation and wire netting which is used for open air ventilation.
10. Provision of bulb fitted above the ground as hoover may be utilized to keep the chicks warm.

Brooding of Chicks

There are two types of brooding viz. Natural brooding and Artificial brooding that can be adopted for backyard poultry farming.

1. Natural brooding

If natural brooding is practiced the local broody hen is used as the indigenous hens are very good sitters. Improved variety of fertile eggs is put into incubation. The hen is provided with nesting materials. Food and water supply is given throughout the incubation period. A broody hen can easily take care for brooding and hatching of 12 to 15 chicks. After hatching the chicks are let loose along with mother for scavenging. There should be provision for separate place inside the shed for young chicks and mother at night.

2. Artificial brooding

In artificial brooding provision of artificial heat is necessary. Artificial heat can be provided with electricity, gas, kerosene, wood, sawdust etc. 'Bukharies' also can be used as a source of artificial heat. Wood, charcoal or sawdust is used in 'Bhukaries' and it is an ideal source of artificial heating when there is acute shortage of electricity, gas and kerosene. The optimum temperature is 95p F in first week and it can be reduced 5p F per week up to 6 week till 70p F. Two watt/ chick heat is required up to 6 week in brooder house. The necessity of light in brooder house is to increase feed consumption for maximum growth in a short period and also to prevent stampeding or piling if scared. Initially (up to 6 weeks) there should be provision for at least continuous light up to 48 hours in brooder house and in growing stage (8 to 18 weeks) light hour is 10 to 12. But in laying period light should be for 15 to 16 hours. The provision for extra light may be in the morning or evening or may be morning and evening both. To prevent direct contact with heat a chick guard made up of card board or metallic guard can be used in brooder house. The height of chick guard is 15 to 18" is placed in circular shape at a distance of 3' away from the hover.

Feeding Management

The feed cost alone is 70% of total expenditure in poultry production. In backyard poultry farming the feed cost is considered to be minimum. Hence, the birds are let loose for scavenging in the open yard and collect the required protein, energy, minerals and vitamins etc from insects, snail, termites, seeds of grasses and weeds, leftover grains, crop residues and household wastes. Feed ingredients like broken rice, ground nut straw, wheat or job's tear grain, rice bean etc also can be given to the birds. During rainy season poultry feed should not be stored more than 1.5 months to avoid fungus growth (Aflatoxicosis). In backyard poultry farming generally two times feeding is practiced; once at morning and another at evening. The space requirement for feeder is 2 to 7 cm at brooding period, 7 to 10 cm during growing stage and 12 to 15 cm / bird at laying stage. The water space should be 0.5 to 1.5 cm during brooding, 1.5 to

2.5 during growing and 2.5 cm during laying period. The birds may be supplied with extra concentrate ration @ 30 to 60 gm/ day/ bird for better performance. The balanced ration may be formulated with appropriate percentage of maize, rice polish, wheat bran, ground nut cake, fish meal, shell grit or lime stone along with salt, minerals and vitamins or may be with locally available ingredients. The poultry feed should contain at least 20% protein in starter level, 16% in grower and 18% in laying stage with energy level (ME) 2800 kcal/kg feed in starter, 2600 kcal/kg at grower and 2650 kcal/kg in layer ration.

Health Care Management

For better health care in backyard poultry farming the birds should be vaccinated against virus diseases in time. The diseases that mostly effect the birds are Ranikhet disease, Marek's Disease, Fowl pox, Gumbroo disease etc. Regular vaccination schedule may be followed in a poultry farm (Table 1). Deworming for internal and external parasites also should be done to maintain a healthy flock. Other diseases that may affect the poultry birds are Coccidiosis, infectious coryza, Salmonellosis etc.

Table 1: Vaccination schedule for poultry birds

Sl. No.	Age of birds	Name of Vaccine	Name of Disease	Doses	Route of Vaccination
1	Day old Chicks	HVT MD Vaccine	Marek's Disease	0.2 ml	S/c or I/m
2	4 to 7 days	F-1/ Lasota	Ranikhet Disease	One drop	Eye or nostril
3	14 to 18 days	Intermediate plus	Gumboro disease	-	Drinking water
4	35 days	F-1/ Lasota	Ranikhet Disease	One drop	Eye and nostril
5	6 to 7 weeks	Chicken embryo adopted	Fowl Pox	0.5 ml	Wing stab method
6	8 to 10 weeks	Strain killed vaccine	Ranikhet Disease	0.5 ml	S/c or I/m

Finger Tips Suggestion for Backyard Poultry Farming

1. Disease free, improved strain, dual purpose poultry birds may be procured for back yard poultry farming.
2. Periodical vaccination should be done on regular basis.
3. Clean drinking water and fungus free feed should be supplied to the birds.
4. The poultry shed should be regularly cleaned and free from moisture and humid condition.
5. Overcrowding should be avoided.

6. If possible there should be separate space for different age group of birds.
7. Sick bird should be immediately separated/ culled from healthy flock.
8. Poultry equipments particularly waterer and feeder should be regularly cleaned and disinfected.
9. There should be restriction for outsider into the poultry shed or farm.
10. The birds should be free from predators and should not be scared by other animals.
11. Before procuring new flocks the shed should be thoroughly cleaned and disinfected.
12. There should be provision for footbath in front of poultry shed.
13. During summer and winter months the shed should be protected from hot or cold wind by hanging curtain around the side of the wall or shed. In summer months water sprinkling also can be done.

Problems in Poultry Industry and Challenges

Poultry development is chiefly confined in cities or nearby cities. At present rural areas are still lacking in basic infrastructure for poultry development and adequate know-how is also not available in the rural areas. In spite of the extensive development in this field, the efficient experts are in shortage. With the result farmers have to depend upon chick suppliers, feed and medicine dealers and such other persons for their technical needs which may result in lesser profits and sometime losses also. A large number of companies have come in the field of production of poultry feed, vaccines, medicines and equipment etc. With the results, it is difficult to select quality product because most of the companies are interested in their own profits and ignoring the interest of the farmers. Due to lack of proper distribution and marketing facilities, farmers have to suffer loss in spite of high production while consumers have to pay higher cost of the poultry produce. The cost of various poultry inputs are increasing day by day with no proportionate increase in poultry products and with the result there is drop in profits.

In present context, it is necessary that the maximum production be done at minimum inputs. Hence, high quality technologies be developed in this field and subsequently be transferred to farmers. A strict quality control measures should be applied at each stage of poultry production. An effective and efficient marketing system should be developed for marketing the poultry products to ensure proper and satisfactory price for their produce.In rural areas the poultry production is practiced as backyard venture. Hence, suitable breed and techniques should be developed so that the poultry farming can be done at very low cost with better efficiency.

CONCLUSION

Domestication of poultry is practiced from ancient time but it is very difficult to say at what time they are first domesticated in the world. Poultry occupies a very important place in all domesticated animals. A large chunk of income from livestock industry is contributed by poultry industry. Hence, its importance is increasing day by day in developed and developing countries. It generates self-employment, provides supplementary income with protein rich food at relatively low cost. The biggest advantage with the poultry is that it can be done as family business where all the members whether male or female, adult or children, old or young can contribute substantially for successfully running of this enterprise. Thus it provides sufficient job opportunities for all the members of family. At present poultry farming is a scientific and organized industry and playing an important role in creating job opportunities. Over and above, poultry farming is quite and eco-friendly activity and can revert the trends of migration of rural youth.

REFERENCES

https://www.poultryworld.net

http://www.dahd.nic.in

Chapter - 52

Factors Dominate on Adoption of Modern Aquaculture Technologies at Fish Farms in Bogra District of Bangladesh

Md. Hashmi Sakib[1] and Md. Safiul Islam Afrad[2]

[1]Lecturer, Department of Agricultural Extension and Rural Development, EXIM Bank Agricultural University, Bangladesh

[2]Professor, Department of Agricultural Extension and Rural Development Bangabandhu Sheikh Mujibur Rahman Agricultural University, Gazipur

ABSTRACT

Bangladesh is a land of rivers, wet lands and tributaries across her deltaic plains in which aquaculture is a potential arena to meet dietary needs of protein. The main objective of this study was to bring out the dominating factors on adoption of modern aquaculture technologies by the fish farmers. The study was conducted in Kahaloo and Sherpurupazilas of Bogra district. Respondent fish farmers were selected using proportionate random sampling technique. Quantitative data were collected using predesigned interview schedule during August to September 2011. Socio-demographic characteristics of the respondents viz. 'use of information sources' 'innovativeness', 'social participation', 'commercialization' and hindrances of adoption i.e. 'lack of technical knowledge', 'unavailability of credits in time', 'lack of quality fish fry', 'comparatively low market price', 'fish fry price is high', 'unavailability of quality fish feed', 'adulteration of fertilizers' were the major dominating factors on adoption of modern aquaculture technologies. 'formulated feed is costly', 'high production cost'

Key words: Adoption, Factors, Aquaculture, Technology

INTRODUCTION

Fisheries are now one of the most important sectors in the Bangladesh economy. In this country, the total fish production was 2701 thousand metric tons (MT) in 2008-09 which contributed 4.18 percent of total gross domestic product (BBS, 2010). During 2009, around 17.3 million US dollars were spent to import fish compared to 1.1 million US dollars in 1999 but in 2007, the total fish export earned was 630.8 million US dollars that decreased to 516.0 million US dollars in 2009 (FAO, 2012). These scenarios indicate that the supply of fish in domestic markets and export earning is decreasing. Hence, the demand of fish in Bangladesh is increasing day by day with her ever growing population. People of the country fulfill their increased demand of fish only through capturing from open water bodies. Thus, fish production of open water bodies showed decreasing trend and availability of some species have greatly reduced. Even though, in the early 1980s, more than 80 percent of animal protein requirements were met from fish (Ahmad and Hassan, 1983). To meet the increased fish demand, Aquaculture is becoming key arena to increase the fish production. In this sector, further technological development can only be a solution of current crisis. In this connection it is essential to ascertain the key factors which are directly or indirectly influencing the adoption of modern aquaculture technologies. By considering above circumstances researcher are motivated to i) find out the farmer's socio-demographic influencing components on adoption of modern aqua cultural technologies, ii) constrain confrontation by the fish farmers regarding adoption of modern aqua cultural technologies.

METHODOLOGY

Kahaloo and Sherpurupazilas of Bogra district were the locale of this study. Bogra district consists of 12 upazilas from which Kahaloo and Sherpur were selected using simple random technique. Kahaloo is located at western part of Bogra district whereas Sherpur is located at the southern part. Six unions viz. Kahaloo, Jamgaon, Malancha, Kusumbi, Bishalpur, Khanpur unions were selected by using simple random technique procedure from a total 18 unions of Kahaloo and Sherpurupazilas. Then a list of fish growers of these six unions was prepared with the help of the Upazila Fisheries Officer (UFO). Total number of fish farmers of these unions was 366. Only heads of these 366 fish farmers constituted the population. Thirty percent of the fish growers were selected from villages of six unions by using proportionate random sampling technique. Resultant 110 fish farmers constituted the sample of this study. In order to collect valid and reliable information from the farmers an interview schedule containing both open and closed form questions was developed considering the objectives of the study.

Ten selected characteristics viz. age, education, family size, farm size, fish farming area, information sources use, annual family income, commercialization, social participation and innovativeness of the respondents were the independent variables. Adoption of aquaculture technologies was the dependent variable of this study which consists of eleven selected technologies viz. (i) pond preparation with lime, (ii) acclimatization of fish fry before stocking, (iii) appropriate stocking density (iv) monosex tilapia cultivation, (v) pangus cultivation, (vi) catfish cultivation, (vii) poly culture of carp, (viii) Thai koi cultivation, (ix) use of formulated feed, (x) use of urea and (xi) use of TSP.

Adoption of aquaculture technologies like, monosex tilapia cultivation, pangus cultivation, catfish cultivation, poly culture of carp, Thai koi cultivation were computed by using following formula of Kashem (2004):

$$\text{Adoption Index} = \frac{\text{Ei}}{\text{Pi}} \times 100$$

Where,

Ej = Extent of adoption (actual area under practice) expressed in terms of summation of obtained adoption score of the practice under study.

Pi = Potential adoption (possible area of practice) expressed in terms of possible maximum obtainable adoption score of the practice under study.

Adoption index of modern aquaculture technologies namely, pond preparation with lime, appropriate stocking density, use of urea, use of TSP, use of formulated feed were computed by following the formula of Sakib and Afrad (2014):

$$\text{Adoption Index} = \frac{\text{Maintained unit}}{\text{Recommended unit}} \times \frac{\text{Covered area}}{\text{Total area}} \times 100$$

On the other hand, by using the formula of Sakib and Afrad (2014), adoption index of acclimatization of fish fry before stocking was computed is given below:

$$\text{Adoption Index} = \frac{\text{Maintained unit}}{\text{Recommended unit}} \times \frac{\text{Covered area}}{\text{Total area}} \times 100$$

Data were collected using predesigned interview schedule during 02 August 2011 to 26 September 2011. The statistical measures such as, number and percentage and rank order were used for describing the overall scenario of the study. One of the Statistical tools i.e. regression was employed in order to explore the contributing socio-demographic factors on adoption of aquaculture technologies.

RESULTS AND FINDINGS

Involvement of some specific socio-demographic contributing factors of the respondents towards adoption of modern aquaculture technologies

With a view to finding out contributing socio-democratic characteristics of the respondent's adoption of modern aquaculture technologies, general linear multiple regression analysis was employed.Tenindependent variables were taken in to account to run the multiple regression analysis. The findings are presented in Table 1.

Table 1: Regression coefficient of farmer's adoption ofmodern aquaculture technologies with their selected characteristics

Farmers characteristics	Unstandardized coefficients	Standardized coefficients	t-value	Level of significance
Constant	24.188		3.232	.002
Age	.030	.032	.375	.708
Education	.057	.027	.230	.819
Use of information sources	.285	.268	2.374	.020
Farm size	-5.165	-.464	-2.014	.047
Fish farming area	6.191	.551	2.516	.014
Annual family income	0.0000028	-.078	-.535	.594
Commercialization	.174	.142	2.035	.045
Social participation	-1.339	-.226	-3.405	.001
Innovativeness	.435	.268	4.361	.000
Knowledge	.978	.559	4.296	.000

R^2 = .72Adjusted R^2 = .69F-value= 22.785P = 0.000

The regression coefficient of seven variables namely, use of information sources, farm size, fish farming area, commercialization, social participation, innovativeness and knowledge were being indicated their significant contribution to the adoption of modern aquaculture technologies. The remaining characteristics had no significant contribution to the respondents' adoption. The R^2 value was 0.72 and the corresponding F value was 22.785, at zero (0.000) level of significance. This R^2 value indicated that 72.00percent of the total variation in the respondents' adoption was explained by the ten variables included in the regression analysis. To ascertain the proper contribution of the characteristics, stepwise multiple regression analysis was run and the findings are presented in Table 2.

Table 2 evinces that seven characteristics viz. use of information sources, farm size, fish farming area, commercialization, social participation, innovativeness and knowledge entered into the stepwise regression model and the contribution has been visualized of five attribute namely, use of information sources, commercialization, social participation, innovativeness and knowledge, all of them was significant at one percent level of significance except commercialization. Hence, whatever variation was in the respondents' knowledge was mainly due to the contribution of these five characteristics. The coefficient of multiple determinations (R^2) for the stepwise regression model was 0.70 which indicated that the independent variables explained 70 percent variation to the farmers' adoption of modern aquaculture technologies. Finally, another linear multiple regression analysis was run involving only the five significantly contributing variables found in the stepwise regression analysis. In this regression analysis, R^2 value was also 0.70 (significant at 0.000 level). This final analysis thus, indicated that 70 percent of the total variation in respondents' adoption of modern aquaculture practices was explained by the mentioned five characteristics (Table 2).

Table 2: Regression coefficient of adoption of modern aquaculturetechnologies by the farmers with their characteristics entered in regression model

Respondents characteristics	Coefficients		t-value	Level of significance
	Unstandardized	**Standardized**		
Constant	26.879		4.793	.000
Use of information sources	.354	.332	3.521	.001
Commercialization	.146	.119	2.075	.040
Social participation	-1.248	-.211	-3.246	.002
Innovativeness	.389	.239	4.076	.000
Knowledge	.893	.510	5.474	.000

$R^2 = 0.700$ Adjusted $R^2 = 0.686$ F-value= 48.550P = 0.000

The unique contribution of each of the five characteristics was determined by taking the changes in R^2 value occurred for enter of a particular variable in the stepwise regression model. Individual contribution of five characteristics has been presented in Table 3.

Table 3: Changes in the multiple R^2 for enter of a variable into the stepwise regression analysis model for adoption of aquaculture technologies

Variable entered	Multiple R	Multiple R^2	Change in R^2	Variation explained percent	level of Significance of F-change
1. Knowledge	0.772	0.596	0.596	59.6	0.000
2. Use of information sources	0.791	0.626	0.030	3.0	0.004
3. Innovativeness	0.810	0.657	0.031	3.1	0.003
4.Social participation	0.829	0.688	0.031	3.1	0.002
5. Commercialization	0.837	0.700	0.012	1.2	0.040

Table 3 reveals that the knowledge contributed 59.60 percent, use of information sources contributed 3.00 percent, innovativeness contributed 3.1 percent, social participation and commercialization contributed 3.1 and 1.20 percent respectively to the explained variance of 70.00 percent.

Problematic Factors were revealed by the respondents regarding adoption of modern aquaculture technologies

Table 4: Constraints of modern aquaculture technologies as perceived by the respondents

Practices	Problems	*F	#P	Rank
Pond preparation with lime	High production cost	95	86.40	2
	Caustic and irritant to human	21	19.10	3
	Lack of technical knowledge	107	97.30	1
Acclimatization of fish fry	Lack of technical knowledge	33	30.00	1
Appropriate stocking	Lack of technical knowledge	32	29.10	3
density maintain	Unavailability of fish fry in time	92	83.60	2
	Unavailability of credits in time	109	99.10	1
Monosex tilapia cultivation	Comparatively low market price	20	18.20	4
	Lack of technical knowledge	27	24.50	3
	High production cost	39	35.50	2
	Lack of quality fish fry	51	46.40	1
	Widely practice of carp culture	17	15.50	5
Pangus cultivation	High production cost	30	27.30	6
	Lack of quality fish fry	34	30.90	5
	Lack of technical knowledge	49	44.50	3

Contd...

	Comparatively low market price	70	63.60	1
	Unavailability of credit in time	43	39.10	4
	Management complexity	55	50.00	2
Catfish cultivation	Cannot cultivate with others fish	36	32.70	6
	Management complexity	52	47.30	4
	Lack of technical knowledge	56	50.90	3
	Lack of quality fish fry	77	70.00	2
	High price of fish fry	78	70.90	1
	Widely practice of carp culture	40	36.40	5
Poly culture of carp	Management complexity	24	21.80	7
	Lack of quality fish fry in time	83	75.50	2
	Competition among different species	54	49.10	6
	Lack of technical knowledge	59	53.60	5
	Sometimes growth become stunted	75	68.20	3
	Unavailability of credit in time	87	79.10	1
	High production cost	68	61.80	4
Thai koi cultivation	Unavailability of credits in time	85	77.30	1
	Lack of quality fish fry	65	59.10	4
	Fish fry price is high	53	48.20	5
	Lack of technical knowledge	70	63.60	3
	Cannot cultivate with others	77	70.00	2
	High production cost	51	46.40	6
Use of formulated feed	Adulterants in feed ingredients	66	60.00	2
into the pond	Unavailability of quality fish feed	96	87.30	1
	Formulated feed is costly	44	40.00	3
Use of urea into the pond	Sometimes algal bloom occurred	34	30.90	3
	Adulteration of fertilizers	85	77.30	1
	Lack of technical knowledge	51	46.40	2
Use of TSP in the pond	Algal bloom occurred	48	43.60	2
	Lack of technical knowledge	85	77.30	1

*F = Number of farmer, #P = Percent

In this study, some very important problematic factors are mentioned by the farmers i.e. 'lack of technical knowledge', 'unavailability of credits in time', 'lack of quality fish fry', 'comparatively low market price', 'fish fry price is high', 'unavailability of quality fish feed', 'adulteration of fertilizers' (Table 4). Moreover 'unavailability of fish fry in time', 'lack of quality fish fry', 'management complexity', 'sometimes growth become stunted', 'formulated feed is costly', 'high production cost', 'cannot cultivate with others', 'management complexity', 'algal bloom occurred', 'unavailability of quality fish feed', 'adulterants in feed ingredients' were the main hindrances of adoption of modern aquaculture technologies as identified by the farmers in the study area. However, the practice wise constrains are presented in the Table 4.

DISCUSSION

In this study socio-demographic factors ("Knowledge", "Use of information sources","Innovativeness") have influence on adoption of modern aquaculture technologies as like as the findings of Singh *et al.* (2012).Pannu *et al.* (2014) found "Heavy initial investment for the construction of *tanka"* "Difficulties of getting loan", "Less subsidy as compared to investment", "Maintenance cost of *tanka* is high", "Rate of interest for loans is high" in his study which is in line with present findings. Besides problematic factors viz. "Credit is not available at proper time", "Credit is not available at proper amount", "High rate of interest on credit" were brought out by Singh *et al.* (2012) are akin to results of present study. Moreover, Singh *et al.* (2012) indicated the problems in his study namely "Erratic price fluctuations","Insufficient loans/ subsidies" have concordance with these results.

CONCLUSION

Adoption of modern aquaculture technologies are influenced by socio-demographic characteristics viz. 'use of information sources' 'innovativeness', 'social participation', 'commercialization' of the respondents. On the other hand some major problems found out by the respondents those are responsible to hinder in case of adoption i.e. 'lack of technical knowledge', 'unavailability of credits in time', 'lack of quality fish fry', 'comparatively low market price', 'fish fry price is high', 'unavailability of quality fish feed', 'adulteration of fertilizers'. 'formulated feed is costly', 'high production cost'

REFERENCES

Ahmad, K. and N. Hassan.1983. Nutrition survey of rural Bangladesh 1981–1982. Dhaka, Institute of Nutrition and Food Science, University of Dhaka.

BBS, 2010.Statistical Yearbook of Bangladesh. Statistics Division, Ministry of Planning, Government of the People's Republic of Bangladesh.

FAO, 2012.Statistical Yearbook. FAO, Rome

Kashem, M. A. 2004. Fundamentals of Extension Education.Lima Printing Press.Mymensingh.

Pannu, H.R., J.P. Lakhera, J.P. Yadav and K.C. Jeengar. 2014. Constraints Perceived in Adoption of Indigenous Water Harvesting Practices. Indian Research Journal of Extension Education, 14 (2) : 120-122.

Sakib, M. H. and M. S. I. Afrad. 2014. Adoption of Modern Aquaculture Technologies by the Fish Farmers in Bogra District of Bangladesh. International Journal of Agriculture Innovations and Research, 3 (2): 414-421.

Singh, I., K.K. Singh and U. S. Gautam.2012. Constraints in Adoption of Soybean Production Technology. Indian Research Journal of Extension Education,Special Issue (Volume II):169-171.

Singh, Y. J., R. Santhakumar, D. K Pandey, H. Bharati and P. DebRoy. 2012. Adoption of Hygienic Fish Handling Practices by Fishermen. Indian Research Journal of Extension Education, 12 (1): 36-38.

CHAPTER - 53

Global Revolution in Hydroponic Farming System : A Lesson From FIJI Island

Anamika and Paras Nath
Department of Agriculture, College of Agriculture, Fisheries, and Forestry
Fiji National University, Koronivia, Suva, Fiji Island

ABSTRACT

Hydroponics is a plant growth system that provides a more precise control of growth media composition. It offers people the ability to grow food in places where traditional agriculture simply isn't possible. As Fiji has a warm tropical climate, the development of the tropical fruit and vegetable industry through hydroponics farming to meet local consumption and export market is a lucrative opportunity. The potential for hydroponics farming is enormous and is in demand to sustain production at reliable and consistent level in Fiji. Potential areas for the setting up of hydroponic farms in Fiji include Savusavu, Nadi, Navua, Sigatoka Valley and Labasa. These areas have established theirlogistical status with the market. A wide range of commodities are grown through hydroponics technology which complements the need to harness the "too hot" and "too wet" weather of tropical Fiji. The pristine environment and availability of resources such as water and other hydroponic farming input are added bonus to establishment of such venture. Hydro nut Hydroponic farming system are constructed as simplest form made of bamboo and coconuts using coconut husk as a growing media and worm casting and humus tea are used as fertilizer. It offers some significant benefits over traditional farming, and as word about these benefits spreads, more people will turn to hydroponics for their agricultural needs.As hydroponics allows produce to be grown locally and requires fewer areas to import their crops, there is a reduction in both price and greenhouse gas emissions due to reduced transportation requirements. This saves time and produces healthier, disease and pest free, lusher plants in about half the time as traditional agriculture farming system.

Key words : Hydroponics, environment, agriculture, traditional farming

INTRODUCTION

Agriculture is the mainstay of Fiji's economy which contributes around 28% to total employment in the formal sector and many more people are employed indirectly in this sector (Nath, 2014). The agriculture sector was once a major stronghold of Fiji's economy which is now the third largest sector contributing $451 million (9%) annually to the nations GDP. More than three-quarters of all Fijian households used to engage in agricultural-related activities, but now many of those workers have switched over to the growing service industry. Sugarcane which used to dominate the agriculture sector now only contributes (0.9%) and has been surpassed by other crops, horticulture, and livestock production and subsistence sector. Having a rich resource base and tropical rain forest climate, Fiji has an advantage in producing a wide variety of tropical fruits and vegetables which are in high demand by local hotel industry. This is a major development in the fresh food market and provides lucrative opportunities to farmers, investors and stakeholders. Potential horticultural commodities for value added processing include; papaya, tomatoes, pineapple, coconut, duruka, mango, chilies, banana, cassava, breadfruit, sweet potatoes, ginger, cocoa, yaqona, taro, cassava and other root crops. Horticulture is one of the largest growing sub-sectors of agriculture and holds potential for continued growth. It is the only sub-sector within agriculture with increasing exports of a large variety of products. In horticulture sector hydroponic system of farming may play a vital role involving traditional and commercial farmers and this is also very good to practice by the urban population as back yard farming to meet their own demand and excess produce may be supplied in the local market. This will ensure family supply of nutritious vegetables at a lower cost, properutilization of backyard space, work for house ladies and additional income to the family.

What is Hydroponics?

The term hydroponics originates from the ancient Greek "hydros," meaning water, and "ponos," meaning labor/work. The concept of soil less gardening or hydroponics has been around for thousands of years. The hanging Gardens of Babylon and the Floating Gardens of China are two of the earliest examples of hydroponics. Scientists started experimenting with soil less gardening around 1950.

Perhaps if we would like to start a garden so that we can grow your own vegetables, but don't have the space in our yard, or we're overwhelmed by pests and insects, we can set up a hydroponics garden in our home and plants can grow readily without a big investment. Hydroponics systems require only around 10 percent of the water that soil-based agriculture requires. This is due to the fact that hydroponic systems allow recycling and reuse of water and nutrient solutions and the fact that no water is wasted and also requires little or no pesticides and only around 25 percent of the nutrients and fertilizers required of soil-based plants. This represents not only a cost savings but also benefits the environment in that no chemicals are being released into the air. Finally, we must consider the environmental impacts of transportation. Plants grown in this manner have direct access to water and nutrients and therefore, are not forced to develop extensive root systems to allow them to find the nutrients they need. This saves time and produces healthier, lusher plants in about half the time as traditional agriculture. Any vegetable, flower and even shrub can be grown without the need to cultivate, weed, and mulch, while watering and fertilizing can be taken care of by automatic system. Hydroponic gardens can be established anywhere, irrespective of the soil or climate.

For the Commercial grower hydroponic systems are ideal, they are highly efficient, requiring a low capital cost while producing high yields with less labour input. These same advantages apply to the home family grower who can cheaply and easily produce quality vegetables and fruit almost all year round. Whatever is our preference, hydroponics will provide us with a challenging and exciting hobby or vocation. Mainly vegetables are grown for commercial purpose such as lettuce, tomato, capsicum, cucumber, herbs as parsley, spinach etc.

Hydroponic Farming

Hydroponic farming is a method of growing plants in nutrient solutions and/or inert media such as hydrocorn, perlite, sand, or other inert media rather than traditional potting mixes or soil. This system provides a precise control of growth media composition. It offers an opportunity to grow food in places where traditional agriculture alone is not possible (Fig. 1).

Fig.1 : Hydroponic system, onion seedlings and trained tomato plants in fruiting stage in Fiji.

Hydroponic Farming in Fiji

Fiji is the largest and most developed of the Pacific Island and may be classified as a lower middle income country. While Fiji is restricted with geographic remoteness small domestic market and vulnerability to extreme weather changes, its constraints are not as severe as those of smaller Pacific Island nations. In Fiji, agriculture is organized more along commercial lines, although the subsistence sector remains important. Large-scale agriculture comprises mainly of palm oil, coconut, sugarcane, ginger, cassava, banana, papaya, green vegetables etc. Typical constraints faced by producers include a shortage of labour, poor quality and availability of planting materials, lack of efficient pest control and monitoring programmes, high post-harvest losses, poor animal health and high cost of purchased feed, and weaknesses in both domestic and export marketing. An impressive quantity and range of traditional food crops through family farming are grown in Fiji. As the population of our planet soars and arable land available for crop production declines, hydroponics will offer us a lifeline of sorts and allow us to produce crops in greenhouses or in multilevel buildings dedicated to agriculture. Already, where the cost of land is at a premium, crops are being produced underground, on rooftops and in greenhouses using hydroponic methods. In Fiji, hydroponics is used as traditional soil-based plant growth and useful in remotes locales,which allows the citizens of the island to enjoy year-round local produce without the expense and delay of importation. Hydro-nut hydroponic farming system are very common in Fiji and constructed as simplest form made of bamboo and coconuts using coconut husk as a growing media and worm casting and humus tea are used as fertilizer at very low investment (Fig.2).

Fig.2 : Hydronut hydroponic system growing lettuce and hot pepper. Hydroponic system made completely of coconut and bamboo using coconut coir as growing media and humus and worm casting tea as fertilizer in the Fiji Islands.

Hydroponics offers people the ability to grow food in places where traditional agriculture simply isn't possible. Hydroponics has been in use for decades and now there is a progressive movement in local village peoples mind. This science allows the people of these areas to enjoy locally grown produce and to expand their food production. Similarly, hydroponics is useful in dense urban areas, where land is at a premium. Potential areas for the setting up the hydroponic farms in Fiji include Savusavu, Nadi, Navua, Sigatoka Valley and Labasa. These are grown throughout the country and are identified as a "hidden strength" of the economy. It is estimated only 40% of energy needs come from locally grown food. Among all grown crops, the world's most nutritious lettuce are sustainably grown with pure Fiji water through hydroponics grow system. The Joes Hydroponic Farm in Suva area of Fiji is growing nine varieties of lettuce such as Green coral, Red coral, Green oak leaf, Red oak leaf, Green butter head, Red butter head, Cos lettuce, Ice burg and Endives, and the other vegetables which are mainly grown are parsley, coriander, tomato etc(Fig.3).

Fig.3 : Hydroponic system: Sustainable hydroponic family farm made out of local materials and resources; 3 tier hydroponic farming (Top tier- lettuce, Middle tier- parsley and Ground tier- coriander) and different varieties of lettuce grown in hydroponic system in Fiji.

Fiji has a warm tropical climate, where development of tropical fruit and vegetable industries through hydroponic farming in order to meet family/local consumption and export market is an emerging priority for lucrative family business. The potential for hydroponic farming is enormous which is being managed by the family and is in demand to sustain production at reliable and consistent level in Fiji. Potential areas for the setting up the hydroponic farms in Fiji include Savusavu, Nadi, Navua, Sigatoka Valley and Labasa. These areas have already developed their logistic status with the market. A wide range of commodities are grown through hydroponic technology, which complements the need to harness the "too hot" and "too wet" weather of tropical Fiji. The pristine environment and availability of resources such as water and other hydroponic farming input are added bonus to establishment of such venture. Hydro-nut hydroponic farming systems are developed as the simplest form made of bamboo and coconuts using coconut husk as a growing media, wherein worm casting and humus tea are used as fertilizer. It offers some significant benefits over traditional farming as such benefits spread to more people to get facilitated to hydroponics for their agricultural needs. As hydroponics allow produce to be grown locally requiring lesser area to grow the crops and help in reducing the import billof their crops, there is also a reduction in both price and greenhouse gas emissions due to reduced transportation requirements. This saves time, labour cost and produces healthier, disease and pest free, lusher plants in about half the time as traditional agriculture farming system.

Principles of Hydroponics

Overview: Hydroponics is simply the growing of plants without soil. Instead, plants are grown in an inert medium (or none at all) and an oxygenated, balanced nutrient solution is delivered to a their root systems

Media: Several hydroponic media are available, including rockwool, geolite, perlite, vermiculite, sphagnum moss and gravel. They retain water well and provide aeration for root systems.

Nutrient Solution: Two types of nutrient solutions are used - grow formula and bloom formula. The grow formula is used for the vegetative growth phase, while the bloom formula is used for the flowering growth phase. These nutrient solutions for all plants, weather grown hydroponically or in soil, need 16 basic elements to grow. Most of us are familiar with the top three that are consumed by plants. These elements are called the macro-elements and they are Nitrogen, Phosphorous and Potassium, they are abbreviated N P K. The other 13 elements are grouped together as the micro-elements. The pH affects the ability of a plant's roots to absorb nutrients. The range in which nutrient absorption is best is from 5.8 to 6.5. To lower pH acidis add. The best acids to use are phosphoric, nitric, and sulfuric acid, these acids disassociate and free up phosphorous, nitrogen and sulfur respectively. Nitrogen, Phosphorous and Sulfur are all elements that plants need for their growth. The bases used in hydroponics are Potassium hydroxide and Sodiumhydroxide. These disassociate and provide Potassium and Sodium, respectively. Main chemicals used for the requirements of macro- and micro nutrients: NPK are directly added in vermiculite whereas micro-elements (Magnisium sulphate, Potassium nitrate, Mono-potassium phosphate, Potassium sulphate, Magnese sulphate, Copper sulphate, Boric acid, Zinc sulphate, Molybdenum etc) are added in water in the main tank.

Setup: A hydroponic system consists of a long bed or trough that is shallow and sloped in a "V" shape. A drain at the bottom connects to a nutrient solution reservoir. A pump forces the nutrient solution up into the bed. The media absorbs the solution, and the plant roots feed on the nutrients. When the pump is turned off, the solution gradually drains back into the reservoir because of gravity.

Maintenance: The only thing need to do is monitor the nutrient level in the hydroponic system topping the level off with water between scheduled nutrient changes. Initially, young plants will not use much nutrient at all, so changing the nutrient once every two weeks is sufficient. Once the plants start consuming the nutrient then once-a-week change is done in order to drain the old nutrient and recharge the system with fresh.

pH: The water pH affects the ability of a plant's roots to absorb nutrients. The range in which nutrient absorption is best is from 5.8 to 6.5.

Types of Hydroponic Systems

1. **Liquid Culture:** The roots are hanging into the nutrient solution which can be either in the form of a liquid or a mist.
2. **Aggregate Culture**: The roots grow into an inert medium such as sand, gravel, Rockwool, perlite, vermiculite, peat moss, foam, coconut coir, etc. and are then irrigated with a complete nutrient solution

Critical Features for Building Hydroponic Systems

1. Water
2. Light
3. Inorganic mineral nutrients
4. Carbon dioxide
5. Oxygen
6. The proper temperature and relative humidity
7. Support systems

Advantage

1. Use of unproductive land soil which is not fit for farming.
2. Crops free from most of the disease caused by soil borne pathogen present in soil.
3. Reduced labor costs as high labor costs are one of the most expensive aspects of running a small family farm.
4. Maintenance of nutrient concentration and pH accurately.
5. No weeding or cultivation is needed.
6. Transplanting of seedlings is easy.

7. Faster germination and maturity rates can be achieved rather than with regular soil.
8. Nutrient solutions are re-circulated so there is no run-off and minerals are conserved, reducing the nutrients used by one-sixth.

REFERENCES

Nath, P. (2014). Horticulture Potential in Fiji Island. Progressive Horticulture. 46(2):175-183.

About the Editors

Dr. M.L. Choudhary, PhD, is currently Vice-chancellor of Bihar Agricultural University in Sabour, Bhagalpur, Bihar, India. He received a master's degree in horticulture from Banaras Hindu University, Varanasi, Uttar Pradash, India, and his PhD in the USA. The government of India deputed him for an advanced study on Hi-Tech Horticulture and Precision Farming in Israel and Chile. In a career spanning 30 years, he has occupied several executive positions, including horticulture commissioner, Ministry of Agriculture, India; chairman, Coconut Development Board, Kochi (Kerala); Ministry of Agriculture, India; national project director, FAO; visiting scientist at Rutgers University, New Jersey, USA; head of the Department of Ornamental Crops, IIHR, Bangalore; and head of the Division in Floriculture at IARI, New Delhi. As horticulture commissioner of the Ministry of Agriculture, India, he conceived, conceptualized, and implemented flagship programs such as the National Horticulture Mission and the National Bamboo Mission and Micro-irrigation. Apart from his professional career, he was also chairman and member secretary of various committees constituted by the government of India as well as state governments. He has guided 18 PhD students and 21 MSc students in the field of horticulture and has published 124 book chapters, 19 books, and 109 research papers of national and international repute. Dr.Choudhary has represented India at various international forums in the capacity of chairman and member. He was also the chairman of the Codex Committee of the Scientific Committee for Organic Standard of Ministry of Commerce. He has been conferred with 15 awards from various scientific and non-scientific organizations and government committees for his outstanding contribution in the field of horticulture/floriculture. He has also been awarded the Fellowship of Horticultural Society of India. Dr.Choudhary in his 30 years of academic, research, and administrative career has visited more than 15 countries to participate in various professional meetings. Recently, he has been conferred with the most coveted Sivashakhti Lifetime achievement award during 6th Indian Horticulture Congress - 2014 (IHCK2014), organized by the

Horticulture Society of India on 7th November, 2014 for his best contribution in the field of Horticulture besides establishing quality education system in the Agricultural University of the state, directing research projects of the University in perspective of changing climatic conditions and easy dissemination of farmers friendly agricultural technology to the farmers of the nation by strengthening extension and training programme.

Aditya is an assistant professor/scientist in the Department of Extension Education, Bihar Agricultural University, India. He acquired a B.Sc. Hons. (Agriculture) from Visva-Bharati University, Santiniketan, India. He received M.Sc. (Extension Education) from Banaras Hindu University, Varanasi, India, with specialization in Information and Communication Technologies (ICTs). He has written several papers related to the use of Information and Communication Technologies (ICTs) in teaching and training. He has been serving as an editorial board member of several journals. He is passionate about educational technology and teaching innovations. He has also published more than 12 peer-reviewed research and review papers and some popular articles, covering aspects of advanced extension practices in particular.

FSC
www.fsc.org
MIX
Papier | Fördert gute Waldnutzung
FSC® C083411

Zeitfracht Medien GmbH
Ferdinand-Jühlke-Straße 7
99095 Erfurt, Deutschland
produktsicherheit@kolibri360.de